Mycorrhizae in Crop Production

T0305813

HAWORTH FOOD & AGRICULTURAL PRODUCTS PRESS®
Crop Science

Plant-Derived Antimycotics: Current Trends and Future Prospects edited by Mahendra Rai and Donatella Mares

Concise Encyclopedia of Temperate Tree Fruit edited by Tara Auxt Baugher and Suman Singha

Landscape Agroecology by Paul A. Wojtkowski

Concise Encyclopedia of Plant Pathology by P. Vidhyasekaran

Molecular Genetics and Breeding of Forest Trees edited by Sandeep Kumar and Matthias Fladung

Testing of Genetically Modified Organisms in Foods edited by Farid E. Ahmed

Fungal Disease Resistance in Plants: Biochemistry, Molecular Biology, and Genetic Engineering edited by Zamir K. Punja

Plant Functional Genomics edited by Dario Leister

Immunology in Plant Health and Its Impact on Food Safety by P. Narayanasamy

Abiotic Stresses: Plant Resistance Through Breeding and Molecular Approaches edited by M. Ashraf and P. J. C. Harris

Teaching in the Sciences: Learner-Centered Approaches edited by Catherine McLoughlin and Acram Taji

Handbook of Industrial Crops edited by V. L. Chopra and K. V. Peter

Durum Wheat Breeding: Current Approaches and Future Strategies edited by Conxita Royo, Miloudi M. Nachit, Natale Di Fonzo, José Luis Araus, Wolfgang H. Pfeiffer, and Gustavo A. Slafer

Handbook of Statistics for Teaching and Research in Plant and Crop Science by Usha Rani Palaniswamy and Kodiveri Muniyappa Palaniswamy

Handbook of Microbial Fertilizers edited by M. K. Rai

Eating and Healing: Traditional Food As Medicine edited by Andrea Pieroni and Lisa Leimar Price

Physiology of Crop Production by N. K. Fageria, V. C. Baligar, and R. B. Clark

Plant Conservation Genetics edited by Robert J. Henry

Introduction to Fruit Crops by Mark Rieger

Generations Gardening Together: Sourcebook for Intergenerational Therapeutic Horticulture by Jean M. Larson and Mary Hockenberry Meyer

Agriculture Sustainability: Principles, Processes, and Prospects by Saroja Raman

Introduction to Agroecology: Principles and Practice by Paul A. Wojtkowski

Handbook of Molecular Technologies in Crop Disease Management by P. Vidhyasekaran

Handbook of Precision Agriculture: Principles and Applications edited by Ancha Srinivasan

Dictionary of Plant Tissue Culture by Alan C. Cassells and Peter B. Gahan

Handbook of Potato Production, Improvement, and Postharvest Management edited by Jai Gopal and S. M. Paul Khurana

Carbon Sequestration in Soils of Latin America edited by Rattan Lal, Carlos C. Cerri, Martial Bernoux, Jorge Etchevers, and Eduardo Cerri

Genetically Engineered Crops: Interim Policies, Uncertain Legislation edited by Iain E. P. Taylor

Mycorrhizae in Crop Production edited by Chantal Hamel and Christian Plenchette

Mycorrhizae in Crop Production

Chantal Hamel, PhD
Christian Plenchette, PhD
Editors

CRC Press is an imprint of the
Taylor & Francis Group, an **informa** business

First published 2007 by CRC Press

Published 2019 by CRC Press
Taylor & Francis Group
6000 Broken Sound Parkway NW, Suite 300
Boca Raton, FL 33487 2742

© 2007 by Taylor & Francis Group, LLC
CRC Press is an imprint of the Taylor & Francis Group, an informa business

No claim to original U.S. Government works

ISBN-13: 978-1-56022-307-8 (pbk)
ISBN-13: 978-1-56022-306-1 (hbk)

This book contains information obtained from authentic and highly regarded sources. Reasonable efforts have been made to publish reliable data and information, but the author and publisher cannot assume responsibility for the validity of all materials or the consequences of their use. The authors and publishers have attempted to trace the copyright holders of all material reproduced in this publication and apologize to copyright holders if permission to publish in this form has not been obtained. If any copyright material has not been acknowledged please write and let us know so we may rectify in any future reprint.

Except as permitted under U.S. Copyright Law, no part of this book may be reprinted, reproduced, transmitted, or utilized in any form by any electronic, mechanical, or other means, now known or hereafter invented, including photocopying, microfilming, and recording, or in any information storage or retrieval system, without written permission from the publishers.

For permission to photocopy or use material electronically from this work, please access www. copyright.com (http://www.copyright.com/) or contact the Copyright Clearance Center, Inc. (CCC), 222 Rosewood Drive, Danvers, MA 01923, 978-750-8400. CCC is a not-for-profit organiza-tion that provides licenses and registration for a variety of users. For organizations that have been granted a photocopy license by the CCC, a separate system of payment has been arranged.

Trademark Notice: Product or corporate names may be trademarks or registered trademarks, and are used only for identification and explanation without intent to infringe.

Visit the Taylor & Francis Web site at
http://www.taylorandfrancis.com

and the CRC Press Web site at
http://www.crcpress.com

Publisher's Note
The publisher has gone to great lengths to ensure the quality of this reprint but points out that some imperfections in the original copies may be apparent.

Cover design by Kerry Mack.

Library of Congress Cataloging-in-Publication Data

Mycorrhizae in crop production / Chantal Hamel, Christian Plenchette, editors.
p. cm.
Includes index.
ISBN: 978-1-56022-306-1 (hard : alk. paper)
ISBN: 978-1-56022-307-8 (soft : alk. paper)
1. Mycorrhizas in agriculture. 2. Mycorrhizal fungi. I. Hamel, Chantal, 1956- II. Plenchette, Christian.

SB106.M83M93 2007
631.4'6—dc22

2007000478

CONTENTS

NOTES FOR PROFESSIONAL LIBRARIANS AND LIBRARY USERS

This is an original book title published by Haworth Food & Agricultural Products Press®, an imprint of The Haworth Press, Inc. Unless otherwise noted in specific chapters with attribution, materials in this book have not been previously published elsewhere in any format or language.

CONSERVATION AND PRESERVATION NOTES

All books published by The Haworth Press, Inc., and its imprints are printed on certified pH neutral, acid-free book grade paper. This paper meets the minimum requirements of American National Standard for Information Sciences-Permanence of Paper for Printed Material, ANSI Z39.48-1984.

DIGITAL OBJECT IDENTIFIER (DOI) LINKING

The Haworth Press is participating in reference linking for elements of our original books. (For more information on reference linking initiatives, please consult the CrossRef Web site at www.crossref.org.) When citing an element of this book such as a chapter, include the element's Digital Object Identifier (DOI) as the last item of the reference. A Digital Object Identifier is a persistent, authoritative, and unique identifier that a publisher assigns to each element of a book. Because of its persistence, DOIs will enable The Haworth Press and other publishers to link to the element referenced, and the link will not break over time. This will be a great resource in scholarly research.

ABOUT THE EDITORS

Chantal Hamel, PhD, is Research Scientist at the Semiarid Prairie Agricultural Research Center of Agriculture and Agri-Food Canada in Swift Current, Saskatchewan, Adjunct Professor in the Department of Soil Science at the University of Saskatchewan, and Chercheuse Associee at the Institut de recherché en biologie vegetale of the Universite de Montreal in Quebec. Dr. Hamel completed her PhD at McGill University in Montreal, on the role of arbuscular mycorrhizae (AM) on nitrogen dynamics in mixed cropping agroecosystems. During her post-doctorate research at the Universite de Montreal, she explored aspects of AM-microbe interactions. As a Research Scientist at the Ministere de l'agriculture des pecheries et de l'amentation du Quebec in St. Hyacinthe, Quebec, and later as Assistant Professor in the Department of Natural Resource Science at McGill, she applied ecological principles to exploit AM in crop production. Dr. Hamel has been Associate Editor of the *Canadian Journal of Soil Science* and the *Canadian Journal of Microbiology*. She has had more than 75 refereed articles published and has contributed to several scientific books.

Christian Plenchette, PhD, is Research Director at the National Institut for Agronomic Research Centre in Dijon, France, where he has worked for 22 years. He is also Associate Professor at the University of Algiers, and Associate Researcher at the Laboratory for Tropical Symbiosis of the Research and Development Institut in Montpellier, France. He previously worked for 12 years at the Soil Service of the Quebec Ministry of Agriculture. His research on arbuscular mycorrhizae (AM) has focused on the mycorrhizal dependency of plants, the mycorrhizal soil infectivity, the role of AM in phosphorus nutrition of plants, and the AM inoculum production. Dr. Plenchette has authored and/or co-authored more than 100 refereed scientific articles and contributed to several scientific books.

Mycorrhizae in Crop Production
© 2007 by The Haworth Press, Inc. All rights reserved.
doi:10.1300/5425_a

CONTRIBUTORS

Alok Adholeya: Centre for Mycorrhizal Research, TERI (The Energy & Resources Institute), DS Block, India Habitat Centre, Lodi Road, New Delhi-110003, India.

Alejandro Alarcón: Microbiología, Edafología, Colegio de Postgraduados, Montecillo, Texcoco, Estado de México. CP 56230. México.

Ronald Ferrera-Cerrato: Microbiología, Edafología, Colegio de Postgraduados, Montecillo, Texcoco, Estado de México. CP 56230. Mexico.

Félix Fernández: Instituto Nacional de Ciencias Agrícolas, Cuba.

Kalyanne Fernández: Instituto Nacional de Ciencias Agrícolas, Cuba.

Mayra E. Gavito: Centre for Ecosystems Research, National Autonomous University of Mexico. Apdo. Postal 27-3. C P 58090 Morelia, Michoacán, Mexico.

John Larsen: Department of Integrated Pest Management, Research Centre Flakkebjerg, Danish Institute of Agricultural Sciences, DK-4200 Slagelse, Denmark.

Aiguo Liu: Pacific Research Center, Agriculture and Agri-Food Canada/Agriculture, et Agroalimentaire Canada, 6947 #7 Highway, P.O. Box 1000, Agassiz, BC, V0M 1A0.

Deepak Pant: Centre for Mycorrhizal Research, TERI (The Energy & Resources Institute), DS Block, India Habitat Centre, Lodi Road, New Delhi-110003, India.

Jesús Pérez-Moreno: Microbiología, Edafología, Colegio de Postgraduados, Montecillo, Texcoco, Estado de México. CP 56230. Mexico.

Mycorrhizae in Crop Production
© 2007 by The Haworth Press, Inc. All rights reserved.
doi:10.1300/5425_b

Sabine Ravnskov: Department of Integrated Pest Management, Research Centre Flakkebjerg, Danish Institute of Agricultural Sciences, DK-4200 Slagelse, Denmark.

Manuel Riera: Centro Universitario Guantánamo, Cuba.

Ramón Rivera: Instituto Nacional de Ciencias Agrícolas, Cuba.

Luis Ruiz: Instituto Nacional de Investigaciones en Viandas Tropicales, Cuba.

Ciro Sánchez: Estación Experimental de Café de Jibacoa, Cuba.

Seema Sharma: Centre for Mycorrhizal Research, TERI (The Energy & Resources Institute), DS Block, India Habitat Centre, Lodi Road, New Delhi-110003, India.

Sujan Singh: Centre for Mycorrhizal Research, TERI (The Energy & Resources Institute), DS Block, India Habitat Centre, Lodi Road, New Delhi-110003, India.

Radha R. Sinha: Department of Biotechnology, Government of India, Ministry of Science & Technology, Block 2 (7th Floor), C.G.O. Complex, Lodi Road, New Delhi-110 003, India.

Jorn Nygaard Sorensen: Department of Horticulture, Research Centre Aarslev, Danish Institute of Agricultural Sciences, DK-5792 Aarslev, Denmark.

Marc St-Arnaud: Institut de recherche en biologie végétale, Jardin botanique de Montréal, 4001 Sherbrooke Street East, Montreal, QC, Canada, H1X 2B2.

Vladimir Vujanovic: AFIF Chair, Applied Microbiology and Bioproducts, Department of Applied Microbiology and Food Science, University of Saskatchewan, Saskatoon, SK, Canada S7N 5A8.

Preface

It has been a revelation that, strictly speaking, most plants do not have roots but rather mycorrhizae, a fact that has had tremendous consequences on the life of plants and the evolution of soil-plant systems. The research on arbuscular mycorrhizal (AM) symbioses has been intensive over the past forty years and we have learned a lot on the physiology, biology, ecology, and genetics of the symbiosis and the fungi involved in it. Most important, it appeared that cropping systems could be more sustainable with the management of AM fungi and reduced reliance on agrochemicals. The extraradical mycelia of AM fungi are an essential link between the plants, which are the consumers, and the soil, which is the provider. They are key organs enhancing plant uptake of nutrients, particularly phosphorus in high P-fixing soils, and consequently reducing crop dependence on fertilizers. They also improve soil quality. Thus, the nature of AM extraradical mycelia must be considered in the design of cropping practices that optimize the contribution of AM fungi to crop production. The nature and role of AM mycelia as plant providers are discussed in Chapters 1 and 2. How AM fungi reduce disease incidence in plants has not been clarified, but what appears clear from the extensive literature review presented in Chapter 3 is that AM fungi do provide an important level of bioprotection to plants.

All research efforts on the study of AM do not translate into biotechnologies for agriculture and forestry in all parts of the world. In developed countries, the availability of agrochemicals at prices that farmers can afford has limited AM-related biotechnologies almost exclusively to soilless horticultural production. Arbuscular mycorrhizal inoculation has more impact in soilless than in field systems where native AM fungi are present. Chapter 4 reports on how AM fungi could be best used in horticultural production.

Mycorrhizae in Crop Production
© 2007 by The Haworth Press, Inc. All rights reserved.
doi:10.1300/5425_c

Incentives to increase fertilizer use efficiency were larger in countries with weaker economies. This was true in Cuba where an important research group has been successful in developing better practices for crop inoculation with effective AM strains since the early 1990s, as reported in Chapter 5. Chapter 6 presents numerous reports, some of which are difficult to access directly, indicating that AM biotechnologies would be advantageously applicable in a large number of tropical crops. Plant nutrition and health and soil quality benefit from AM in tropical seedlings. Chapter 7 summarizes the mycorrhizal research conducted in India, where the use of AM inoculants is rapidly expanding. India has been a leader in the development of AM technologies for crop production, where AM inoculants are used not only in crop production but are also very useful in soil rehabilitation.

Evidence of negative impacts of human activities, including crop production, on the environment and climate of the Earth are presented in Chapter 8, in a warning call reestablishing the need for AM agricultural research and development in wealthier countries. This book was prepared to serve as the basis for a second round of research efforts to improve cropping systems' sustainability throughout the world.

The editors would like to thank all reviewers for their precious contribution to the scientific quality of the texts. They are: Y. Dalpé, F. A. de Souza, D. D. Douds, E. Dumas-Gaudot, E. George, M. Giovannetti, N. Goicoechea, M. Gryndler, A. Fitter, J. A. Fortin, K. Hanson, Y. Kapulnic, R. Linderman, M. Ryan, and F. Selles. The contribution of the staff of the Agriculture and Agri-Food Canada Translation and Revision Services of the Communication and Consultations Team is also gratefully acknowledged. These people have made reading the following chapters enjoyable. We also thank Michelle Waldron, Administrative Assistant, and Dr. Amarjit S. Basra, former editor-in-chief FPP, and all other staff at The Haworth Press who made this publication possible.

Chapter 1

Extraradical Arbuscular Mycorrhizal Mycelia: Shadowy Figures in the Soil

Chantal Hamel

Arbuscular mycorrhizal (AM) fungi have coevolved with plants and soils for over 400 million years to become part of the root system of a very large number of terrestrial plant species (Taylor et al., 1995) and the "backbone" of the soil (Barea, Azcón, and Azcón-Aguilar, 2002). At some point in evolution, other types of mycorrhizal associations appeared under the selection pressure of their environment (Read and Perez-Moreno, 2003). While certain plant species, particularly primary colonizing species, are known to be non-mycorrhizal (Janos, 1980), it is a fact that most of today's terrestrial plant species use leaves to fulfill their C needs and mycorrhizae to take up water and nutrients.

It might be wise to capitalize on the mechanisms developed by plants to better use environmental resources for tomorrow's agricultural production. Plant production biotechnology allows remarkable achievements and can bring us closer to agricultural sustainability. While the genes that code for the fungal part of mycorrhizae are not contained in a plant seed, several genes that control AM fungi development and mycorrhizae function are. Although it is challenging, it will be effective in the long run to build on mycorrhizae in the development of highly performing plants and cropping systems, because they are the result of a long natural selection in the plant-soil system.

Mycorrhizae in Crop Production
© 2007 by The Haworth Press, Inc. All rights reserved.
doi:10.1300/5425_01

An investment in the improvement of AM symbioses in crop production will certainly yield economically and environmentally high-performance crop plants and cropping systems. Furthermore, effective enhancement of water utilization by crops is a feature provided by mycorrhizal fungi which is difficult to reproduce with current technologies.

What the cropping systems of tomorrow will look like is still unclear, but what is clear is that to effectively manage the AM symbiosis, we must first understand its nature and function. The extraradical phase of AM fungi, a key component of the performance of the AM symbiosis, is still poorly understood. In the early 1980s, Bethlenfalvay, Bayne, and Pacovsky (1983) gave us an insight into the importance of AM extraradical hyphae in an experiment using a range of P levels to create various mycorrhizal associations, from parasitic to mutualistic to parasitic, between soybean [Glycine max (L.) Merrill.] and *Glomus fasciculatum* (Thaxt.) Gerd. & Trappe emend. C. Walker & Koske. In this experiment, the greatest ratio of extra- to intraradical mycelium, which was found at the intermediate P levels, was coincident with plant growth enhancement. The importance of the extraradical hyphae of mycorrhizae has since been repeatedly demonstrated. Although AM-derived effects on plant growth cannot be entirely attributed to the function of the extraradical AM mycelium, the latter is clearly very important. The extraradical phase of AM fungi is involved in water and nutrient uptake by most plants, and is at the forefront of microbe-plant interactions and soil aggregate stabilization, a process that has considerable bearing on aeration, water relations (Augé et al., 2001; Rillig and Steinberg, 2002), and soil quality (Lupwayi et al., 2001). This chapter presents our current understanding of the nature and development of the extraradical AM mycelium. The reader is referred to Chapter 2 of this book for information on the role of the AM mycelium in nutrient uptake.

NATURE OF THE EXTRARADICAL ARBUSCULAR MYCORRHIZAL MYCELIUM

An extraradical AM mycelium appears as a simple structure made of coenocytic multinuclear hyphae. These hyphae grow three-dimensionally through the soil while older hyphal segments are being emptied from their cytoplasm. Hyphal regions isolated from the rest of the mycelium by evacuated hyphae may reconnect to the network by means of anastomosis formation, at least in the Glomeraceae and Gigasporaceae. Figure 1.1 shows numerous lateral hyphae linking, through anastomoses, runner hyphae of *Glomus intraradices* N.C. Schenck & G.S. Sm. traveling through the bulk soil, while other lateral hyphae produce spores singly or in small groups and, in this case, some infrequent branched absorbing structures. These structures of *G. intraradices* extraradical mycelium were also observed in vitro in monoxenic culture with carrot (*Daucus carota* L.) roots (Bago, Azcón-Aguilar, Goulet, and Piché, 1998; Bago, Azcón-Aguilar, and Piché, 1998). Thinner, lower-order hyphae were found to anastomose and

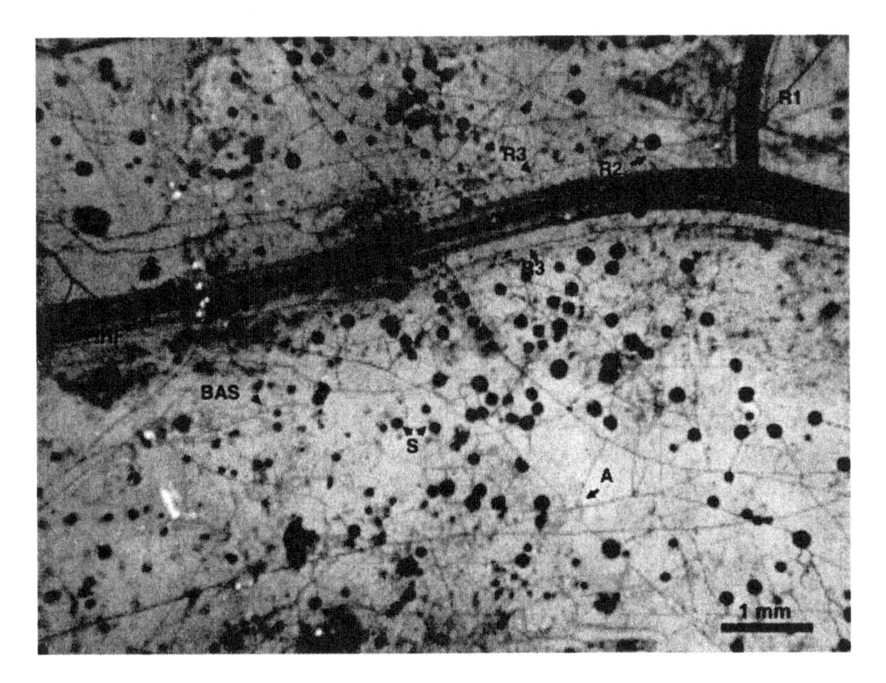

FIGURE 1.1. Network of *Glomus intraradices* N.C. Schenck & G.S. Sm. hyphae associated with alfalfa (*Medicago sativa* L.) roots. R1, runner hyphae extruding from the roots toward remote soil areas; R2, one of the lateral hyphae sent toward the root surface at intervals by hyphae running parallel to roots; R3, hyphae running parallel to the root; A, point of anastomosis; S, spores; BAS, branched absorbing structure; IHF, infective hyphal fan and appressoria formation on the root surface.

to bear auxiliary cells, and branched absorbing structures and to undergo swelling in *Scutellospora reticulata* (Koske, D.D. Mill. & C. Walker) C. Walker & F.E. Sanders grown in monoxenic culture with Ri T-DNA transformed carrot roots (de Souza and Declerck, 2003). It appears from the observation of these mycelia that anastomosis formation is both a healing process involved in the reconnection of broken, evacuated, or obstructed hyphae and a way to ensure good resource distribution in the network.

Anastomosis: A Determinant of AM Fungal Population Composition in Cropping Systems?

In vitro and in vivo experiments have revealed that numerous anastomoses were seen linking hyphae of compatible *Glomus* isolates into a network (Giovannetti, Azzolini, and Citernesi, 1999; Giovannetti et al., 2001). Anastomoses bridge hyphal sections and allow for the transport of nuclei and cytoplasmic elements from one part of the mycelium to another. The extraradical AM mycelium is able to form numerous anastomoses. Giovannetti et al. (2001) found 0.46 anastomoses mm^{-1} and 0.51 anastomoses mm^{-1} of *Glomus mosseae* (T.H. Nicolson & Gerd.) Gerd. & Trappe hyphae developing on *Prunus cerasifera* and *Thymus vulgaris*, respectively. Anastomoses also occur between mycelia originating from different plant species, suggesting that a mycelial network is an indefinitely large structure, even under mixed plant communities (Giovannetti et al., 2004). Protoplasmic continuity and the presence of nuclei in the hyphal connections confirmed the viability of anastomosed hyphae. Anastomosis formation has not been described between different AM fungal species and was not seen in all AM fungal isolates studied. It was observed in mycelia from the same or different spores of the same isolate of three *Glomus* species (Giovannetti et al., 1999), but no anastomosis was formed between the hyphae of *Gigaspora rosea* T.H. Nicolson & N.C. Schenck or *Scutellospora castanea* C. Walker, even when the hyphae originated from the same spore, suggesting that different taxa have different growth habits. The observation of precontact tropism, protoplasm retraction from the tip and septum formation in the incoming hyphae, in incompatible reactions between isolates of *G. mosseae* from different geographical areas (Giovannetti et al., 2003) suggests, interestingly, that specific recognition signals may be involved in anastomosis formation.

Anastomosis formation could have ramifications in agricultural fields. It could be useful in speeding up the reestablishment of fragmented AM hyphal networks in soil after tillage operations, crack formation in drying soil, or grazing by microarthropods, for example. In disturbed systems, the capacity of species to form anastomoses may be an important competitive advantage for the AM fungal species which possess this ability. Without this ability to reconnect fragments, AM hyphae that are far from roots, their C source, may become exhausted and disappear. Hyphal networks formed by isolates that can form anastomoses are likely to be large and to extend over a whole root system and even beyond, to link the root systems of several plants. The networks of fungal isolates that are unable to form anastomoses are likely to be smaller in size and to extend over only part of a root system.

A differential capacity of isolates from different genera to form anastamoses could possibly explain why the *Glomus* species were more abundant than non-*Glomus* species in tilled agricultural soils (Jansa, Mozafar, Anken et al., 2002; Jansa et al., 2003). Chlamydospores of *Glomus* and *Acaulospora* spp. were prevalent in soil shortly after the establishment of perennial asparagus plantations, while *Gigaspora* spp. predominated in asparagus plantations over twelve years of age (Wacker, Safir, and Stephenson, 1990). According to Bianciotto et al. (2004), anastomoses have been observed mainly in *Glomus* spp., although they were also reported in *Scutellospora reticulata* (de Souza and Declerck, 2003), *Gi. margarita,* and *Gi. rosea* (de la Providencia et al., 2005). The ability to form anastomoses may not be a characteristic of AM fungal genera, however, as suggested by the results of a study on the impact of cultivation on AM fungal biodiversity (Hamel et al., 1994). Both *Gi. margarita* and *Glomus caledonium* (T.H. Nicolson & Gerd.) Trappe & Gerd. disappeared after the onset of cultivation of an old meadow, an event more likely to occur with AM fungal species unable to reconnect through anastomosis formation. At the same time, *Scutellospora aurigloba*-like and *G. clarum* appeared, while eight other *Glomus* species appeared to be relatively unaffected. Tillage was reported to favor *G. mosseae, Glomus* sp 6, *Scutellospora pellucida* (T.H. Nicolson &

N.C. Schenck) C. Walker & F.E. Sanders (Menéndez, Scervino, and Godeas, 2001), and species of the *Glomus etunicatum* group (Douds et al., 1995), while no-tillage appeared to favor *Acaulospora denticulata* Sieverd. & S. Toro, *Entrophospora* spp., *Glomus* spp. 1-5, *Glomus aggregatum* N.C. Schenck & G.S. Sm., *Glomus microaggregatum* Koske, Gemma & P.D. Olexia, *Glomus coremioides* (Berk & Broome) D. Redecker & J.B. Morton (=*Sclerosystis coremioides*) (Menéndez, Scervino, and Godeas, 2001), and species of the *Paraglomus occultum* (C. Walker) J.B. Morton & D. Redecker (=*Glomus occultum*) group (Douds et al., 1995; Galvez et al., 2001). Merryweather and Fitter (1998) report the particular sensitivity of *Acaulospora* morphotypes to soil disturbance although, in their study, all AM populations were negatively affected. Clearly, more study is required on the formation of anastomoses in AM fungi mycelium. It is possible that the ability to form anastomoses is an isolate-specific feature, or a universal phenomenon that can be inhibited by certain environmental conditions, such as those prevailing in vitro, and that some AM fungal isolates are somehow more susceptible to disturbance although they form anastomoses. Whether soil disturbance selects for AM fungal taxa with anastomosis formation ability still needs to be demonstrated.

AM Mycelial Network Heterogeneity

A number of studies suggest the occurrence of genetic (Sanders, Clapp, and Wiemken, 1996) and phenotypic (Bever and Morton, 1999; Redecker, 2002) variations within populations of AM fungi. Recombination is a rare event in these presumably asexual and seemingly haploid (Hijri and Sanders, 2004; but see Pawlowska and Taylor, 2004) organisms, and the many nuclei contained in each AM fungal individual seem to have genetically diverged from each other through mutations to form a population of multiple genomes (Kuhn, Hijri, and Sanders, 2001; Sanders, Koch, and Kuhn, 2003; Vandenkoornhuyse, Leyval, and Bonnin, 2001). The multigenomic nature of AM networks could cause considerable spatial heterogeneity in each network. It is likely that heterogeneity in the mycelial network develops with its expansion. Specific nuclei assemblages may be segregated by chance in different parts of the network or selected by small-scale variations in the soil environment, giving rise to phenotypic variation within a network. If this is true, variation in topography or drainage, for example, may modify the soil environment and select for particular fungal traits in localized areas of a large network. Nuclei were found to be regularly spaced at 36 μm from each other within cytoplasm-filled AM hyphae sections (Bago et al., 2002). Thus, one centimeter of active hyphae contains approximately 300 nuclei; the size of these AM hyphal networks has not been measured but it is theoretically as large as the land area covered by host plants in AM species, especially if anastomosis formation occurs (Giovannetti et al., 2004). Thus, there is considerable room for heterogeneity in the AM mycelial networks developing in a farmer's field. Soil tillage disturbs the soil and we can foresee that it may increase genetic diversity by breaking AM mycelial networks into isolates, or in contrast, homogenize the AM mycelial networks of isolates able to form anastomoses. Koch et al. (2004) found phenotypic and genotypic variations among the isolates of a population of *G. intraradices* in soils under either tillage or no-tillage management. The results of this recent study, showing a higher level of variability at the genotype level than in quantitative traits, suggest that a high level of genetic redundancy in AM fungi can buffer the effect of high-genetic variability among isolates. The practical consequences of tillage on AM mycelial network heterogeneity are still unclear.

Formation of anastomoses could facilitate the distribution of endophytes in AM mycelial networks. The endophytic bacteria *Candidatus Glomeribacter gigasporarum* (Bianciotto et al., 2003) and strains of *Burkholderia* (Levy et al., 2003) were observed in spores and hyphae of a number of *Gigaspora* and *Scutellospora* spp. The role of these endophytes still needs to be clarified, but some evidence suggests that they could be involved in nutrient exchange between the partners in this consequently tripartite AM symbiosis (Minerdi, Bianciotto, and Bonfante, 2002) and stimulate AM spore germination (Bianciotto et al., 2004). The vertical transmission of *Candidatus G. gigasporum* through the generation of vegetative spores in *Gigaspora margarita* W.N. Becker & I.R. Hall grown in vitro on root culture has shown that the endobacteria can travel within hyphae to forming spores. Endobacteria were described in *Gigaspora* and *Scutellospora* spp., which belong to the Gigasporaceae, a family where anastomosis formation is reported. Endobacteria existing in AM fungal species with anastomosis formation capacity could possibly move from one mycelium to another via hyphal bridges.

The extraradical mycelium appears to have a simple structure, but various parts of this mycelium may be morphologically, structurally, and functionally different. Bonfante-Fasolo and Grippiolo (1982) observed a gradual thinning of the extra- to intraradical hyphae, changes in ultrastructural architecture of hyphal wall, and different responses to cytochemical reactions, suggesting a gradual simplification of the wall components from the extraradical to the intraradical hyphae. It would also probably be worthwhile to compare the structural architecture of the branched absorbing structures and the runner hyphae making up the extraradical phase of AM mycelia.

AM Mycelium Differential Morphologies and Responses: A Matter of Spatial Organization

The occurrence of different morphologies and the observation of different hyphal behaviors within localized portions of the extraradical AM mycelium suggest both the specialization of some segments and probably the occurrence of differential plant-fungus exchanges of signal molecules at different locations along the roots. For example, three types of coarse running hyphae are recognized. Some incoming running hyphae undergo branching in the vicinity of a root surface, before contacting the root, as if they were responding to a signal emitted by the root (Smith and Read, 1997). Such a differential morphological response to root vicinity was strongly expressed by the hyphae of *G. mosseae* from germinated spores (Giovannetti et al., 1993). Other coarse hyphae run along the roots, sending lateral hyphae toward the root surface at irregular intervals (R2 in Figure 1.1). Coarse hyphae running along the roots' surface appear to be involved in the infection of elongating roots. Still other coarse hyphae come out of infection points heading straight off to remote areas of the soil (R1 in Figure 1.1). These hyphae grow into the bulk soil at a high elongation rate. While creating mycorrhizae from spores on carrot root culture, these fast growing hyphae protruding through the root surface indicate that a mycorrhizal association was formed. Friese and Allen (1991) have described these growth patterns from root observation chambers. In vitro cultivation of mycorrhizae has also permitted observation of the development of these AM fungal structures (Fortin et al., 2002).

These differential growth patterns could be explained by differential signaling within the mycorrhizae. Different signals between the root and the AM fungi could be exchanged depending on the specific physiological status of different root and hyphal segments at given times. Conversely, the same message from one organism could trigger different responses in the recipient organism depending on its physiological status. The localization in space and time of fungi-root signal exchanges or an impact of recipient physiological status on the response triggered could explain the simultaneous attraction to root (R2 in Figure 1.1), parallel development of root and hyphal sections (R3 in Figure 1.1), and repulsion from root demonstrated by AM hyphae associated with a zone of a host root system (e.g., R1 in Figure 1.1). The co-occurrence of such contrasting behavior is required for the network to be fully functional. Infection events, which are transient and thus must be reiterated throughout the life span of a symbiosis, are favored by attraction of the fungi to roots. Proximity and parallel growth of hyphae and roots may synchronize mycorrhizal root infection spread and root growth. Finally, efficient foraging for water and nutrients by AM hyphae may take place away from the root, outside of its surrounding depletion zone.

The AM symbiosis is the result of a controlled infection expressed through modification of root metabolism to control the extent of fungal development as well as its growth pattern. The involvement of plant produced flavonoids in recognition and development of the AM symbioses has been proposed (Lum and Hirsch, 2002). AM fungi were found to react when exposed to flavonoid compounds (Bécard, Douds, and Pfeffer, 1992; Douds, Nagahashi, and Abney, 1996; Tsai and Phillips, 1991), which were identified as signal molecules in other symbiotic systems. A host-plant produced signaling factor triggering a presymbiotic response in AM fungi, namely germ tube branching and nuclear division, however, was not flavonoidic in nature (Buée et al., 2000). Although the nature of signaling factors involved in the AM symbiosis is unknown, it appears that diffusible root-produced compounds are involved in AM hyphal growth control. This is further suggested by the fact that the growth and sporulation of the AM mycelium is much more prolific when the latter is located in a compartment other than that of the root (St-Arnaud et al., 1996) (Figure 1.2). Thus, it is very likely that the development pattern of the AM hyphal network is affected by the conditions affecting diffusion in soil, that is, temperature, distance and path tortuosity, and the presence of functional groups on the soil solid surfaces that could adsorb signal molecules. According to the laws of diffusion, AM hyphae would be sent foraging at greater distance from the root in a warm, light soil and at shorter distance in a cool, clay-rich soil, if the production of signal molecules is the same in both soils. Such behaviors would maximize the efficiency of nutrient recovery from soil as fine textured soils are more fertile.

In addition to the involvement of compounds diffusible in water, it appears that some volatile compounds may also be involved in chemotactic responses of AM fungi. The germ tube of *Gigaspora gigantea* (T.H. Nicolson & Gerd.) Gerd. & Trappe responded to maize root-produced volatile compounds (Suriyapperuma and Koske, 1995). In-

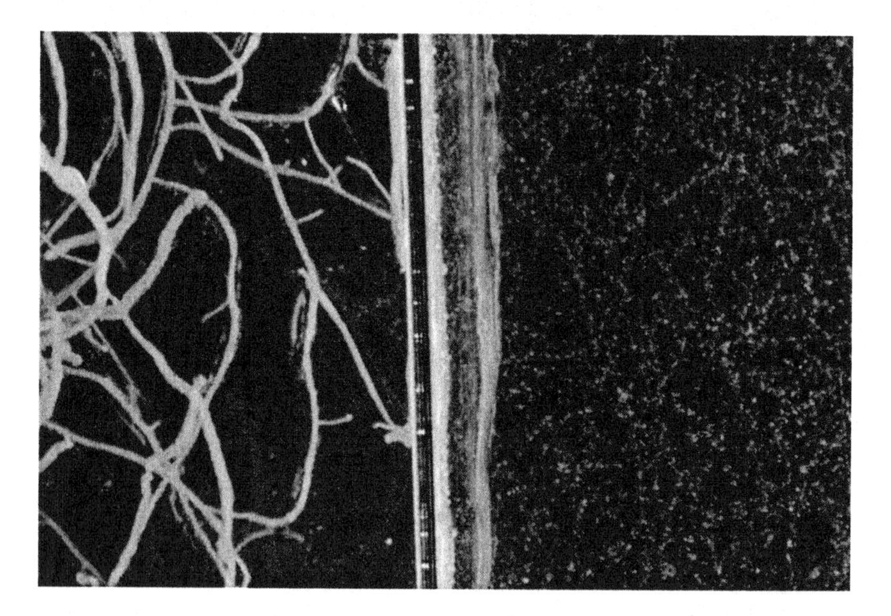

FIGURE 1.2. Differential growth of *Glomus intraradices* N.C. Schenck & G.S. Sm. in root-containing and in root-free Petri dish compartments. This differential growth pattern may indicate the repression of *G. intraradices* growth by roots (From M. St-Arnaud et al., 1996 with permission).

terestingly, the strength of the germ tube response to the presence of root-produced volatiles differed with maize cultivars and depended also on the P level of the growth medium. Some level of specificity is to be expected in the response of different AM fungal species to plant produced signals, as it would reflect the subtle plant-fungus specificity observed in the AM symbiosis (e.g., Helgason et al., 2002; Klironomos, 2003; Sylvia et al., 2003).

AM Mycelium Dynamics

The extraradical AM mycelium is a dynamic structure which can spread in soil at a rate of 0.2 mm day^{-1} to 2.5 mm day^{-1} (Smith and Read, 1997). Schubert et al. (1987) observed that the mycelium of *Glomus clarum* T.H. Nicolson & N.C. Schenck associated with *Trifolium repens* increased steadily, reaching maximum length eight weeks after inoculation. They observed that hyphal viability decreased from 100 percent to less than 15 percent during this time, starting in the second week following inoculation. This concurs with the results of a recent study reporting the dynamic nature of extraradical AM hyphae; in *Glomus* spp., the extraradical AM hyphal turnover rate has been estimated as five to six days in the absence of predators (Staddon et al., 2003). Branched absorbing structures and arbuscules were reported to have a similar "life span" in vitro (Fortin et al., 2002). The rapid disappearance of cytoplasm and hyphal structure was sometimes seen as the result of apoptotic processes within AM fungi (Fortin et al., 2002). However, hyphal turnover in AM fungi may not be apoptotic, that is, related to programed cell death. Filamentous fungi are special organisms known to move their cytoplasm within their hyphal tubes to more favorable areas (Klein and Paschke, 2004). For example, the nutrient level can be reduced, or metabolites excreted by the hyphae could reduce the suitability of a soil microsite after some time. While sections of hyphal tubes are indeed abandoned, the living portion of the fungi contained within the cell membrane simply moves from one place to another within the tubular cell wall. A one-week "life span" of extraradical hyphal segments coupled with constant growth would maximize the exploitation of the soil volume in time and space. Another potential benefit of "travelling" AM fungi could be the avoidance of parasite buildup that would likely occur around a stationary fungus. The cost to rapid hyphal turnover could be considerable, although it is mitigated by the withdrawal of cytoplasm from the old hyphal segments. The cost-effectiveness of AM hyphae turnover has not been evaluated. In any case, it appears to be effective, overall, since it exists.

Extraradical AM hyphae influence their biological environment directly and indirectly. The hyphosphere, the zone of soil immediately surrounding AM hyphae, is colonized by selected rhizosphere bacteria (Vancura et al., 1990). Since many of these bacteria require amino acid and growth factors presumably provided by AM fungi, the size of these bacterial populations may fluctuate with the appearance and disappearance of AM hyphae in the soil volume. The impact of AM fungi on the overall microbial community, however, is less clear. AM fungi were found to have little effect or negative impact on total soil bacterial number and activity. In a study (Olsson et al., 1996), bacterial number as measured by direct count of viable bacteria, and bacterial activity as measured by ^{3}H-labeled thymidine incorporation, were highest in cucumber root compartments but were not affected by AM mycelium, after 30 days of growth. After harvest, disconnected AM hyphae still did not have a significant impact on bacterial activity, and did not affect bacterial phospholipid fatty acids (PLFA). In contrast, Marschner and Baumann (2003) found a significant influence of AM fungi on the soil bacterial 16S rDNA community composition. In another study, also involving direct bacterial counts and ^{3}H-labeled thymidine incorporation (Christensen and Jakobsen, 1993), the presence of AM fungi decreased the rate of bacterial DNA synthesis and bacterial biomass. Discrepancy in results may depend on differences in scale in the various studies reported. An increase in the number of a few cultivable bacteria associated with AM hyphae, for example, may go unnoticed in the bulk soil bacterial population. While numerous researchers report changes in the quality of the soil microbial community (Marschner and Baumann, 2003) and variations in the size of specific microbial populations (e.g., Amora-Lazcano, Vazquez, and Azcón, 1998; Andrade, Linderman, and Bethlenfalvay, 1998; Andrade et al., 1998; Edwards, Young, and Fitter, 1998; Filion, St-Arnaud, and Fortin, 1999; Green et al., 1999; St-Arnaud et al., 1997), the impact of AM fungi on soil microorganisms is not well understood and no general trend emerges from the published literature. Complex competitive interactions between AM fungi, plants, and soil microorganisms, in conjunction with the influence of resource availability on the system, determine the fate of soil microbial populations. This system is very complex, and it is unrealistic to expect an absolute mycorrhizal effect on soil microorganisms. Interested readers will find more information on the subject of microbial interaction with AM fungi, as reviewed by Hodge (2000).

AM fungal mycelium and spores are also the source of a cell surface glycoprotein, glomalin, which is believed to improve soil physical quality. This glycoprotein and its immunoreactive degradation products appear to be recalcitrant to decomposition in soil, where they accumulate, as determined by their long-lasting detection with a monoclonal antibody. Glomalin is sometimes (Franzluebbers, Wright, and Stuedemann, 2000) but not always related to total soil organic matter levels (Rillig et al., 2001). It seems, nevertheless, that glomalin could contribute significantly to soil organic matter. First, AM fungi are abundant in soil, accounting for approximately 25 percent of total soil microbial biomass (Hamel, Neeser et al., 1991; Olsson et al., 1999) and up to 90 percent of root biomass (Hamel, 2004). Second, they have a fast turnover rate of 5 to 7 days (Fortin et al., 2002; Staddon et al., 2003), and their glomalin seems to somehow bypass the microbial processing imposed on most plant residues, contributing directly to the stable pool of soil organic matter. Carbon dating indicated that the Bradford-reactive-soil-protein (BRSP) pool extracted from tropical soils, had a minimum residence time of 6 to 42 years. This supports the hypothesis of an important contribution of glomalin-related soil proteins to the stable soil organic matter pool (Rillig et al., 2001). The abundance of BRSP was estimated to reach over 60 mg cm^{-3} of soil (approximately 2.7 percent) in tropical forests, and the C and N it contained amounted to approximately 4 to 5 percent of total soil C and N. Glomalin was found to be closely related to stable soil aggregate formation (Wright and Upadhyaya, 1998), except in highcarbonate soils where soil aggregate stability depends on carbonates rather than on soil organic materials (Franzluebbers et al., 2000). Thus, AM fungi hyphal networks seem to possess the ability to modify the physical quality of their habitat, through their important contribution to soil organic matter buildup and stabilization of soil aggregates. This environmental impact of extraradical AM mycelia may, in turn, impact on other soil microorganisms.

SPECIFIC ASSOCIATIONS WITH SPECIFIC OUTCOMES

It is important to bear in mind that the term "arbuscular mycorrhizae" refers to a group of associations taking place between genetically diverse plants and fungal species, the extent and efficiency of which varies with environmental conditions (see Chapter 5) as well as partners' genotypes. As such, the effect of the symbiosis on plant uptake of nutrients and water may vary. For example, drought tolerance in chili ancho pepper was provided by *Glomus* species, but not by *G. fasciculatum* (Davies et al., 2002). Plants colonized by a mix of AM fungi from a semiarid area had a higher level of drought tolerance than those colonized by *G. intraradices* from a culture collection (Augé et al., 2003), indicating a selective effect of the environment on fungal performance. Nutrient uptake and translocation efficacy is also influenced by the fungal isolate. For example, Frey and Schuëpp (1993) found that *G. intraradices, Acaulospora laevis* Gerd. & Trappe, and *Gi. margarita* associated with maize differed in their ability to take up and translocate $^{15}NH_4SO_4$ across a 40 μm nylon barrier. *Glomus versiforme* (P. Karst.) S.M. Berch produced a larger extraradical mycelium than *G. aggregatum* and *G. intraradices* on four apple rootstocks grown in high-P soil (644 kg ha^{-1} Bray extractible P) (Morin et al., 1994). Mycorrhizal efficiency, as expressed by plant growth enhancement, was related to external hyphal development.

AM fungi play numerous roles in the soil-plant system. They absorb water and nutrients, stimulate plant growth through hormone production, provide protection against plant pathogens, and improve soil structural quality. Different fungal species, and therefore different mycelial networks, probably have a different ability to perform different tasks. In a field situation, several AM fungal species are usually present. While some AM fungi can inhibit the positive effect of a co-occurring species, presumably through interspecific competition, others can be without effect (McGonigle, Yano, and Shinhama, 2003) or have positive effects. The specific beneficial influence of a species may be complemented by that of other co-occurring species (Hart and Klironomos, 2002).

The targeted selection of AM fungal isolates for inoculation purposes is desirable to optimize the success of specific applications, such as the production of drought tolerant transplants or transplants with improved ability to use soil nutrients. However, this task is complicated by the fact that the host plant is also a determinant of mycorrhizal development and efficacy. It would be important to consider the improvement of crop mycorrhizal efficacy in plant breeding (Tawaraya, 2003). Mycorrhizal efficacy of crops may not be a priority in an agriculture based on abundant fertilizer supply. But it would certainly be an asset for the implementation of sustainable and environmentally sound cropping systems.

The roots of mycorrhizal plants are connected to networks consisting of the hyphae of AM fungal isolates extending over a distance in soil. Different AM hyphal networks are likely to coexist in most units of surface soil volume, as several fungal species can associate with a given host plant species. One AM fungus can colonize the roots of many plant species, and conversely, the roots of one plant species can be connected to several AM networks. Because specific plant-AM fungus interactions have a differential impact on the development of the fungi and the plants involved, and thus on their competitivity in a biodiverse environment, the coexistence of AM hyphal networks probably has an important bearing on plant ecology in mixed stands. According to the model proposed by Bever and collaborators (Bever, Westover, and Antonovics, 1997; Bever, Pringle, and Schultz, 2002), differential affinities and influences between soil microorganisms associated with plants modifies plant-to-plant interactions in a population. AM fungi, because of their intimate relationship with plants, are particularly influential on these interactions. Figure 1.3 explains the possible impact of AM fungi on the interaction between two plants of different species. A plant X influences the AM networks associated with its root system, which in turn influences plant X productivity and competitive ability against a neighboring plant Y. The same pattern of mutual influence (indicated by vertical arrows pointing up and down) drives the growth of plant Y and its associated fungi. In the example of Figure 1.3, plant X and Y share three AM hyphal networks. Plant X differentially favors the development of fungus A and plant

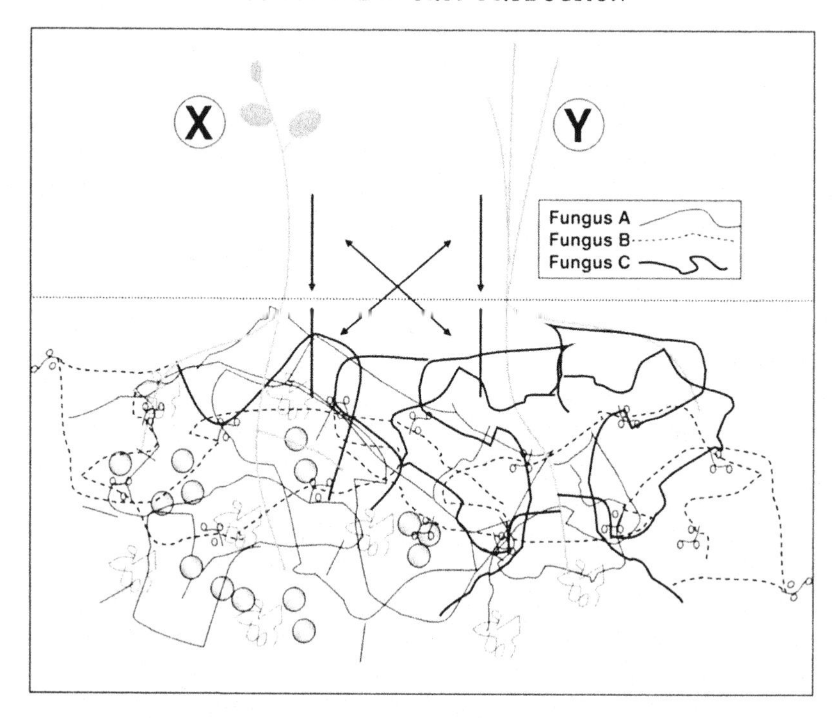

FIGURE 1.3. Graphical representation of the way the different affinities that may exist between AM fungi and plant species in a mixed stand may modify plant-to-plant interactions. *Arrows* represent the influence of organisms on one another. Different plants may enhance the development of some AM fungal species that grow less well on other plant species. The fungal species whose development is favored most by a plant will have a larger influence on neighboring plants. Thus, plants in a mixed stand have direct influence on each other and an indirect influence via their specific impact on AM fungi development.

Y favors the development of fungus C, while the development of fungus B is influenced to the same extent by plant X and plant Y. The promotion of fungus C by plant Y increases the influence of this fungus on plant X, and conversely the promotion of fungus A by plant X increases the influence of this fungus on plant Y. We can also see from Figure 1.3 that the uneven distribution of the hyphae of fungi A and C is a source of soil heterogeneity. AM fungi are important components of soil quality; consequently their uneven distribution creates heterogeneity in the soil, increases the number of available niches over a given area, and in this way favors biodiversity (Reynolds et al., 2003). High AM fungal richness has increased plant biodiversity, nutrient capture, and productivity (van der Heijden et al., 1998; see also Cuenca et al., 2004).

Networks of extraradical mycelium link plants of the same or different species. Mycorrhizae-assisted transfer of C, N, and P between plants has been well demonstrated using isotopic techniques. This topic was reviewed by Simard, Jones, and Durall (2002). In almost all reports, the C transferred from one AM plant to a neighboring AM plant could not reach the shoot but rather remained in roots, presumably in fungal structures. Pfeffer et al. (2004) showed in vitro that C flow was unidirectional from the plant to the fungus and while C could move within the fungus from one metabolically active root to another, it remained within the fungus. N also appears to move unidirectionally, but from the fungus to the plant. N-transfer between mycorrhizal plants occurs indirectly through N release from roots to soil and subsequent mycorrhizae-assisted N-uptake (George, Marschner, and Jakobsen, 1995; Hamel, Barrantes-Cartín et al., 1991). Thus, the nursing effect of a larger plant on smaller plants connected through hyphal links, which would be expected if resources moved from source to sink as had been hypothesized, may never happen. Accordingly, AM fungi had no influence on the belowground interaction between a N and C stressed soybean seedling and an older nonstressed soybean plant, in an experiment using pots compartmentalized by screen barriers of varying permeability (Franson et al., 1994). The disturbance of extraradical hyphal links between nurse *Calamagrostis epigejos* plants and seedlings of the same species had no effect on seedling growth in industrial waste substrates (Malcova, Albrechtova, and Vosátka, 2001), an observation that does not support the role of AM hyphae in intraspecific interactions. Although N-transfer between mycorrhizal plants is indirect and proceeds through release in soil and subsequent uptake, the phenomenon of a common network is likely to be a very important factor in plant dynamics. Research results have shown that even though a network is shared between plants, the presence of this AM network gives more competitive ability to one plant than another. It favors the plant that benefits most from the symbiosis, that is, the plant with highest mycorrhizal dependency (Hamel, 1992; van der Heijden, 2002). While N-transfer between plants may have an obvious impact on the proportion of plant species in pastures of mixed hay fields including legumes, it may also have a bearing on the competitive interactions between nonlegumes and crops and weeds. The outcome of such interaction depends on the quality of specific AM-crops and AM-weeds relationships. Specificity in the AM symbiosis was recently reviewed by Sanders (2002).

INFLUENCES ON THE DEVELOPMENT OF THE EXTRARADICAL MYCELIUM IN CROPPING SYSTEMS

The extraradical mycelium of AM fungi is a normal absorbing organ in most plants. This component of AM symbiosis is therefore very important for normal plant growth. The AM extraradical mycelium is responsible for a large portion of nutrient uptake in plants, and in some cases, could even be responsible for 100 percent of plant P uptake. Smith, Smith, and Jakobsen (2004) proposed that the AM symbiosis may, in specific fungus-plant associations, shut off the P absorption mechanism in roots. The AM fungi have evolved in soil-plant systems with a certain level of plant diversity under relatively undisturbed conditions. In contrast, agricultural soils tend to support low levels of plant biodiversity and are highly disturbed notably by tillage operations. Thus, mycorrhizal efficacy may be influenced by different factors in agricultural soils. This section examines the influences shaping the development of the extraradical AM fungal mycelium in agricultural soils.

The Influence of the Crop

In agricultural fields under monoculture of annual crops a strong seasonal cycle is imposed on soil biological activity as these succeeding crops are the main providers of C to underground life. AM fungi, which are obligate biotrophs and need the support of a living host plant to exist, are also subjected to the seasonal cycle imposed by crop succession, and consequently the development of the AM fungi indigenous to these soils is closely linked to plant phenology (Bohrer, Friese, and Amon, 2004; Kabir, O'Halloran, Fyles et al., 1998), in addition to being influenced by plant nutrition. This phenomenon has a particularly significant bearing under monocropping systems since the development of all plants in a field is synchronized. The abundance of AM hyphae in soil typically increases during the development of roots and photosynthetic tissues up to the plant reproduction stage and decreases thereafter, until the next cropping cycle. Kabir, O'Halloran, Fyles et al. (1998) observed that the proportion of viable hyphae also increased during vegetative growth up to flowering, in maize, and decreased thereafter. This can be explained by the fact that grain crops reallocate their resources to seed production, at the expense of source elements, after their vegetative and generative stages (Peltonen- Sainio, 1999). This reallocation of resources in the plant is followed by senescence and death. A similar development pattern, where resources are first allocated to source elements followed by a shift toward allocation to sinks and finally by leaf senescence, is also seen in crops such as potato (Vos, 1999). Whether the AM hyphal development pattern follows that of a source element involved in the production of yield in tuber and root crops remains to be seen.

The fungal component of the crop root system remains alive in soil after crop senescence and can colonize the roots of the following crop, regardless of whether the latter is planted immediately after harvest of the first crop, under warm climates, or after overwintering for several months, under cool climates. The rate of mycorrhizal development in the new crop depends on the abundance of AM mycelium and propagules left by the preceding crop and on the soil conditions. In annual cropping systems, seedlings become colonized and connected to AM hyphal networks as they send their roots in the vicinity of active AM mycelium or germinating propagules. Active mycelium, root fragments, spores, and even vesicles can serve as propagules. Klironomos and Hart (2002) have shown, however, that while their isolates of the genera *Glomus* and *Acaulospora* could produce infection from spores, root pieces, and extraradical hyphae, *Gigaspora* and *Scutellospora* isolates could not produce infection from extraradical hyphae and produced limited infection from root fragments. Isolates with less flexibility in their propagation strategy would be at a disadvantage in disturbed environments where plant colonization must be reestablished every spring. The different propagation strategies in different AM fungi appear to be a likely explanation for the scarcity of *Gigaspora* and *Scutellospora* in agricultural fields, where mycorrhizal associations have to be reinitiated each year.

Influences on Propagule Survival Between Crops and Reinfection

The importance of the different AM propagules as sources of infection depends on their abundance in soil. Several factors can affect propagules survival in agricultural fields. Winter can be an important factor under cool-temperate climates. The AM mycelium in Canadian maize fields retained infectivity in frozen soil over winter (Addy, Miller, and Peterson, 1997; Kabir, O'Halloran, and Hamel, 1997). Spores, in contrast, did not seem to be an effective inoculum after the winter (Addy et al., 1997). The persistence of AM propagules may also vary with species. The inoculums of different AM fungi responded differently to wetting and drying cycles. The infectivity of *Acaulospora laevis* Gerd. & Trappe and *Glomus monosporum* Gerd. & Trappe increased, that of *Scutellospora calospora* (T.H. Nicolson & Gerd.) C. Walker & F.E. Sanders was not affected by wetting and drying, while that of mycorrhizal root fragments of *Glomus invermaium* Hall and *Scutellospora calospora* (T.H. Nicolson & Gerd.) C. Walker & F.E. Sanders decreased and increased, respectively (Braunberger, Abbott, and Robson, 1996). *G. invermaium* hyphae were no longer infective after a wetting and drying period. The ability of AM propagules of different species to survive between cropping seasons appears to be a key determinant of the structure of the mycorrhizal population selected by the production of annual agricultural plants.

The timing of mycorrhizal development in spring seems to be determined by soil temperature. In cool-temperate climates, air temperature increases faster than soil temperature in spring, and at seeding soil temperature is often low. Few studies have tested the impact of soil temperature on mycorrhizae formation. Inhibition of AM development at temperatures allowing plant growth was repeatedly reported. AM colonization was strongly repressed at a soil temperature of 15°C in soybean and sorghum (*Sorghum bicolor* L.) roots (Zhang et al., 1995), and inhibited at 10°C in barley (*Hordeum vulgare* L.) (Baon, Smith, and Alston, 1994), and sorghum (Liu, Wang, and Hamel, 2003). Both root and *G. intraradices* hyphae on carrot root culture grew at 10°C, although at a reduced rate as compared to higher temperatures, suggesting that low soil temperatures specifically inhibit the formation of symbiosis. Winter wheat did not form mycorrhizae after seeding in fall, when soil temperatures were low, although mycorrhizae developed in spring (Mohammad, Pan, and Kennedy, 1998). Pea plants did not develop extraradical AM hyphae at 10°C but did at 15°C (Gavito, Schweiger, and Jakobsen, 2003). The mycelium developed at 15°C could take up and translocate ^{33}P. In another study, *G. intraradices* increased ^{32}P in leek (*Allium porrum* L.) over a non-mycorrhizal control both at 23°C and 15°C, but not at 0°C (Wang et al., 2002).

The formation of mycorrhizae at low soil temperatures may also vary with the fungal species involved and this variation may stem from adaptation of isolates to the specific climatic conditions prevailing where they live. For example, Baon et al. (1994) found reduced colonization by *G. intraradices* in barley roots held at 15°C, but Volkmar and Woodbury (1989) found no negative effect of low root zone temperature on barley root colonization by AM species indigenous to the Canadian prairie. Even higher colonization levels were found at 12°C than at 16°C or 20°C when the inoculum was placed 5 cm below the soil surface rather than dispersed throughout the soil. Similarly, AM fungi indigenous from a Denmark field soil were more active at translocating ^{33}P to pea at 15°C than a *G. caledonium* isolate from the BEG collection of mycorrhizal fungi (Gavito et al., 2003). Interestingly, Daft, Chilvers, and Nicolson (1980) reported mycorrhizal development in dormant bulbous spring-flowering forest floor plants, English bluebells (*Endymion non-scriptus* L.), during winter months, when soil temperature was close to 5°C. While the growth habit of a plant may dictate the rules for its mycorrhizal development, it seems that the origin of the host plant is unlikely to be responsible for the inhibition of mycorrhiza formation in cool soils, as mycorrhizae formation were repressed at 15°C-16°C in three cool season species, *Medicago truncculata, Trifolium subterraneum,* and barley (Smith and Bowen, 1979), just as it was in two species of tropical origin, soybean (Zhang et al., 1995) and sorghum (Liu et al., 2003), when their root zone temperature was maintained at 15°C. It makes sense that selection plays in favor of the AM fungal isolates that are more effective at forming symbioses under low soil temperature conditions, in agricultural soils under cool-temperate climates, since spring after spring, the cold tolerant isolates would be favored. Thus, the problem with delayed mycorrhizae formation in agricultural soil might not be a problem in normal springs, at least when indigenous AM fungal species are involved.

Influence of Soil Tillage

In soils under no-tillage management, active mycelial networks will rapidly become connected to new roots. Studies conducted in Canada have shown the importance of mycelial networks as AM propagules in agricultural fields (Miller, 2000). It seems, however, that fragmentation of the mycelial networks by tillage per se does not reduce soil AM infectivity, although it may reduce translocation of nutrients to plants. Rather, mycelium fragmentation would reduce the length of time hyphal fragments can survive before the next crop is planted (Kabir, O'Halloran, and Hamel, 1997, 1999). The impact of tillage on AM mycelium may therefore result in delayed AM colonization, particularly when there is a fallow period or winter in between two crops, in soil with medium to low levels of inoculum. The impact of tillage, however, may not be seen in soils where the level of AM inoculum is saturating (Jasper, Abbott, and Robson, 1991; McGonigle and Miller, 2000). The ability to survive soil disturbance appears to depend on the AM fungal taxonomic group. Glomeraceae seem to tolerate soil disturbance better than other groups, and observed tillage-related shifts in AM fungal population structure indicate the proliferation of members of this group in agricultural fields (Jansa, Mozafar, Banke, et al., 2002; Jansa et al., 2003), although variation may occur at the species level (Galvez et al., 2001; Douds et al., 1995).

Even when there is no difference in root colonization between plants produced under no-tillage and conventional tillage systems, plants under no-tillage usually benefit from the presence of a non-disrupted AM mycelial network and usually have better early P uptake. The detrimental effects of soil disturbance on early plant P uptake is well documented (Galvez et al., 2001; Gavito and Miller, 1998; Goss and de Varennes, 2002; Kabir, O'Halloran, Fyles et al., 1998; Miller, 2000; Mozafar et al., 2000). Soil disturbance-induced reduction in plant P uptake was shown to vary with the AM fungal species indicating a differential sensitivity of AM fungal species to soil disturbance (McGonigle et al., 2003). *G. mosseae* stimulated maize P uptake to the same extent whether or not the soil was disturbed before seeding.This was not the case for *G. aggregatum,* or *Gi. margarita*. It remains that under conventional tillage management, the viability of the extraradical AM hyphae in spring may be reduced, root colonization may be delayed; early P uptake by crops and yields may also be reduced (Kabir, O'Halloran, Fyles et al., 1998). Although AM-enhanced P uptake at early plant development stage is often seen under no-tillage, positive impacts on yield have been rare. Miller (2000) explained that in order for better P nutrition to lead to better yield, the negative impact of no-tillage on other soil variables, which limit yield, must be eliminated. No-tillage soils are typically more compact and take more time to warm up in spring, which is an important consideration under cool climates. Although it has proven difficult to isolate the impact of AM fungi in a field situation, it appears that consequences of good mycorrhizae development on crop production can be significant, especially in legumes where the AM and diazotrophic symbiosis often interact synergistically. For example, soybean colonization by both AM fungi and *Bradyrhizobium japonicum* developed faster as shown by a root nodule biomass almost four times larger in soybean grown in undisturbed soil and a higher level of mycorrhizal colonization of roots which was maintained beyond pod-filling stage (Goss and de Varennes, 2002). Germinated spores of four AM species were found to synthesize a diffusible factor capable of triggering the expression of an early nodulin gene in *Medicago truncatula* Gaertn. (Kosuta et al., 2003). This indicates that the response of legume N_2-fixation could be attributed to the production of enhancing signal molecules by AM fungi, in addition to improved P nutrition (Stribley and Snellgrove, 1986).

Reduced tillage, in the form of discing the soil to prepare the seed bed in spring, in contrast to mouldboard ploughing in the fall and discing in the spring, has had a reduced impact on mycorrhizae development in maize. This moderate impact on mycorrhizal development can be attributed to the placement of the extraradical AM mycelium in the soil. Kabir, O'Halloran, Widden et al. (1998) found that most of this mycelium is located in the top 20 cm of soil, in a maize field at grain-filling stage (Figure 1.4). A depth of 20 cm corresponds

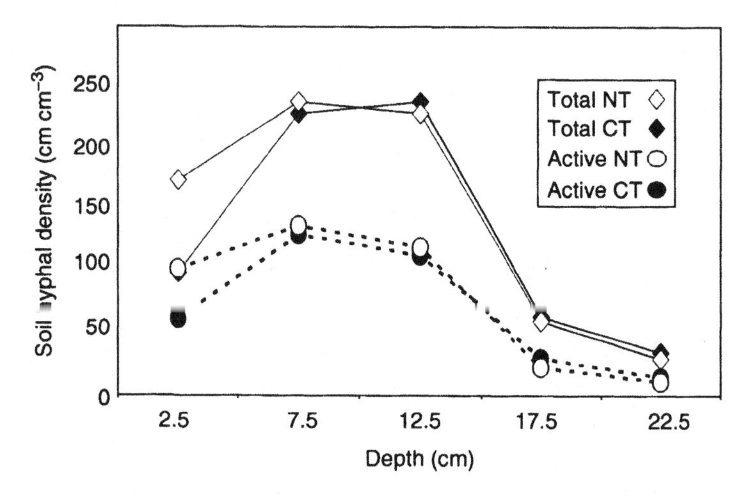

FIGURE 1.4. Distribution of total and active AM hyphae in the soil profile, as influenced by conventional tillage (CT), which involves moldboard ploughing to a depth of 20 cm in fall and disking in spring, and no-tillage (NT) soil management systems.

to ploughing depth. Hence, the AM mycelial networks are thoroughly disturbed by moldboard ploughing, while much of the networks and their connections to old root systems would remain intact after a discing operation, which works the soil at shallower depths. Mycelium distribution was similar in tilled and no-tillage soils, except that hyphal density was higher in the uppermost layer of no-tillage soil. Higher hyphae density was attributed to the more stable soil moisture and temperature conditions under no-tillage, where accumulating organic matter on the soil surface acts as protective mulch. Higher densities might also be due to an organic matter-related AM fungal growth stimulation in the top soil layer.

Other Influences

The stimulation of AM fungi by organic matter has been reported. Compost amendment significantly increased the number of AM fungi infective propagules in the root zone of inoculated shrubs, planted in desertified soil (Palenzuela et al., 2002). Cellulose (Gryndler et al., 2002) and chitin (Gryndler et al., 2003) stimulated sporulation and extraradical hyphae length development. Autoclaved mycelium of *Fusa-*

rium oxysporum also stimulated AM fungi growth. In these studies, it is difficult to conclude that the observed effects are those of the organic matter per se rather than the effect of some "mycorrhizae-helper bacteria" that use these organic compounds to multiply. For example, *Bacillus thuringiensis* (Vivas, Marulanda, Gomez, et al., 2003) and *Bacillus* spp. (Vivas, Marulanda, Ruiz-Lozano, et al., 2003) enhanced both AM fungi development and the benefit of the symbiosis. The saprophytic bacteria *Agrobacterium radiobacter* also stimulated root colonization by and metabolic activity of the extraradical mycelium of *G. fasciculatum* (Vosátka et al., 1995). In a study in which water extract of composted grape pomace enhanced onion (*Allium cepa* L.) root colonization by *G. intraradices* in a low P soil, it was unclear whether the stimulating effect of the extract was attributable to a P input or to a microbial effect (Linderman and Davis, 2001). N and P fertilization can increase AM fungal biomass in nutrient limited soil, especially in isolates adapted to high soil fertility. Some AM fungi genotypes are better adapted to high soil fertility than others (Treseder and Allen, 2002). It must be emphasized, however, that it is the high levels of soil-available nutrients, particularly P, which usually limit AM development in farmlands in the Western world. Other factors that may limit AM development in agricultural soils are inadequate soil pH (van Aarle, Olsson, and Söderström, 2002; van Aarle, Söderström, and Olsson, 2003) and high salinity (Carvalho et al., 2003). However, under most field conditions, AM fungi develop and it seems that the management of the AM symbiosis has potential in crop production. Although we can produce high yields without consideration of AM symbiotic development, and hence, we may not need its input for this specific purpose, we may want to include AM management in crop production to reduce fertilizer inputs and the impact of agriculture on environmental quality, to improve water use efficiency (Augé, 2001), and to improve soil quality (Jeffries et al., 2003).

CONCLUSIONS

The extraradical mycelium of AM fungi is a key feature of AM symbiosis. It is largely responsible for the uptake function of mycorrhizae and for translocation of nutrients and water from the soil to the plant. Furthermore, the extraradical AM mycelium is an important component of soil quality. While we have a relatively good understanding of the general functioning of this mycelium, especially if the fungal species used as test models are representative of the whole group, it would appear that much work remains to be done on the physiology of the AM root and on the ecology of these fungi. The study of mycorrhizal ecology is the important challenge we are now facing. It may yield additional information leading to the design of cropping practices that best valorize the contribution of mycorrhizae to sustainable crop production. Progress in AM physiology and genetics would provide information required by plant breeders to produce plant cultivars with efficient mycorrhizae. The requirements for AM fungi development are generally compatible with conditions required for the growth of the plants with which they have coevolved. AM extraradical extensions to roots are an integral part of most plant root systems and a feature that has persisted for 400 million years, presumably due to its efficiency. It makes sense to consider mycorrhizae in the future of crop plants. Although working with two organisms simultaneously is more complicated, in the long run the results should be rewarding.

REFERENCES

Addy, H.D., M.H. Miller, and R.L. Peterson (1997). Infectivity of the propagules associated with extraradical mycelia of two AM fungi following winter freezing. *New Phytologist* 135:745-753.

Amora-Lazcano, E., M.M. Vazquez, and R. Azcón (1998). Response of nitrogen-transforming microorganisms to arbuscular mycorrhizal fungi. *Biology and Fertility of Soils* 27:65-70.

Andrade, G., R.G. Linderman, and G.J. Bethlenfalvay (1998). Bacterial associations with the mycorrhizosphere and hyphosphere of the arbuscular mycorrhizal fungus *Glomus mosseae*. *Plant and Soil* 202:79-87.

Andrade, G., K.L. Mihara, R.G. Linderman, and G.J. Bethlenfalvay (1998). Soil aggregation status and rhizobacteria in the mycorrhizosphere. *Plant and Soil* 202:89-96.

Augé, R.M. (2001). Water relations, drought and vesicular-arbuscular mycorrhizal symbiosis. *Mycorrhiza* 11:3-42.

Augé, R.M., J.L. Moore, K.H. Cho, J.C. Stutz, D.M. Sylvia, A.K. Al-Agely, and A.M. Saxon (2003). Relating foliar dehydration tolerance of mycorrhizal *Phaseolus vulgaris* to soil and root colonization by hyphae. *Journal of Plant Physiology* 160:1147-1156.

Augé, R.M., A.J.W. Stodola, J.E. Tims, and A.M. Saxton (2001). Moisture retention properties of a mycorrhizal soil. *Plant and Soil* 230:87-97.
Bago, B., C. Azcón-Aguilar, A. Goulet, and Y. Piché (1998). Branched absorbing structures (BAS): A feature of the extraradical mycelium of symbiotic arbuscular mycorrhizal fungi. *New Phytologist* 139:375-388.
Bago, B., C. Azcón-Aguilar, and Y. Piché (1998). Architecture and developmental dynamics of the external mycelium of the arbuscular mycorrhizal fungus *Glomus intraradices* grown under monoxenic conditions. *Mycologia* 90:52-62.
Bago, B., P.E. Pfeffer, W. Zipfel, P. Lammers, and Y. Shachar-Hill (2002). Tracking metabolism and imaging transport in arbuscular mycorrhizal fungi. Metabolism and transport in AM fungi. *Plant and Soil* 244:189-197.
Baon, J.B., S.E. Smith, and A.M. Alston (1994). Phosphorus uptake and growth of barley as affected by soil temperature and mycorrhizal infection. *Journal of Plant Nutrition* 17:479-492.
Barea, J.M., R. Azcón, and C. Azcón-Aguilar (2002). Mycorrhizosphere interactions to improve plant fitness and soil quality. *Antonie Van Leeuwenhoek International Journal of General and Molecular Microbiology* 81:343-351.
Bécard, G., D.D. Douds, and P.E. Pfeffer (1992). Extensive *in vitro* hyphal growth of vesicular-arbuscular mycorrhizal fungi in the presence of CO_2 and flavonols. *Applied and Environmental Microbiology* 58:821-825.
Bethlenfalvay, G.J., H.G. Bayne, and R.S. Pacovsky (1983). Parasitic and mutualistic associations between a mycorrhizal fungus and soybean: The effect of phosphorus on host plant-endophyte interactions. *Physiologia Plantarum* 57:543-548.
Bever, J.D. and J. Morton (1999). Heritable variation and mechanisms of inheritance of spore shape within a population of *Scutellospora pellucida*, an arbuscular mycorrhizal fungus. *American Journal of Botany* 86:1209-1216.
Bever, J.D., A. Pringle, and P.A. Schultz (2002). Dynamics within the plant—Arbuscular mycorrhizal fungal mutualism: Testing the nature of community feedback. In Diversity of arbuscular mycorrhizal fungi and ecosystem functioning, in *Mycorrhizal Ecology*, M.G.A. van der Heijden and I.R. Sanders (eds.). Berlin: Springer- Verlag, pp. 267-292.
Bever, J.D., K.M. Westover, and J. Antonovics (1997). Incorporating the soil community into plant population dynamics: The utility of the feedback approach. *Journal of Ecology* 85:561-573.
Bianciotto, V., A. Genre, P. Jargeat, E. Lumini, G. Bécard, and P. Bonfante (2004). Vertical transmission of endobacteria in the arbuscular mycorrhizal fungus *Gigaspora margarita* through generation of vegetative spores. *Applied and Environmental Microbiology* 70:3600-3608.
Bianciotto, V., E. Lumini, P. Bonfante, and P. Vandamme (2003). *Candidatus Glomeribacter gigasporarum* gen. nov., sp. nov., an endosymbiont of arbuscular mycorrhizal fungi. *International Journal of Systematic and Evolutionary Microbiology* 53:121-124.
Bohrer, K.E., C.F. Friese, and J.P. Amon (2004). Seasonal dynamics of arbuscular mycorrhizal fungi in differing wetland habitats. *Mycorrhiza* 14:329-337.

Bonfante-Fasolo, P. and R. Grippiolo (1982). Ultrastructural and cytochemical changes in the wall of a vesicular-arbuscular mycorrhizal fungus during symbiosis. *Canadian Journal of Botany* 60:2303-2312.

Braunberger, P.G., L.K. Abbott, and A.D. Robson (1996). Infectivity of arbuscular mycorrhizal fungi after wetting and drying. *New Phytologist* 134:673-684.

Buée M., M. Rossignal, A. Jauneau, R. Ranjeva, and G. Bécard (2000). The pre-symbiotic growth of arbuscular mycorrhizal fungi is induced by a branching factor partially purified from plant root exudates. *Molecular Plant-Microbe Interaction* 13:693-698.

Carvalho, L.M., P.M. Correia, I. Cacador, and M.A. Martins-Loucao (2003). Effects of salinity and flooding on the infectivity of salt marsh arbuscular mycorrhizal fungi in *Aster tripolium* L. *Biology and Fertility of Soils* 38:137-143.

Christensen, H. and I. Jakobsen (1993). Reduction of bacterial growth by a vesicular-arbuscular mycorrhizal fungus in the rhizosphere of cucumber (*Cucumis sativus* L.). *Biology and Fertility of Soils* 15:253-258.

Cuenca, G., Z. De Andrade, M. Lovera, L. Fajardo, and E. Meneses (2004). The effect of two arbuscular mycorrhizal inocula of contrasting richness and the same mycorrhizal potential on the growth and survival of wild plant species from La Gran Sabana, Venezuela. *Canadian Journal of Botany* 82:582-589.

Daft, M.J., M.T. Chilvers, and T.H. Nicolson (1980). Mycorrhizas of the Liliiflorae. I. Morphogenesis of *Endymion non-scriptus* (L.) Garcke and its mycorrhizas in nature. *New Phytologist* 85:181-189.

Davies, F.T., V. Olalde-Portugal, L. Aguilera-Gomez, M.J. Alvarado, R.C. Ferrera-Cerrato, and T.W. Boutton (2002). Alleviation of drought stress of Chile ancho pepper (*Capsicum annuum* L. cv. San Luis) with arbuscular mycorrhiza indigenous to Mexico. *Scientia Horticulturae* 92:347-359.

de la Providencia, I.E., F.A. de Souza, F. Fernandez, N.S. Delmas, and S. Declerck (2005). Arbuscular mycorrhizal fungi reveal distinct patterns of anastomosis formation and hyphal healing mechanisms between different phylogenic groups. *New Phytologist* 165:261-271.

de Souza, F.A. and S. Declerck (2003). Mycelium development and architecture, and spore production of *Scutellospora reticulata* in monoxenic culture with Ri T- DNA transformed carrot roots. *Mycologia* 95:1004-1012.

Douds, D.D., Jr., L. Galvez, R.R. Janke, and P. Wagoner (1995). Effect of tillage and farming system upon populations and distribution of vesicular-arbuscular mycorrhizal fungi. *Agriculture, Ecosystems and Environment* 52:111-118.

Douds, D.D., Jr., G. Nagahashi, and G.D. Abney (1996). The differential effects of cell wall-associated phenolics, cell walls, and cytosolic phenolics of host and non-host roots on the growth of two species of AM fungi. *New Phytologist* 133:289-294.

Edwards, S.G., J.P.W. Young, and A.H. Fitter (1998). Interactions between *Pseudomonas fluorescens* biocontrol agents and *Glomus mosseae*, an arbuscular mycorrhizal fungus, within the rhizosphere. *FEMS Microbiology Letters* 166:297-303.

Filion, M., M. St-Arnaud, and J.A. Fortin (1999). Direct interaction between the arbuscular mycorrhizal fungus *Glomus intraradices* and different rhizosphere microorganisms. *New Phytologist* 141:525-533.

Fortin, J.A., G. Bécard, S. Declerck, Y. Dalpé, M. St-Arnaud, A.P. Coughlan, and Y. Piché (2002). Arbuscular mycorrhiza on root-organ cultures. *Canadian Journal of Botany* 80:1-20.

Franson, R.L., C. Hamel, D.L. Smith, and G.J. Bethlenfalvay (1994). Below-ground interactions between a seedling soybean and pre-established soybean plant with and without mycorrhizal fungi. 1. Plant biomass, root growth, and mycorrhizal colonization. *Agriculture Ecosystems and Environment* 49:131-138.

Franzluebbers, A.J., S.F. Wright, and J.A. Stuedemann (2000). Soil aggregation and glomalin under pastures in the Southern Piedmont USA. *Soil Science Society of America Journal* 64:1018-1026.

Frey, B. and H. Schüepp (1993). Acquisition of nitrogen by external hyphae of arbuscular mycorrhizal fungi associated with *Zea mays* L. *New Phytologist* 124:221-230.

Friese, C.F. and M.F. Allen (1991). The spread of VA mycorrhizal fungal hyphae in the soil: Inoculum types and external hyphal architecture. *Mycologia* 83:409-418.

Galvez, L., D.D. Douds, Jr., L.E. Drinkwater, and P. Wagoner (2001). Effect of tillage and farming system upon VAM fungus populations and mycorrhizas and nutrient uptake of maize. *Plant and Soil* 228:299-308.

Gavito, M.E. and M.H. Miller (1998). Early phosphorus nutrition, mycorrhizae development, dry matter partitioning and yield of maize. *Plant and Soil* 199: 177-186.

Gavito, M.E., P. Schweiger, and I. Jakobsen (2003). P uptake by arbuscular mycorrhizal hyphae: Effect of soil temperature and atmospheric CO_2 enrichment. *Global Change Biology* 9:106-116.

George, E., H. Marschner, and I. Jakobsen (1995). Role of arbuscular mycorrhizal fungi in uptake of phosphorus and nitrogen from soil. *Critcal Reviews in Biotechnology* 15:257-270.

Giovannetti, M., D. Azzolini, and A.S. Citernesi (1999). Anastomosis formation and nuclear and protoplasmic exchange in arbuscular mycorrhizal fungi. *Applied and Environmental Microbiology* 65:5571-5575.

Giovannetti, M., P. Fortuna, A.S. Citernesi, S. Morini, and M.P. Nuti (2001). The occurrence of anastomosis formation and nuclear exchange in intact arbuscular mycorrhizal networks. *New Phytologist* 151:717-724.

Giovannetti, M., C. Sbrana, L. Avio, A.S. Citernesi, and C. Logi (1993). Differential hyphal morphogenesis in arbuscular mycorrhizal fungi during pre-infection stages. *New Phytologist* 125:587-593.

Giovannetti, M., C. Sbrana, L. Avio, and P. Strani (2004). Patterns of below-ground plant interconnections established by means of arbuscular mycorrhizal networks. *New Phytologist* 164:175-181.

Giovannetti, M., C. Sbrana, P. Strani, M. Agnolucci, V. Rinaudo, and L. Avio (2003). Genetic diversity of isolates of *Glomus mosseae* from different geographic areas detected by vegetative compatibility testing and biochemical and molecular analysis. *Applied and Environmental Microbiology* 69:616-624.

Goss, M.J. and A. de Varennes (2002). Soil disturbance reduces the efficacy of mycorrhizal associations for early soybean growth and N_2 fixation. *Soil Biology and Biochemistry* 34:1167-1173.

Green, H., J. Larsen, P.A. Olsson, D.F. Jensen, and I. Jakobsen (1999). Suppression of the biocontrol agent *Trichoderma harzianum* by mycelium of the arbuscular mycorrhizal fungus *Glomus intraradices* in root-free soil. *Applied and Environmental Microbiology* 65:1428-1434.

Gryndler, M., J. Jansa, H. Hrselová, I. Chvatalova, and M. Vosátka (2003). Chitin stimulates development and sporulation of arbuscular mycorrhizal fungi. *Applied Soil Ecology* 22:283-287.

Gryndler, M., M. Vosátka, H. Hrselová, I. Chvatalova, and J. Jansa (2002). Interaction between arbuscular mycorrhizal fungi and cellulose in growth substrate. *Applied Soil Ecology* 19:279-200.

Hamel, C. (1992). Mycorrhizal effects on ^{15}N-transfer from legume to grass intercrops, plant growth and interspecific competition. *Dissertation Abstracts International. B, Sciences and Engineering* 53:1117B.

Hamel, C. (2004). Impact of arbuscular mycorrhizal fungi on soil microorganisms, N and P cycling and crop growth. *Canadian Journal of Soil Science* 84:383-395.

Hamel, C., U. Barrantes-Cartín, V. Furlan, and D.L. Smith (1991). Endomycorrhizal fungi in nitrogen transfer from soybean to maize. *Plant and Soil* 138:33-40.

Hamel, C., Y. Dalpé, C. Lapierre, R.R. Simard, and D.L. Smith (1994). Composition of the vesicular-arbuscular mycorrhizal fungi population in an old meadow as affected by pH, phosphorus and soil disturbance. *Agriculture Ecosystems and Environment* 49:223-231.

Hamel, C., C. Neeser, U. Barrantes-Cartín, and D.L. Smith (1991). Endomycorrhizal fungal species mediate ^{15}N transfer from soybean to corn in non-fumigated soil. *Plant and Soil* 138:41-47.

Hart, M.M. and J.N. Klironomos (2002). Diversity of arbuscular mycorrhizal fungi and ecosystem functioning. In *Mycorrhizal Ecology,* M.G.A. van der Heijden and I.R. Sanders (eds.). Berlin: Springer-Verlag, pp. 225-242.

Helgason, T., J.W. Merryweather, J. Denison, P. Wilson, J.P.W. Young, and A.H. Fitter (2002). Selectivity and functional diversity in arbuscular mycorrhizas of cooccurring fungi and plants from a temperate deciduous woodland. *Journal of Ecology* 90:371-384.

Hijri, M. and I.R. Sanders (2004). The arbuscular mycorrhizal fungus *Glomus intraradices* is haploid and has a small genome size in the lower limit of eukaryotes. *Fungal Genetics and Biology* 41:253-261.

Hodge, A. (2000). Microbial ecology of the arbuscular mycorrhiza. *FEMS Microbiology Ecology* 32:91-96.

Janos, D.P. (1980). Tropical succession. *Biotropica* 12 (supplement no. 2):1-95.

Jansa, J., A. Mozafar, T. Anken, R. Ruh, I.R. Sanders, and E. Frossard (2002). Diversity and structure of AMF communities as affected by tillage in a temperate soil. *Mycorrhiza* 12:225-234.

Jansa, J., A. Mozafar, S. Banke, B.A. McDonald, and E. Frossard (2002). Intra- and intersporal diversity of ITS rDNA sequences in *Glomus intraradices* assessed by cloning and sequencing, and by SSCP analysis. *Mycological Research* 106: 670-681.

Jansa, J., A. Mozafar, G. Kuhn, T. Anken, R. Ruh, I.R. Sanders, and E. Frossard (2003). Soil tillage affects the community structure of mycorrhizal fungi in maize roots. *Ecological Applications* 13:1164-1176.

Jasper, D.A., L.K. Abbott, and A.D. Robson (1991). The effect of soil disturbance on vesicular-arbuscular mycorrhizal fungi in soils from different vegetation types. *New Phytologist* 118:471-476.

Jeffries, P., S. Gianinazzi, S. Perotto, K. Turnau, and J.M. Barea (2003). The contribution of arbuscular mycorrhizal fungi in sustainable maintenance of plant health and soil fertility. *Biology and Fertility of Soils* 37:1-16.

Kabir, Z., I.P. O'Halloran, and C. Hamel (1997). Overwinter survival of arbuscular mycorrhizal hyphae is favored by attachment to roots but diminished by disturbance. *Mycorrhiza* 7:197-200.

Kabir, Z., I.P. O'Halloran, and C. Hamel (1999). Combined effects of soil disturbance and fallowing on plant and fungal components of mycorrhizal corn (*Zea mays* L.). *Soil Biology and Biochemistry* 31:307-314.

Kabir, Z., I.P. O'Halloran, J.W. Fyles, and C. Hamel (1998). Dynamics of the mycorrhizal symbiosis of corn (*Zea mays* L.): Effects of host physiology, tillage practice and fertilization on spatial distribution of extra-radical mycorrhizal hyphae in the field. *Agriculture Ecosystems and Environment* 68:151-163.

Kabir, Z., I.P. O'Halloran, P. Widden, and C. Hamel (1998). Vertical distribution of arbuscular mycorrhizal fungi under corn (*Zea mays* L.) in no-till and conventional tillage systems. *Mycorrhiza* 8:53-55.

Klein, D.A. and M.W. Paschke (2004). Filamentous fungi: The indeterminate lifestyle and microbial ecology. *Microbial Ecology* 47:224-235.

Klironomos, J.N. (2003). Variation in plant response to native and exotic arbuscular mycorrhizal fungi. *Ecology* 84:2292-2301.

Klironomos, J.N. and M.M. Hart (2002). Colonization of roots by arbuscular mycorrhizal fungi using different sources of inoculum. *Mycorrhiza* 12:181-184.

Koch, A.M., G. Kuhn, P. Fontanillas, L. Fumagalli, J. Goudet, and I.R. Sanders (2004). High genetic variability and low local diversity in a population of arbuscular mycorrhizal fungi. *Proceedings of the National Academy of Sciences of the United States of America* 101:2369-2374.

Kosuta, S., M. Chabaud, G. Lougnon, C. Gough, J. Dénarié, D.G. Barker, and G. Bécard (2003). A diffusible factor from arbuscular mycorrhizal fungi induces symbiosis-specific MtENOD11 expression in roots of *Medicago truncatula*. *Plant Physiology* 131:952-962.

Kuhn, G., M. Hijri, and I.R. Sanders (2001). Evidence for the evolution of multiple genomes in arbuscular mycorrhizal fungi. *Nature* 414:745-748.

Levy, A., B.J. Chang, L.K. Abbott, J. Kuo, G. Harnett, and T.J.J. Inglis (2003). Invasion of spores of the arbuscular mycorrhizal fungus *Gigaspora decipiens* by *Burkholderia* spp. *Applied and Environmental Microbiology* 69:6250-6256.

Linderman, R.G. and E.A. Davis (2001). Vesicular-arbuscular mycorrhiza and plant growth response to soil amendment with composted grape pomace or its water extract. *HortTechnology* 11:446-450.

Liu, A., B. Wang, and C. Hamel (2003). Arbuscular mycorrhiza colonization and development at suboptimal root zone temperature. *Mycorrhiza* 14:93-101.

Lum, M.R. and A.M. Hirsch (2002). Roots and their symbiotic microbes: Strategies to obtain nitrogen and phosphorus in a nutrient-limiting environment. *Journal of Plant Growth Regulation* 21:368-382.

Lupwayi, N.Z., M.A. Arshad, W.A. Rice, and G.W. Clayton (2001). Bacterial diversity in water-stable aggregates of soils under conventional and zero tillage management. *Applied Soil Ecology* 16:251-261.

Malcova, R., J. Albrechtova, and M. Vosátka (2001). The role of the extraradical mycelium network of arbuscular mycorrhizal fungi on the establishment and growth of *Calamagrostis epigejos* in industrial waste substrates. *Applied Soil Ecology* 18:129-142.

Marschner, P. and K. Baumann (2003). Changes in bacterial community structure induced by mycorrhizal colonisation in split-root maize. *Plant and Soil* 251: 279-289.

McGonigle, T.P. and M.H. Miller (2000). The inconsistent effect of soil disturbance on colonization of roots by arbuscular mycorrhizal fungi: A test of the inoculum density hypothesis. *Applied Soil Ecology* 14:147-155.

McGonigle, T.P., K. Yano, and T. Shinhama (2003). Mycorrhizal phosphorus enhancement of plants in undisturbed soil differs from phosphorus uptake stimulation by arbuscular mycorrhizae over non-mycorrhizal controls. *Biology and Fertility of Soils* 37:268-273.

Menéndez, A.B., J.M. Scervino, and A.M. Godeas (2001). Arbuscular mycorrhizal populations associated with natural and cultivated vegetation on a site of Buenos Aires province, Argentina. *Biology and Fertility of Soils* 33:373-381.

Merryweather, J.W. and A.H. Fitter (1998). Patterns of arbuscular mycorrhiza colonisation of the roots of *Hyacinthoides non-scripta* after disruption of soil mycelium. *Mycorrhiza* 8:87-91.

Miller, M.H. (2000). Arbuscular mycorrhizae and the phosphorus nutrition of maize: A review of Guelph studies. *Canadian Journal of Plant Science* 80:47-52.

Minerdi, D., V. Bianciotto, and P. Bonfante (2002). Endosymbiotic bacteria in mycorrhizal fungi: From their morphology to genomic sequences. *Plant and Soil* 244:211-219.

Mohammad, M.J., W.L. Pan, and A.C. Kennedy (1998). Seasonal mycorrhizal colonization of winter wheat and its effect on wheat growth under dryland field conditions. *Mycorrhiza* 8:139-144.

Morin, F., J.A. Fortin, C. Hamel, R.L. Granger, and D.L. Smith (1994). Apple rootstock response to vesicular-arbuscular mycorrhizal fungi in a high phosphorus soil. *Journal of the American Society for Horticultural Science* 119:578-583.

Mozafar, A., T. Anken, R. Ruh, and E. Frossard (2000). Tillage intensity, mycorrhizal and nonmycorrhizal fungi, and nutrient concentrations in maize, wheat, and canola. *Agronomy Journal* 92:1117-1124.

Olsson, P.A., E. Bååth, I. Jakobsen, and B. Söderström (1996). Soil bacteria respond to presence of roots but not to mycelium of arbuscular mycorrhizal fungi. *Soil Biology and Biochemistry* 28:463-470.

Olsson, P.A., I. Thingstrup, I. Jakobsen, and F. Bååth (1999). Estimation of the biomass of arbuscular mycorrhizal fungi in a linseed field. *Soil Biology and Biochemistry* 31:1879-1887.

Palenzuela, J., C. Azcón-Aguilar, D. Figueroa, F. Caravaca, A. Roldan, and J.M. Barea (2002). Effects of mycorrhizal inoculation of shrubs from Mediterranean ecosystems and composted residue application on transplant performance and mycorrhizal developments in a desertified soil. *Biology and Fertility of Soils* 36:170-175.
Pawlowska, T.E. and J.W. Taylor (2004). Organization of genetic variation in individuals of arbuscular mycorrhizal fungi. *Nature* 427:733-737.
Peltonen-Sainio, P. (1999). Growth and development of oat with special reference to source-sink interaction and productivity. In *Crop Yield: Physiology and Processes,* D.L. Smith and C. Hamel (eds.). Berlin, Heidelberg: Springer-Verlag, pp. 39-66.
Pfeffer, P.E., D.D. Douds, H. Bucking, D.P. Schwartz, and Y. Shachar-Hill (2004). The fungus does not transfer carbon to or between roots in an arbuscular mycorrhizal symbiosis. *New Phytologist* 163:617-627.
Read, D.J. and J. Perez-Moreno (2003). Mycorrhizas and nutrient cycling in ecosystems—a journey towards relevance? *New Phytologist* 157:475-492.
Redecker, D. (2002). Molecular identification and phylogeny of arbuscular mycorrhizal fungi. *Plant and Soil* 244:67-73.
Reynolds, H.L., A. Packer, J.D. Bever, and K. Clay. (2003). Grass roots ecology: Plant-microbe-soil interactions as drivers of plant community structure and dynamics. *Ecology* 84:2281-2291.
Rillig, M.C. and P.D. Steinberg (2002). Glomalin production by an arbuscular mycorrhizal fungus: A mechanism of habitat modification? *Soil Biology and Biochemistry* 34:1371-1374.
Rillig, M.C., S.F. Wright, K.A. Nichols, W.F. Schmidt, and M.S. Torn (2001). Large contribution of arbuscular mycorrhizal fungi to soil carbon pools in tropical forest soils. *Plant and Soil* 233:167-177.
Sanders, I.R. (2002). Specificity in the arbuscular mycorrhizal symbiosis. In *Mycorrhizal Ecology: Ecological Studies,* M.G.A. van der Heijden and I.R. Sanders (eds.). Berlin: Springer-Verlag, pp. 415-437.
Sanders, I.R., J.P. Clapp, and A. Wiemken (1996). The genetic diversity of arbuscular mycorrhizal fungi in natural ecosystems—a key to understanding the ecology and functioning of the mycorrhizal symbiosis. *New Phytologist* 133:123-134.
Sanders, I.R., A. Koch, and G. Kuhn (2003). Arbuscular mycorrhizal fungi: Genetics of multigenomic, clonal networks and its ecological consequences. *Biological Journal of the Linnean Society* 79:59-60.
Schubert, A., C. Marzachi, M. Mazzitelli, M.C. Cravero, and P. Bonfante-Fasolo (1987). Development of total and viable extraradical mycelium in the vesicular-arbuscular mycorrhizal fungus *Glomus clarum* Nicol. & Schenck. *New Phytologist* 107:183-190.
Simard, S.W., M.D. Jones, and D.M. Durall (2002). Carbon and nutrient fluxes within and between mycorrhizal plants. Diversity of arbuscular mycorrhizal fungi and ecosystem functioning. In *Mycorrhizal Ecology,* M.G.A. van der Heijden and I.R. Sanders (eds.). Berlin: Springer-Verlag, pp. 33-74.

Smith, S.E. and G.D. Bowen (1979). Soil temperature, mycorrhizal infection and nodulation of *Medicago truncatula* and *Trifolium subterraneum*. *Soil Biology & Biochemistry* 11:469-473.

Smith, S.E. and D.J. Read (1997). *Mycorrhizal Symbiosis*, London: Academic Press, 605 pp.

Smith, S.E., F.A. Smith, and I. Jakobsen (2004). Functional diversity in arbuscular mycorrhizal (AM) symbioses: The contribution of the mycorrhizal P uptake pathway is not correlated with mycorrhizal responses in growth or total P uptake. *New Phytologist* 162:511-524.

Staddon, P.L., C.B. Ramsey, N. Ostle, P. Ineson, and A.H. Fitter (2003). Rapid turnover of hyphae of mycorrhizal fungi determined by AMS microanalysis of C-14. *Science* 300:1138-1140.

Stribbley, D.P. and R.C. Snellgrove (1986). Vesicular-arbuscular mycorrhizas and the phosphorus physiology of plants. *Journal of the Science of Food and Agriculture* 37:11-12.

St-Arnaud, M., C. Hamel, B. Vimard, M. Caron, and J.A. Fortin (1996). Enhanced hyphal and spore production of the arbuscular mycorrhizal fungus *Glomus intraradices* in an *in vitro* system in the absence of host roots. *Mycological Research* 100:328-332.

St-Arnaud, M., C. Hamel, B. Vimard, M. Caron, and J.A. Fortin (1997). Inhibition of *Fusarium oxysporum* f.sp. *dianthi* in the non-VAM species *Dianthus caryophyllus* by co-culture with *Tagetes patula* companion plants colonized by *Glomus intraradices*. *Canadian Journal of Botany* 75:998-1005.

Suriyapperuma, S.P. and R.E. Koske (1995). Attraction of germ tubes and germination of spores of the arbuscular mycorrhizal fungus *Gigaspora gigantea* in the presence of roots of maize exposed to different concentrations of phosphorus. *Mycologia* 87:772-778.

Sylvia, D.M., A.K. Alagely, M.E. Kane, and N.L. Philman (2003). Compatible host/mycorrhizal fungus combinations for micropropagated sea oats—I. Field sampling and greenhouse evaluations. *Mycorrhiza* 13:177-183.

Tawaraya, K. (2003). Arbuscular mycorrhizal dependency of different plant species and cultivars. *Soil Science and Plant Nutrition* 49:655-668.

Taylor, T.N., W. Remy, H. Hass, and H. Kerp (1995). Fossil arbuscular mycorrhizae from the Early Devonian. *Mycologia* 87:560-573.

Treseder, K.K. and M.F. Allen (2002). Direct nitrogen and phosphorus limitation of arbuscular mycorrhizal fungi: A model and field test. *New Phytologist* 155: 507-515.

Tsai, S.M. and D.A. Phillips (1991). Flavonoids released naturally from alfalfa promote development of symbiotic *Glomus* spores *in vitro*. *Applied and Environmental Microbiology* 57:1485-1488.

van Aarle, I.M., P.A. Olsson, and B. Söderström (2002). Arbuscular mycorrhizal fungi respond to the substrate pH of their extraradical mycelium by altered growth and root colonization. *New Phytologist* 155:173-182.

van Aarle, I.M., B. Söderström, and P.A. Olsson (2003). Growth and interactions of arbuscular mycorrhizal fungi in soils from limestone and acid rock habitats. *Soil Biology and Biochemistry* 35:1557-1564.

Vancura, V., M.O. Orozco, O. Grauová, and Z. Prikryl (1990). Properties of bacteria in the hyphosphere of a vesicular-arbuscular mycorrhizal fungus. *Agriculture Ecosystems and Environment* 29:421-427.
Vandenkoornhuyse, P., C. Leyval, and I. Bonnin (2001). High genetic diversity in arbuscular mycorrhizal fungi: Evidence for recombination events. *Heredity* 87: 243-253.
van der Heijden, M.G.A. (2002). Arbuscular mycorrhizal fungi as a determinant of plant diversity: In search of underlying mechanisms and general principles. In *Mycorrhizal Ecology: Ecological Studies,* M.G.A. van der Heijden and I.R. Sanders (eds.). Berlin: Springer-Verlag, pp. 243-265.
van der Heijden, M.G.A., J.N. Klironomos, M. Ursic, P. Moutoglis, R. Streitwolf-Engel, T. Boller, A. Wiemken, and I.R. Sanders (1998). Mycorrhizal fungal diversity determines plant biodiversity, ecosystem variability and productivity. *Nature* 396:69-72.
Vivas, A., A. Marulanda, M. Gomez, J. M. Barea, and R. Azcón (2003). Physiological characteristics (SDH and ALP activities) of arbuscular mycorrhizal colonization as affected by *Bacillus thuringiensis* inoculation under two phosphorus levels. *Soil Biology and Biochemistry* 35:987-996.
Vivas, A., A. Marulanda, J.M. Ruiz-Lozano, J.M. Barea, and R. Azcón (2003). Influence of a *Bacillus* sp on physiological activities of two arbuscular mycorrhizal fungi and on plant responses to PEG-induced drought stress. *Mycorrhiza* 13: 249-256.
Volkmar, K.M. and W. Woodbury (1989). Effects of soil temperature and depth on colonization and root and shoot growth of barley inoculated with vesicular-arbuscular mycorrhizae indigenous to Canadian prairie soil. *Canadian Journal of Botany* 67:1702-1707.
Vos, J. (1999). Potato. In *Crop Yield: Physiology and Processes,* D.L. Smith and C. Hamel (eds.). Berlin, Heidelberg: Springer-Verlag, pp. 39-66.
Vosátka, M., H. Vejsadová, M. Gryndler, and H. Hrselová (1995). Effect of *Agrobacterium radiobacter* on vesicular-arbuscular-mycorrhizal infection and external mycelium of maize. *Folia Microbiologica* 40:100-103.
Wacker, T.L., G.R. Safir, and S.N. Stephenson (1990). Evidence for succession of mycorrhizal fungi in Michigan asparagus fields. *Acta Agriculturae ISHS* 271: 273-278.
Wang, B., D.M. Funakoshi, Y. Dalpé, and C. Hamel (2002). Phosphorus-32 absorption and translocation to host plants by arbuscular mycorrhizal fungi at low root-zone temperature. *Mycorrhiza* 12:93-96.
Wright, S.F. and A. Upadhyaya (1998). A survey of soils for aggregate stability and glomalin, a glycoprotein produced by hyphae of arbuscular mycorrhizal fungi. *Plant and Soil* 198:97-107.
Zhang, F., C. Hamel, H. Kianmehr, and D.L. Smith (1995). Root-zone temperature and soybean [Glycine max. (L.) Merr.] vesicular-arbuscular mycorrhizae: Development and interactions with the nitrogen fixing symbiosis. *Environmental and Experimental Botany* 35:287-298.

Chapter 2

Soil Nutrient and Water Providers: How Arbuscular Mycorrhizal Mycelia Support Plant Performance in a Resource-Limited World

Aiguo Liu
Christian Plenchette
Chantal Hamel

Arbuscular mycorrhizal (AM) fungi are an essential component of natural soil-plant systems. While it is possible to design systems devoid of AM fungi, as in greenhouse production, for example, it is clear that cropping systems that fully utilize AM symbiosis are the most efficient, sustainable, and environmentally sound large-scale methods for food production. The contribution of AM fungi to soil structural quality—which promotes water infiltration, limits erosion, and creates soil heterogeneity that is conducive to biodiversity—is well recognized. Most important, AM fungi can extract soil nutrients and water efficiently, allowing good crop yields to be produced from soils with limited fertility. Thus, AM fungi are essential for the design of cropping systems that have a low impact on the environment. Advances in molecular biology techniques have made it possible to develop diagnostic tools to estimate the potential contribution of AM fungi in a soil to a given crop yield. The determination of soil mycorrhizal potential as a component of routine soil testing will lead to more accurate fertilizer recommendations and the safe reduction of

Mycorrhizae in Crop Production
© 2007 by The Haworth Press, Inc. All rights reserved.
doi:10.1300/5425_02

fertilizer inputs. The production of good yields with reduced fertilizer use will translate into reduced nutrient seepage from farmland into the environment. A good understanding of how AM fungi contribute to plant nutrient uptake will facilitate the calibration of fertilizer recommendations in accordance with this novel soil testing approach as well as the design of efficient and environmentally sound cropping systems. This chapter summarizes the current state of knowledge of nutrient and water uptake by AM fungal mycelia.

PHOSPHORUS UPTAKE

P in Soil

Phosphorus is not a rare element in the Earth's crust: it is 11th in order of abundance. However, its quantity in soil varies considerably depending on the nature of the parent material, the degree of weathering, and the extent to which P has been lost through leaching and runoff. There is therefore much variability in the soil P content reported by different authors. Tisdale, Nelson, and Beaton (1985) and Barber (1995) report that the P content in surface soils varies between 200-1,000 mg P kg^{-1} and 200-5,000 mg P kg^{-1} respectively, while Stevenson and Cole (1986) report values from as little as 100 mg P kg^{-1} in sandstone to over 2,000 mg P kg^{-1} in high-phosphate limestone.

Although phosphorus is relatively abundant in soil, total soil P content bears little relation to the amount of P that is available to plants (Stevenson and Cole, 1986). In soil, phosphorus is mainly chemically fixed, sorbed to clay, or linked with Al, Fe, or Ca in often highly insoluble compounds or bound in organic complexes; P is unavailable to plants in both circumstances, as plants absorb only phosphorus anions from the soil solution. Phosphorus is certainly one of the least available plant nutrients in the rhizosphere, and phosphorus deficiency is one of the major limitations in crop production in numerous parts of the world, particularly in the tropics.

As a result of the strong P fixation power of soils, the size of the bioavailable P pool is not sufficient to ensure maximum crop yields (Fardeau, 1996). Also, soil solution, which usually has an average concentration of 0.05 mg L^{-1} (Tisdale, Nelson, and Beaton, 1985), rarely exceeds 0.31 mg L^{-1} in non-fertilized soils (Bieleski, 1973). Soil solution P is in equilibrium with P sorbed to the solid phase. This equilibrium is regulated by the buffering power of the soil (Barber, 1980). Soil solution P is rapidly absorbed by plants, and a P depletion zone, as large as the length of the root hairs, arises around the roots (Lewis and Quirk, 1967; Nye and Tinker, 1977). To replenish this zone, P ions need to be desorbed and transported toward the roots. Two mechanisms are involved in depletion zone replenishment: P diffusion and mass flow. The presence of P depletion zones in the soil around roots shows that P uptake by plants is faster than its supply by diffusion or mass flow. Because of the low concentration of P in solution, mass flow, that is, the quantity of ions transported by the water absorbed by plants, provides less than 5 percent of plant P requirements. On the scale of a growing period diffusion occurs over only a few millimeters, since the diffusion constant of P ions in soil is very low ($D = 10^{(-8-11)}$ cm^2/s) (Barber, Walker, and Vasey, 1963).

To meet their P requirements, plants need to grow more roots that are able to intercept a maximum of P ions. It is well-known that better-rooted species have the ability to take up more P. On one hand, there is the soil P supply, which can be described as the quantity of P ions able to leave the soil. On the other, there is the plant with its P requirements, which can be described as the quantity of P ions that may enter the plant. In an ideal situation, these two quantities would be equal. In fact, this is rarely the case, and fertilizer needs to be applied to supplement soil P supply capacity. In most cases, however, only 15 percent or less of P fertilizers are used by crops during the year of application, the remaining part being sorbed, fixed, or retrograded (Morel, 1988).

This panoramic view of the behavior of phosphorus will help us understand the role of extraradical arbuscular mycorrhizal mycelia.

Absorption

Plant species fulfill their P requirements using different strategies: rapid root growth, root elongation, root hair proliferation, and modifications of rhizosphere conditions (Smith, Dickson, and Smith, 2001). AM roots, in addition, have external mycelia that can be considered an extension of the roots. Mycorrhizal plants have been shown to absorb more P than non-mycorrhizal plants (Gerdemann, 1964; Daft and Nicolson, 1966; Baylis, 1967). Evidence that AM roots are more efficient at P uptake is provided by the higher P concentration in AM plant tissues. It has been shown that AM roots have an increased efficiency of absorption per unit of root tissue and that they are capable of maintaining this rate of absorption over longer periods than non-mycorrhizal roots are capable of doing (Smith et al., 1986). Non-mycorrhizal plants absorb rhizospheric P rapidly, and a P depletion zone occurs around the roots. Studies using ^{32}P have demonstrated the ability of external mycelia to absorb P ions at a distance well beyond the P depletion zone (Pearson and Jakobsen, 1993). The efficiency of external hyphae at acquiring P from soil solution with a low P concentration is certainly related to their small diameter (<10 μm) compared to that of roots or root hairs, which reduces the distance of P ion diffusion and the formation of P depletion zones. The alleviation of P depletion zones allows for continuous uptake during the growing period. Moreover, the thin hyphae proliferate extensively and penetrate smaller pores of soil than roots do, with the result that the volume of soil prospected by a root system is larger in mycorrhizal plants than in non-mycorrhizal plants.

As they grow, roots and extraradical AM hyphae intercept soil volume that contains orthophosphate ions, which are taken up by the roots and hyphae. With their high surface-to-volume ratio and high turnover rate AM hyphae efficiently increase the volume of soil that is exploited during plant development. Hyphae take up orthophosphate ions and translocate most of this P to their host plant. It has been hypothesized that mycorrhizal plants can access pools of soil P that are unavailable to non-mycorrhizal plants. Tracer studies have shown no differences in the specific activity of mycorrhizal and non-mycorrhizal plants despite a large difference in P uptake indicating that, while all the plants were obtaining their P from the same soil P pools, the mycorrhizal plants were extracting P more effectively (Plenchette and Morel, 1994; Morel and Plenchette, 1994). Not only are external hyphae extremely efficient absorbing organs, they also contribute to the P nutrition of plants by increasing the soil P supply in two other ways. It has been shown that mycorrhizae can acidify the rhizosphere through increased proton efflux and pCO_2 enhancement (Rigou and Mignard, 1994). Of course, these mechanisms act in calcareous soils and release P that is bound to calcium. In acid soils, P is mainly bound to Fe and Al, and chelating agents excreted in greater quantities by mycorrhizal roots can also release some bioavailable P and contribute to P supply enhancement (Haselwandter, 1995). Phosphatases are also excreted in greater quantities in the mycorrhizosphere and contribute to the mobilization of organic P (Tarafdar and Marschner, 1994; Koide and Kabir, 2000). The equilibrium between P in soil solution and P in solid phase is controlled by the buffering power of the soil. Greater P uptake by mycorrhizal plants induces faster replenishment of the P depletion zone and therefore increases the soil P supply, since mycorrhizae are particularly efficient in high-P-fixing soils (Plenchette and Fardeau, 1988).

Epstein (1972) showed that there are two uptake systems for P: a low-affinity uptake system operating at a high concentration of soil solution (>1 μM), and a high-affinity system requiring energy when the concentration of the soil solution is low (<1 μM). Both systems have also been shown to operate with mycorrhizal plants (Schweiger and Jakobsen, 1999).

Orthophosphate uptake by AM hyphae operates through energy-dependent systems, as the large concentration gradient that exists between the low concentration of the soil solution and the high concentration in the fungal cytoplasm needs to be overcome. AM fungal hyphae have been found to possess high-affinity transport systems for orthophosphate uptake (Ezawa, Smith, and Smith, 2002) that allow them to take up orthophosphate at soil solution concentrations below the threshold concentration required for P uptake by most plant roots (Mosse, Hayman, and Arnold, 1973). Within AM hyphae orthophosphate is rapidly transformed into polyphosphate, a transformation that helps maintain effective hyphal orthophosphate uptake. In P-starved *Glomus intraradices,* N.C. Schenck & G.S. Sm. polyphosphate accumulation was detectable five hours after P application (Viereck, Hansen, and Jakobsen, 2004). In P-starved *Archeospora leptoticha,* hyphal polyphosphate content doubled within one to three hours after P application (Ezawa et al., 2004). The proportion of hyphae containing polyphosphate increased from 25 percent to a maximum of 50 percent; it is possible that 50 percent was the proportion of active hyphae in the system. AM hyphae turn over rapidly (Staddon et al., 2003), as a result of which a large proportion of the network is normally inactive (Kabir et al., 1998; Schubert et al., 1987).

Transport

Within hyphae, polyphosphates as well as orthophosphates and some organic phosphates appear to be packaged into vacuoles that serve as storage spaces (spherical vacuoles) or vehicles (tubular vacuoles) (Ashford and Allaway, 2002). It appears that P is translocated within these vehicle vacuoles along extraradical AM hyphae (Uetake et al., 2002), which are non-septate, up to the arbuscules (Ezawa et al., 2004). Evidence obtained in ectomycorrhizal systems suggests that these P-filled tubular membranes move by means of cytoplasmic streaming and ultimately reach the arbuscules, where the P can be broken down enzymatically and moved across the symbiotic interface into the host cells (Ashford and Allaway, 2002).

P unloading at the plant-fungus interface could be coupled with hexose acquisition by the hyphae. These hexoses are transformed into storage lipids and translocated throughout the AM hyphal network as triacylglycerols in bodies. Multiphoton microscopy has revealed these bodies, called oleosomes, as they move bi-directionally within AM hyphal tubes (Bago et al., 2002). The motion of oleosomes, similar to that of tubular membranes but in contrast to the movement of nuclei, seems related to cytoplasmic streaming. The higher abundance of oleosomes, observed in AM hyphae close to plant roots than in hyphae in distal areas, suggests that stored lipids are being used as they travel away from their source. Using ^{13}C-labeling in a compartmented-pot experiment, Gavito and Olsson (2003) measured less photosynthesis-derived ^{13}C in AM hyphae in nutrient-rich compartments than in hyphae growing in hyphal compartments filled with washed quartz sand. This finding supports the role of oleosomes as providers of the energy required for hyphal uptake of nutrients. The low concentration of oleosomes in the branched absorbing structures is consistent with the presumed absorbing role of these hyphae.

Some reports attribute 70 to 80 percent of plant P to AM fungal uptake and translocation (Smith and Read, 1997), while others report a smaller contribution, such as 38 percent in wheat (*Triticum aestivum* L.) (Yao et al., 2001). Using ^{33}P, Smith, Smith, and Jakobsen (2004) found that *G. intraradices* delivered 100 percent of the P in three plant species with varying mycorrhizal dependency, while *Glomus caledonium* (T.H. Nicolson & Gerd.) Trappe & Gerd. and *Gigaspora rosea* T.H. Nicolson & N.C. Schenck delivered lesser amounts. In this experiment, the amounts of AM fungi-derived P in plants were unrelated to colonization, growth, and P response.

The factors related to the importance of AM contribution to plant P uptake have been well reviewed by Smith and Read (1997). P inflow in host plants is often related to the density in soil of the associated extraradical AM hyphae, which represent the fungal portion of mycorrhizal uptake capacity, and sometimes to root colonization, the bottleneck through which P is transferred to plants. The size of the extraradical AM network and the extent of root colonization depend on the fungal and plant species as well as on environmental conditions. The rate of P transfer from AM mycelia to root cells varies with the fungal or plant species. The ability of the root itself to take up P is an obvious determinant of plant response to AM colonization. This ability varies with the root transporters' affinity for orthophosphate (Harrison and van Buuren, 1995) as well as with the extent of root development. The general postulate that AM fungi are more important to plant species with thick, unbranched roots bearing few or short root hairs than to species with finely branched roots bearing long and abundant root hairs has usually proven to be true. P availability is known to control the extent of arbuscular mycorrhizae and root development. High P availability inhibits mycorrhizal development and stimulates root growth. The percentage of viable hyphae in a network, a factor that certainly affects the network's function, was shown to decline with soil depth (Kabir et al., 1998) and may well be affected by other environmental conditions such as soil pH or aeration or by the abundance of organisms, such as collembolans, that graze on fungi.

NITROGEN UPTAKE

It has only been recently realized that AM fungi could have a significant impact on N uptake in non-legume plant species. The role and impact of extraradical AM mycelia in the uptake of N are not well defined. N uptake by mycorrhizae is much more complex than P uptake. Both plant roots and AM hyphae can take up N easily. Plant-available N exists in more than one form, and N nutrition has been shown to influence the development of both AM mycelia and roots. Furthermore, these effects are modulated by the level of soil-available

P and the effect of P nutrition on plant physiology. And, of course, the contribution of AM fungi to plant N nutrition also depends on the genotypes of the plants and fungi involved in the symbiosis. A careful examination of the literature on the effects of extraradical AM mycelia on plant N nutrition, including the effect of available N levels on the development and uptake capacity of these mycelia, may help clarify this question.

Inorganic N Ion Uptake and Translocation

Two main forms of mineral N exist in soil. Both forms, ammonium (NH_4^+) and nitrate (NO_3^-) can be used by plants, but the NH_4^+ ion binds to and is retained on soil particles, which generally have a negative charge. Therefore, NH_4^+ is less readily available than NO_3^-, which is not retained on cation exchange sites. NO_3^- is more soluble and more mobile in soil and is therefore the main form of N assimilated by plants (Jansson and Persson, 1982). This is particularly true in agricultural soils where NO_3^--N predominates, as NH_4^+ is rapidly nitrified at pH near neutrality.

It has recently been shown that AM fungi can contribute significantly to plant N nutrition in soil where NO_3^- N dominates, by increasing plant NO_3^- uptake, the percentage of plant N derived from fertilizer, and plant N use efficiency (Azcón, Ruiz-Lozano, and Rodríguez, 2001). The effects of AM fungi were unexpected given the mobile nature of the NO_3^- ion. There should be little benefit in having an extensive root system for the uptake of nutrients that are readily mobile in soil such as NO_3^-. In contrast to P, which moves in soil mainly by diffusion, N is largely carried by mass flow (Table 2.1), and transpiration can bring to the roots large amounts of N in particular

TABLE 2.1. Relative importance of root interception, diffusion, and mass flow in the supply of nutrients to maize (Calculated from Jungk, 1996, p. 531).

	Proportion (%) of the nutrients supply by		
	Root interception	Mass flow	Diffusion
Nitrogen	1.1	78.9	20.0
Phosphorus	2.5	5.0	92.5
Potassium	2.1	17.9	80.0

NO_3^-, which is more readily moved than NH_4^+ (Jungk, 1996). Furthermore, the solubility of NO_3^- is such that its diffusion rate is much faster than that of P, and NO_3^- depletion zones are unlikely to form around plant roots (Figure 2.1). Consequently, fairly homogenous levels of NO_3^- are expected in the root zone, and the root densities needed to draw soil-available N are much lower than those required for adequate plant P nutrition. Theoretically, the presence of AM hyphae should not improve the N nutrition of plants growing where NO_3^- predominates, as there is little impediment to NO_3^- uptake by the roots unless the effect of AM hyphae is not that of a NO_3^- transporter. Contrary to common belief, Azcón, Ruiz-Lozano, and Rodríguez (2001) have clearly shown using ^{15}N techniques, that AM fungi may greatly improve plant NO_3^- uptake and use efficiency. This may simply demonstrate that at least some plants need mycorrhizae to function properly.

Germinating spores of eight AM species belonging to four genera have been shown to possess NO_3^- reductase activity (Tilak and

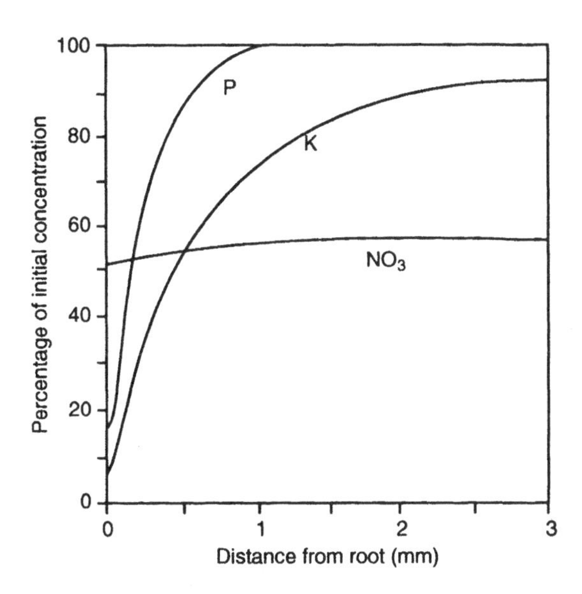

FIGURE 2.1. Calculated concentration gradients around a plant root in soil when NO_3^-, $H_2PO_4^-$ and K^- are carried mainly by diffusion (Redrawn from Jungk, 1996, p. 538).

Dwivedi, 1990), suggesting that the ability to use NO_3^- is a general trait of AM fungi. AM fungi have been found to metabolize NH_4^+ as well, through glutamine synthetase activity (Smith et al., 1985; Toussaint, St-Arnaud, and Charest, 2004). In vitro, *G. intraradices* was shown to take up NO_3^- and NH_4^+ to convert these N forms into arginine, which was translocated to intraracidal AM hyphae (Govindarajulu et al., 2005). Arginine could then be broken down by urease and ornithine aminotransferase, and the ammonia liberated passed into plant tissues through ammonia channels. In a similar in vitro system, Toussaint, St-Arnaud, and Charest (2004) observed translocation of ^{15}N from $^{15}NH_4$ but not from $^{15}NO_3$ to the carrot (*Daucus carota* L.) root culture supporting it. *G. intraradices* hyphae took up and translocated to the plant ^{15}N from NO_3^- and NH_4^+ sources equally well in compartmentalized growth containers (Johansen, Jakobsen, and Jensen, 1993). These results suggest that NH_4^-N could be assimilated preferentially by AM fungi, which—at least sometimes—also have the ability to use NO_3^-N. Less energy expenditure is involved in the use of NH_4^-N, as it is already in a reduced form. However, it appears that NH_4^+ may have a direct toxic effect on extraradical AM hyphal length development, reducing the mycorrhizal contribution to plant N nutrition (Hawkins and George, 2001). NH_4^+ may also have a direct deleterious effect on plant growth and nutrition, at least in some species. Consequently, results from experiments involving NH_4^+ as the sole N source should be regarded with caution. Figure 2.2 summarizes the interaction between roots, AM hyphae, and soil N as it relates to N uptake.

Mechanisms for Plant N Uptake Enhancement

Some studies have shown that AM mycelia can absorb and translocate large amounts of N to the host plant. In an experiment under controlled conditions, 30 percent of the N measured in maize (*Zea mays* L.) was attributed to the action of AM hyphae (George, Marschner, and Jakobsen, 1995). Longer extraradical mycelia have been associated with better NH_4^+ and NO_3^- uptake and translocation to the host plant (Azcón, Ruiz-Lozano, and Rodríguez, 2001; Hawkins and George, 2001), but the mycorrhizal effects on NO_3^- acquisition have sometimes been attributed to better plant P nutrition, as NO_3^- reduction has a high requirement for adenosine triphosphate because it is

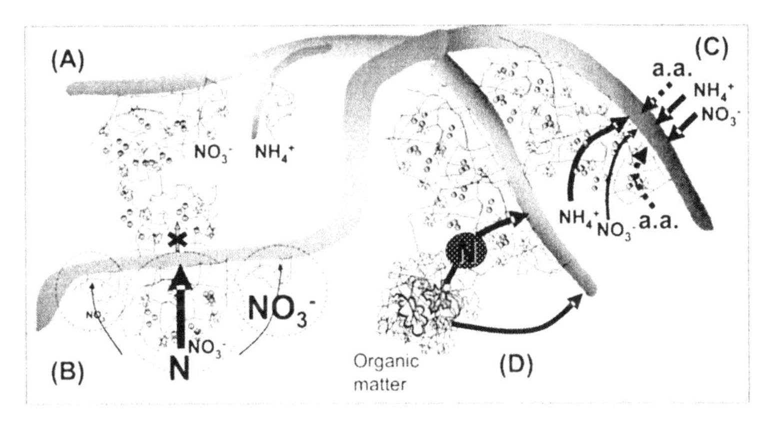

FIGURE 2.2. Graphical summary of the process of N uptake from different sources by arbuscular mycorrhizae. The size of the arrow is proportional to N uptake and a dashed line indicates that we know little about the importance of the N uptake pathway. (A) NH_4^- reduces extraradical AM mycelium development as compared to NO_3^-N (Hawkins and George, 2001). (B) Levels of N supply that are too high (Johansen et al., 1994; Hawkings and George, 2001; Azcón et al., 2003) or too low (Liu et al., 2000) reduce extraradical AM development and N uptake by plants. (C) The ability of plant roots and AM hyphae to absorb NH_4^+, NO_3^- (Hawkins and George, 2001) and amino acids (a.a.) (Hawkins, Johansen, and George, 2000) was shown; it seems that AM hyphae preferentially use NH_4^+ as compared to NO_3^- (Toussaint et al., 2004). (D) AM hyphae were seen to proliferate preferentially in the vicinity of organic matter patches (Hodge et al., 2001), compete effectively against soil microorganisms for the uptake of the N released through mineralization (Hodge et al., 2001; Feng et al., 2002), translocate N from organic sources to the plant (Ames, Reid et al., 1983; Hodge et al., 2001), stimulate mineralization (Hodge, 2001; Hodge et al., 2001), and improve N uptake by roots (Hodge, 2003a,b).

an energy intensive process. No unique pattern for the impact of AM fungi on host-plant N nutrition emerges from the literature at this time. AM fungal development can be limited by N availability, as N is also essential for fungal growth (Treseder and Allen, 2002). High soil N levels can reduce the development of the extraradical mycelia of AM fungi, and this in turn reduces the ability of the mycorrhizae to take up nutrients, including mineral N (Azcón, Ambrosano, and Charest, 2003; Hawkins and George, 2001; Johansen, Jakobsen, and Jensen,1994). It seems that N levels that are too low are no better than levels that are too high, as both situations reduce hyphal development and nutrient uptake (Liu, Hamel, Hamilton, and Smith, 2000). The observation of plant N deficiency associated with reduced hyphal length and ^{15}N uptake by hyphae (Hawkins and George, 2001) supports this hypothesis. A further complication stems from the fact that the effect of N fertilization seems to be differentially expressed depending on soil P level, AM fungal species (Azcón, Ambrosano, and Charest, 2003), and plant genotype (Liu, Hamel, Hamilton, and Smith, 2000). It has been postulated that reduced N uptake by mycorrhizal plants at unsuitable soil N levels can also be attributed to an AM fungi-induced modification of plant ionic regulation, leading to changes in root-shoot transport dynamics, or to a modification of the rhizosphere environment with an impact on soil microorganisms (Azcón, Ambrosano, and Charest, 2003). What is clear, however, is that inadequate levels of N or P can reduce the uptake of nutrients, including N, P, K, Ca, Fe, Cu, Zn, and Mn, in mycorrhizal plants (Azcón, Ambrosano, and Charest, 2003; Liu, Hamel, Hamilton, Ma, and Smith, 2000).

Inorganic N Sources and AM Fungi

Large amounts of N are stored in soil organic matter, and much of the N used by crops can be derived from crop residues or from organic amendments such as animal manure. A study by Frey and Schuëpp (1993) revealed no evidence for organic N use by *G. intraradices*. However, contrasting results have also been obtained. A ^{15}N method was used to determine whether mycorrhizal and non-mycorrhizal cotton plants take up N from fertilizer and from the soil N pool in the same proportion (Feng et al., 2002). Mycorrhizal colonization increases the availability of soil N, as demonstrated by the higher A_N value of mycorrhizal cotton plants (*Gossipium arboreum* L.) grown in 60 Co-γ-irradiated loam soil in the greenhouse in comparison with the value of the non-mycorrhizal controls. In an earlier experiment, also conducted in the greenhouse, a higher A_N value was also found in mycorrhizal sorghum (*Sorghum bicolor* L.) plants, but only at intermediate rates of fertilization, that is, where plant growth and N demand were largest (Ames, Porter et al., 1983). Since most soil N is in organic forms, these results suggest that AM fungi increase organic N availability.

Both plant roots and AM hyphae can take up amino acids, at least to some extent. The ability of the extraradical mycelia of AM fungi to take up and translocate the amino acids glycine and glutamine has been demonstrated in vitro (Hawkins, Johansen, and George, 2000). The uptake rate of these organic forms of N by the fungal hyphae, approximately 4 μmol g hyphae^{-1} h^{-1} in a 4 mM solution, is in the same range as that of inorganic N. Mycorrhizal roots take up approximately 40 nmol N g root^{-1} h^{-1} and 20 nmol N g root^{-1} h^{-1} in 0.2 mM NO_3^- and 2.0 mM NO_3^- solution respectively, as compared to approximately 1 nmol g root^{-1} h^{-1} for non-mycorrhizal roots. Furthermore, using compartmented growth containers and ^{13}C- and ^{15}N-labeled leaf material, it has been shown that AM fungi can not only increase the recovery of N from organic residues (Hodge, 2003b) but also enhance organic matter mineralization (Hodge, 2001; Hodge, Campbell, and Fitter, 2001). It has been observed that the extraradical hyphae of *Glomus hoi* proliferate abundantly in a residue patch (Hodge, Campbell, and Fitter, 2001). In that experiment, hyphae proliferation in the compartment that contained residues was much greater than proliferation in an adjacent, non-inoculated plant compartment, suggesting that the AM fungus was foraging for something other than a host plant. It is also possible that hyphae development in the adjacent, non-amended compartment was nutrient-limited, while the hyphae growing in the organic patch compartment were using nutrients released by decomposers. The nutrient needs of plants are well understood, but those of AM fungi are rarely questioned (Treseder and Allen, 2002). It has been concluded that AM hyphae, which are in close proximity to mineralizing microorganisms, compete efficiently with those organisms for the N released by mineralization. This is consistent with the conclusions of Feng et al. (2002): AM hyphae appear to be effective competitors for the release of nutrients from organic matter mineralization.

The influence of AM fungi on plant uptake of N from organic sources is still unclear. Extraradical AM hyphae have been shown to take up glycine in vitro (Hawkins, Johansen, and George, 2000) but not in soil (Hodge, 2001). AM fungi have improved plant uptake of N from complex organic materials (Hodge, 2003b), and the decomposition of organic residues can be enhanced by the presence of AM fungi (Hodge, 2001; Hodge, Campbell, and Fitter, 2001). This may indicate that, although plant roots may not always need AM fungi to take up glycine, they may sometimes benefit from the enhancing effect of AM fungi on organic residue decomposition. This impact of

AM fungi on organic matter decomposition may be important, as suggested by a study in which plant N uptake from an organic source was enhanced by AM fungi in *Brassica napus,* a plant that is considered a non-host species (Hodge, 2003a).

The capture of N from organic sources has sometimes been related to root length (Hodge, 2003a,b), but it has also been related to extraradical hyphae length when root growth has been contained in divided growth containers (Hodge, 2001; Ames, Reid et al., 1983), indicating that extraradical mycelia do translocate N from organic sources to the host plant. This function, however, may be insignificant when the organic N source is not located in a root exclusion zone and when the competition for N between plants and microorganisms is low.

Extraradical AM hyphae can enhance N mineralization and compete effectively for uptake and translocation of N to the host crop. In light of the previous text, it appears that better N nutrition may explain earlier results from a field experiment conducted in soil where N had been immobilized by straw addition. A strong mycorrhizal effect, which doubled maize biomass production in that field, could not be replicated with P fertilizer applications (Hamel and Smith, 1991).

N Uptake Under Dry Conditions

On a more specific note, the ability of AM plants to extract more soil water may lead to better N nutrition in dry seasons or dry areas. The scarcity of soil water effectively limits the mobility of soil nutrients, including nitrogen. Subramanian and Charest (1998) found higher activities of nitrate reductase, glutamine synthetase, and glutamate synthase in mycorrhizal maize roots than in non-mycorrhizal roots after a period of water stress. Total amino acids, soluble proteins, and total N content were also higher in AM maize, suggesting better N nutrition for mycorrhizal maize under water stress.

UPTAKE OF OTHER NUTRIENTS

AM fungi can enhance plant acquisition of potassium, calcium, magnesium, sulfur, zinc, copper, and iron (Clark, 1997; Clark and Zeto, 2000; Liu, Hamel, Hamilton, Ma, and Smith, 2000; Liu et al., 2002; Persad-Chinnery and Chinnery, 1996). However, reduced uptake of Mn and/or other plant nutrients by AM plants has also been reported (Posta, Marschner, and Römheld, 1994; Weissenhorn et al., 1995). The effect of AM fungi on the uptake of nutrients other than N and P has been variable, and the reason for this variability is unclear. Experiments have often been conducted under conditions of deficiency of P and other nutrients where non-AM plants are smaller than AM plants, and it is difficult in these conditions to define the impact of AM fungi on nutrient uptake. The following sections review the literature on the uptake of nutrients other than N and P.

Uptake of K, Ca, Mg, and S

Although plants require K, Ca, Mg, and S in relatively large amounts, these nutrients are rarely deficient in soil. Only a few experiments have therefore been designed to study the role of AM fungi in plant K, Ca, Mg, and S uptake, with the result that little is known on the subject. There are few reports available, and they are sometimes inconsistent. The concentrations of K, Ca, and Mg in AM plants have either increased or decreased as compared with those in non-AM plants (Buttery et al., 1988; El-Shanshoury et al., 1989; Kothari, Marschner, and Römheld, 1990; Medeiros, Clark, and Ellis, 1994; Pinochet et al., 1997). In general, the results have shown that, when the availability of these nutrients is low, AM fungi enhance their uptake, and when the availability of these nutrients is high, there is no difference between AM and non-AM plants.

Enhanced uptake of K, Ca, Mg, and S in AM plants has been demonstrated in different experiments. The use of labeled Ca and S (^{45}Ca and $^{35}SO_4$) has shown that AM hyphae can absorb and transport Ca and S to the host plant (Rhodes and Gerdemann 1978; Cooper and Tinker, 1978). The capacity of extraradical AM mycelia to transport K, Ca, and S has also been demonstrated under controlled conditions (Marschner and Dell, 1994). The importance of the AM fungi contribution to the uptake of these nutrients varies, however, because these nutrients differ in general availability and properties. The contribution of the symbiosis to plant uptake of K, a plant nutrient with low mobility in soil, is likely to be more important than its contribution to Ca, Mg, or S uptake. Ca and Mg are usually abundant and the S ion is highly mobile in soil, with the result that plants are unlikely to depend on AM extensions for the uptake of these nutrients. However, when the Ca, Mg, or SO_4 availabilities are low (e.g., Ca in acid soil), the contribution of AM fungi to plant Ca, Mg, and SO_4 acquisition can become significant. AM fungi effectively enhance plant uptake of nutrients in poor soils (Siqueira and Saggin, 1995; Clark and Zeto, 1996). For example, Clark (1997) reports that AM plants enhance the uptake of Ca, Mg, and K in an acid soil, in which Ca, Mg, and K levels are deficient. In another experiment, Raju et al. (1990) found enhanced uptake of S in AM sorghum plants compared to non-AM plants. Studying the effects of excess Mn on mineral uptake in sorghum, Medeiros, Clark, and Ellis (1994) found that AM plants had a higher Ca concentration in shoots and a higher Mg concentration in roots than non-AM plants did.

Liu et al. (2002) found that AM fungi increase field-grown plant acquisition of K, Ca, and Mg in soils where the availabilities of K, Ca, and Mg, as well as P, are low. They also found that a correlation between the abundance of extraradical hyphae and the concentrations of Ca and Mg in maize shoots existed only in soils where the available Ca or Mg was relatively low (Liu et al., 2002). The concentration of K in maize shoots was positively correlated with extraradical hyphal length in both soils. These correlations suggest that plant K, Ca, and Mg can be taken up and transported by extraradical hyphae and that, in nutrient-limiting situations, the more developed the hyphal network, the more K, Ca, and Mg the AM plants take up. Because high soil P availability limits extraradical hyphal development, the contribution of AM fungi to the uptake of these nutrients is reduced with increasing P fertilization (Azcón, Ambrosano, and Charest, 2003; Liu et al., 2002).

Uptake of Cu and Zn

It has commonly been reported that AM fungi increase plant Cu and Zn uptake (Bürkert and Robson, 1994; Clark and Zeto, 1996; Marschner and Dell, 1994; Liu, Hamel, Hamilton, Ma, and Smith, 2000). A good example of this is the study by Pacovsky, Paul, and Bethlenfalvay (1986), in which more Cu and Zn were measured in mycorrhizal soybean plants than in P-supplemented non-mycorrhizal soybean plants of the same size, in soil with low levels of Cu and Zn.

^{65}Zn was used by Bürkert and Robson (1994) to show that hyphae of AM fungi can absorb and transport labeled Zn to the host plants. Using pots divided into root and hyphal compartments with mesh of different dimensions, it was possible to conclude that the extraradical AM hyphal contribution to Cu uptake by *Trifolium repens* could reach 62 percent of total uptake (Li, Marschner, and Römheld, 1991). Marschner and Dell (1994) report that *Glomus mosseae* contributes from 16 to 25 percent of total Zn uptake by maize grown in calcareous soil and from 52 to 62 percent of total Cu uptake by clover plants grown in the same soil, when extraradical AM hyphae have access to a volume of soil from which roots were excluded. Thus, better uptake of these elements by AM plants has been attributed to the more thorough exploitation of the soil volume by mycorrhizal roots.

Soil usually contains low levels of available Cu and Zn, and Cu and Zn ions have very low mobility in soil. Thus, similar to what is seen with P, zones of Cu and Zn depletion readily develop around plant roots, and the AM-enhanced uptake of Cu and Zn by plants seems to rely on the ability of the AM roots to exploit a larger volume of soil than roots alone. As such, P fertilization, which reduces AM mycelium development, reduces the uptake of Zn and Cu. Interactions between P and Zn or Cu have been observed by Gardiner and Christensen (1991) using pear (*Pyrus communis* L.), by Lambert, Baker, and Cole (1979) using maize and soybean (*Glycine max* (L.) Merr.), and by Liu, Hamel, Hamilton, Ma, and Smith (2000) using maize. In all cases, AM colonization and concentrations of Cu and Zn decreased with P fertilization in AM plants. The proposed mechanism is that high P likely suppresses the development of AM fungi and thus reduces the contribution to the uptake of Cu and Zn in AM plants (Persad-Chinnery and Chinnery, 1996).

Liu, Hamel, Hamilton, Ma, and Smith (2000) explain the greater uptake of Cu and Zn by maize grown under low levels of P and micronutrients by the greater development of extraradical mycelia. AM fungi extraradical mycelia and root colonization were reduced by approximately 20 percent with micronutrient application in a nutrient poor soil-sand mix. This indicates that high levels of both P and micronutrients inhibit mycorrhizal development, resulting in the reduced uptake ability of extraradical hyphae. In this experiment, the difference between the levels of Zn and Cu uptake between AM and non-AM maize decreased with the increasing micronutrient level, becoming nonsignificant at the high level of application.

The availability of P, micronutrients, and also N in soil influences AM development in plants (differential AM development, as seen previously). Reduced AM development may, in turn, affect the ability of plants to take up nutrients. The impact of N and P fertility on AM development, and consequently, on nutrient acquisition in mycorrhizal lettuce has recently been assessed (Azcón, Ambrosano, and Charest, 2003). At low P and N levels, AM colonization increased specific root absorption of Cu and Zn, as well as that of other nutrients. At high N and P levels, however, AM development was reduced, and furthermore, AM plants took up less Zn and other nutrients (N, K, Fe, and S) per mg of root compared to non-AM plants; in other words, AM colonization reduced the ability of roots to take up these nutrients. This suggests that the occurrence of AM colonization has an impact on plant physiology in a way that is not yet understood. Interestingly, this reduced nutrient uptake in AM lettuce did not result in reduced growth.

The contribution of extraradical AM mycelia to plant uptake of nutrients with low availability is an important component of the AM effect on plant uptake. However, evidence suggests that there are physiological differences between AM and non-AM plants that result in reduced uptake of Zn and Cu by AM plants under conditions of abundance of these nutrients (El-Kherbawy et al., 1989; Weissenhorn et al., 1995). Notwithstanding a physiological effect of AM colonization on nutrient uptake, it appears that there is not much benefit in having AM root extensions for the uptake of a nutrient that is readily available and abundant.

AM fungi reduce plant acquisition of Zn and other metals compared to non-AM plants when the plants are grown in soils containing high toxic levels of these elements (Heggo, Angle, and Chaney, 1990). This suggests that AM fungi may take up nutrients selectively depending on the availability of the specific nutrient in the soil and the plant's requirements. How this occurs is not clear. In order to explain the reduction of Zn toxicity by mycorrhizal fungi, Denny and Wilkins (1987) have proposed that hyphal cell wall and/or plant-fungus interfaces may sorb Zn, and in this way, reduce the severity of its toxicity; this mechanism was deduced from research on ectomycorrhizal fungi. Persad-Chinnery and Chinnery (1996) suggest that the mechanism proposed by Denny and Wilkins (1987) may also be applicable to AM fungi, since the uptake of nutrients by both ectomycorrhizal and AM fungi involves extraradical hyphae. However, there are important differences in morphology and function between arbuscular mycorrhizae and ectomycorrhizae, and the mechanisms for AM-mediated reduction in metal uptake by plants under toxic conditions remain to be experimentally demonstrated.

Uptake of Mn and Fe

Only a few studies have looked at Fe and Mn acquisition by AM plants. AM plants generally have lower Mn content than non-AM plants, although cases of enhanced Mn uptake have also been reported (Al-Karaki and Clark, 1999; Clark, Zobel, and Zeto, 1999; Pirazzi, Rea, and Bragaloni, 1999). AM fungi have enhanced plant acquisition of Fe in alkaline soil, where Fe availability was low, but not in acid soil, where Fe was more available (Clark and Zeto, 1996).

Extraradical AM mycelia enhance the absorbing capacity of plant roots. Therefore, AM plants are intuitively expected to take up more nutrients than non-AM plants. However, reduced Mn uptake in AM plants has been reported in most studies. The reduction in the concentration of a nutrient in an AM plant as compared to a non-AM plant can sometimes be attributed to a dilution effect related to the increased biomass of the AM plant. However, experimental results cannot always be explained by a dilution effect (Nielsen and Jensen, 1983). The AM-related reduction in plant Mn acquisition has been attributed to enhanced Mn oxidation in the rhizosphere (Arines, Porto, and Vilarino, 1992; Posta, Marschner, and Römeld, 1994). The availability of Mn and Fe in soil depends on both soil pH value and soil oxidation-reduction potential. The reduced forms of these elements are more available to plants (Marschner, 1988). Kothari, Marschner, and Römheld (1991) report that Mn acquisition is decreased in AM plants. AM fungi have been found to reduce the activity and number of Mn-reducing bacteria (Posta, Marschner, and Römeld, 1994) or to increase the number of Mn-oxidizing bacteria in the rhizosphere (Arines, Porto, and Vilarino, 1992), thereby indirectly reducing soil oxidation-reduction potential and the availability of Mn and Fe in the mycorrhizosphere. Another explanation, proposed by Bethlenfalvay and Franson (1989), is that the greater root exudates found in non-AM plants are causally related to the increased acquisition of Mn in non-AM plants. These exudates can mobilize and chelate available forms of manganese, such as MnO_2, leading to increased levels of available Mn in the rhizosphere of non-AM plants and to higher uptake. It appears that AM fungi can indirectly reduce plant acquisition of Mn by somehow creating a rhizosphere environment in which Mn is less available. In other words, the reduced uptake of Mn in AM plants may be attributed to the reduced availability of Mn in the rhizosphere, which is likely due to the indirect effects of AM fungi.

Mn and Fe share some chemical properties. Both nutrients may exist in different oxidation states in soil, and both are more soluble and available in their reduced form and more soluble at low pH. Why, then, do AM fungi reduce plant Mn uptake in most soil conditions but increase Fe uptake in alkaline soil and reduce Fe uptake in acid soil? Liu, Hamel, Hamilton, Ma, and Smith (2000) propose the simultaneous occurrence of two opposite AM influences on Mn and Fe dynamics. AM fungi may reduce Mn and Fe availability by increasing the soil oxidation-reduction potential or by reducing the production of nutrient-mobilizing root exudates in the rhizosphere, while their hyphae also enhance the ability of roots to exploit the resources of the soil. The overall influence of AM fungi on plant acquisition of Mn or Fe depends on which of these two influences dominates under a given set of conditions. This would explain how AM fungi can both increase plant Fe uptake when micronutrient abundance or availability (in alkaline soil) is low, and decrease Mn uptake when micronutrient levels are high (Clark and Zeto, 1996; Liu, Hamel, Hamilton, Ma, and Smith, 2000).

AM fungi not only enhance plant acquisition of P, K, Ca, Mg, Cu, and Zn when the level of these nutrients is low but also reduce the toxicity of Mn and/or other metals when the concentration of these metals is too high (Clark, Zobel, and Zeto, 1999; Medeiros, Clark, and Ellis, 1994). Since Mn toxicity often occurs in acid soil, an explanation for the reduced Mn acquisition in AM plants grown in acid soil is that AM plants may have a greater ability than non-AM plants to resist acquisition and/or increase their protection against the toxicity of Mn (Bethlenfalvay and Franson, 1989; Medeiros, Clark, and Ellis,

1994; Azaizeh et al., 1995; Clark, Zobel, and Zeto, 1999). This suggests that AM plants can regulate their acquisition of K, Ca, Mg, Fe, and Mn according to the availability of these nutrients in the soil. Whether mycorrhizal plants enhance or even reduce the uptake of these nutrients depends on the levels of these nutrients. When the availability of these nutrients is high or even toxic, AM plants do not enhance or even reduce the uptake of these nutrients; however, when the availability of these nutrients in the soil is low, mycorrhizal plants enhance the uptake of these nutrients. Unfortunately, the regulation mechanisms involved are not clear.

Water Uptake

The impact of AM fungi on water uptake is briefly summarized as the effect of AM fungi on plant water relations and has recently been reviewed by Augé (2001) and Ruiz-Lozano (2003). Several mechanisms are involved in the reduction of water stress in mycorrhizal plants. Many mechanisms are related to the physiology of the host plant, but some are also related to the extraradical AM mycelia. AM hyphae can absorb water, as demonstrated by the difference in transpiration flux in AM clover (*Trifolium pratense* L.) and leek (*Allium porrum* L.) plants with and without extraradical hyphae (Hardie, 1985).

Some studies have concluded that the enhanced drought tolerance provided by AM fungi is probably due to drought avoidance rather than to a change in the ability of leaves to withstand dehydration (Augé et al., 2001; Davies et al., 2002). The enhanced ability of mycorrhizal plants to absorb water is related to the length of their extraradical mycelia. Marulanda, Azcón, and Ruiz-Lozano (2003) found that different AM fungal species depleted soil water to different degrees. For example, 0.6 percent of soil water (volumetric) was used by *G. mosseae*–colonized lettuce, while 0.95 percent of soil water was used by *G. intraradices*–colonized lettuce plants of the same size. Soil water uptake was related to the abundance of soil hyphae. Path analysis has revealed that the density of the extraradical AM hyphae in soil best explains the variation in both leaf and soil lethal water potential (Augé et al., 2003). The variability explained by hyphal length is greater than that explained by the level of root mycorrhizal colonization, suggesting that, in this case at least, water extraction from the soil is the limiting factor and that there is no bottleneck in the arbuscules for the transport of water from the fungus to the host plant.

The better hydraulic conductance of mycorrhizal plants can be attributed to a number of factors. AM hyphae can increase soil root contact during drying. Upon drying, the soil often retracts and roots may shrink by as much as 30 percent, creating gaps between the soil and the plant's roots (Brady and Weil, 2001). The mycorrhizal hyphae are well anchored in the soil matrix around and in the roots and may help prevent or reduce the extent of gap formation, complementing the action of the root hairs. Extraradical AM hyphae may also improve the capacity of a root system to extract soil water by giving it access to micropore water. Because of their small diameter, hyphae can enter pores that are too small for root hairs to access. Furthermore, AM hyphae proliferate well beyond the limit of root hairs, giving plants access to more water-filled pores.

Lastly, AM mycelia could improve soil water-holding capacity through enhanced soil aggregation (Tisdall, 1991; Augé et al., 2001). Improved soil structure generally improves soil moisture-retention properties related to the amount of mesopores and smaller pores. As a result, mycorrhizal soil may hold more water at a given soil water potential than non-mycorrhizal soil (Augé et al., 2001), and plants growing in arbuscular mycorrhizae–colonized soil have access to a larger reservoir of soil water. Mycorrhizal soil, however, has recently been found to hold a slightly reduced amount of soil-available water at field capacity and to hold soil water more tightly as compared to non-mycorrhizal soil (Augé, 2004). More research needs to be done on the impact of AM hyphae on the availability of soil water to plants. In particular, the impact of AM fungi on soil hydraulic properties and on the movement of water in soil upon drying needs to be determined.

CONCLUSION

Numerous studies have demonstrated the beneficial role that AM fungi play in plant nutrient and water uptake. While the mechanisms of AM mycelium uptake of P are well understood, the often conflicting results of studies targeting other nutrients have underlined the occurrence of nutrient interactions that influence mycelium development, and consequently, alter the impact of AM fungi on nutrient uptake. The development of AM mycelia is largely controlled by soil fertility, with the result that, in a fertile soil, the contribution of AM fungi to plant nutrition is reduced. Studies conducted under controlled conditions using a single species of AM fungi have generated most of our knowledge about AM-assisted plant nutrient and water uptake. According to these studies, the contribution of AM fungi to plant nutrition is highly sensitive to soil fertility conditions. In the field, however, AM fungal biodiversity should make the mycorrhizal system more robust than it may appear from studies under controlled conditions. AM fungal species are distinct and respond differently to environmental conditions, including soil fertility, and when soil conditions are less favorable to the development of some AM species, other AM species are likely to prevail. While this requires further demonstration, little doubt remains at this time about the important role of AM fungi in the uptake of nutrients by plants.

REFERENCES

Al-Karaki, G.N. and R.B. Clark (1999). Varied rates of mycorrhizal inoculum on growth and nutrient acquisition by barley grown with drought stress. *Journal of Plant Nutrition* 22:1775-1784.

Ames, R.N., L.K. Porter, T.V. St. John, and C.P.P. Reid (1983). Nitrogen sources and "A" values for vesicular-arbuscular and non-mycorrhizal sorghum grown at three rates of ^{15}N-ammonium sulfate. *New Phytologist* 96:555-563.

Ames, R.N., C.P.P. Reid, L.K. Porter, and C. Cambardella (1983). Hyphal uptake and transport of nitrogen from two ^{15}N-labelled sources by *Glomus mosseae*, a vesicular-arbuscular mycorrhizal fungus. *New Phytologist* 95:381-396.

Arines, J., M.E. Porto, and A. Vilarino (1992). Effect of manganese on vesicular-arbuscular mycorrhizal development in red clover plants and on soil Mn-oxidizing bacteria. *Mycorrhiza* 1:127-131.

Ashford, A.E. and W.G. Allaway (2002). The role of the motile tubular vacuole system in mycorrhizal fungi. *Plant and Soil* 244:177-187.

Augé, R.M. (2001). Water relations, drought and vesicular-arbuscular mycorrhizal symbiosis. *Mycorrhiza* 11:3-42.

Augé, R.M. (2004). Arbuscular mycorrhizae and soil/plant water relations. *Canadian Journal of Soil Science* 84:373-381.

Augé, R.M., J.L. Moore, K.H. Cho, J.C. Stutz, D.M. Sylvia, A.K. Al-Agely, and A.M. Saxon (2003). Relating foliar dehydration tolerance of mycorrhizal *Phaseolus vulgaris* to soil and root colonization by hyphae. *Journal of Plant Physiology* 160:1147-1156.

Augé, R.M., A.J.W. Stodola, J.E. Tims, and A.M. Saxton (2001). Moisture retention properties of a mycorrhizal soil. *Plant and Soil* 230:87-97.
Azaizeh, H.A., H. Marschner, V. Römheld, and L. Wittenmayer (1995). Effects of a vesicular-arbuscular mycorrhizal fungus and other soil microorganisms on growth, mineral nutrient acquisition and root exudation of soil-grown plants. *Mycorrhiza* 5:321-327.
Azcón, R., E. Ambrosano, and C. Charest (2003). Nutrient acquisition in mycorrhizal lettuce plants under different phosphorus and nitrogen concentration. *Plant Science* 165:1137-1145.
Azcón, R., J.M. Ruiz-Lozano, and R. Rodríguez (2001). Differential contribution of arbuscular mycorrhizal fungi to plant nitrate uptake (^{15}N) under increasing N supply to the soil. *Canadian Journal of Botany* 79:1175-1180.
Bago, B., P.E. Pfeffer, W. Zipfel, P. Lammers, and Y. Shachar-Hill (2002). Tracking metabolism and imaging transport in arbuscular mycorrhizal fungi. Metabolism and transport in AM fungi. *Plant and Soil* 244:189-197.
Barber, S.A. (1980). Soil plant interactions in the phosphorus nutrition of plants. In *The Role of Phosphorus in Agriculture,* F.E. Khaswhneh, E.E. Sample, and E.J. Kamprath (eds.). Madison, Wisconsin, USA: ASA, CSSA, SSSA, pp. 591-615.
Barber, S.A. (1995). *Soil nutrient bioavailability. A mechanistic approach.* 2nd edition. Wiley and Sons, New York, NY.
Barber, S.A., J.M. Walker, and E.H. Vasey (1963). Mechanisms for the movement of plant nutrients from the soil and fertilizer to the plant root. *Journal of Agriculture and Food chemistry* 11:204-207.
Baylis, G.T.S. (1967). Experiments on the ecological significance of phycomycetous mycorrhizas. *New Phytologist* 66:213-243.
Bethlenfalvay, G.J. and R.L. Franson (1989). Manganese toxicity alleviation by mycorrhizae in soybean. *Journal of Plant Nutrition* 12:953-970.
Bieleski, R.L. (1973). Phosphate pools, phosphate transport and phosphate availability. *Annual Review of Plant Physiology* 24:225-252.
Brady, N.C. and R.R. Weil (2001). *The nature and properties of soils.* Upper Saddle River, NJ: Prentice Hall.
Buttery, B.R., S.J. Park, W.I. Findlay, and B.N. Dhanvantari (1988). Effects of fumigation and fertilizer on growth, yield, chemical composition, and mycorrhizae in white bean and soybean. *Canadian Journal of Plant Science* 68:677-686.
Bürkert B. and Robson A. (1994). ^{65}Zn uptake in subterranean clover (*Trifolium subterraneum* L.) by three vesicular-arbuscular mycorrhizal fungi in a root-free sandy soil. *Soil Biology and Biochemistry* 26:1117-1124.
Clark, R.B. (1997). Arbuscular mycorrhizal adaptation, spore germination, root colonization, and host plant growth and mineral acquisition at low pH. *Plant and Soil* 192:15-22.
Clark, R.B. and S.K. Zeto (1996). Mineral acquisition by mycorrhizal maize grown on acid and alkaline soil. *Soil Biology and Biochemistry* 28:1495-1503.
Clark, R.B. and S.K. Zeto (2000). Mineral acquisition by arbuscular mycorrhizal plants. *Journal of Plant Nutrition* 23:867-902.

Clark, R.B., R.W. Zobel, and S.K. Zeto (1999). Effects of mycorrhizal fungus isolates on mineral acquisition by *Panicum virgatum* in acidic soil. *Mycorrhiza* 9:167-176.
Cooper, K.M. and P.B. Tinker (1978). Translocation and transfer of nutrients in vesicular-arbuscular mycorrhizas. II. Uptake and translocation of phosphorus, zinc, and sulphur. *New Phytologist* 81:43-52.
Daft, M.J. and T.H. Nicolson (1966). Effect of Endogone mycorrhiza on plant growth. *New Phytologist* 65:343-350.
Davies, F.T., V. Olalde-Portugal, L. Aguilera-Gomez, M.J. Alvarado, R.C. Ferrera-Cerrato, and T.W. Boutton (2002). Alleviation of drought stress of Chile ancho pepper (*Capsicum annuum* L. cv. San Luis) with arbuscular mycorrhiza indigenous to Mexico. *Scientia Horticulturae* 92:347-359.
Denny, H.J. and D.A. Wilkins (1987). Zinc tolerance in *Betula spp.* IV. The mechanism of ectomycorrhizal amelioration of zinc toxicity. *New Phytologist* 106: 545-553.
El-Kherbawy, M., J.S. Angle, A. Heggo, and R.L. Chaney (1989). Soil pH, rhizobia, and vesicular-arbuscular mycorrhizae inoculation effects on growth and heavy metal uptake of alfalfa (*Medicago sativa* L.). *Biology and Fertility of Soils* 8: 61-65.
El-Shanshoury, A.R., M.A. Hassan, and B.A. Abdel-Ghaffar (1989). Synergistic effect of vesicular-arbuscular mycorrhizas and *Azotobacter chroococcum* on the growth and the nutrient contents of tomato plants. *Phyton-Horn* 29:203-212.
Epstein, E. (1972). *Mineral nutrition of plants: Principles and perspectives.* New York: John Wiley and Sons, Inc.
Ezawa, T., T.R. Cavagnaro, S.E. Smith, F.A. Smith, and R. Ohtomo (2004). Rapid accumulation of polyphosphate in extraradical hyphae of an arbuscular mycorrhizal fungus as revealed by histochemistry and a polyphosphate kinase/luciferase system. *New Phytologist* 161:387-392.
Ezawa, T., S.E. Smith, and F.A. Smith (2002). P metabolism and transport in AM fungi. *Plant and Soil* 244:221-230.
Fardeau J.C. 1996. Dynamics of phosphate in soils. An isotopic outlook. *Fertilizer Research* 45:91-100.
Feng, G., F.S. Zhang, X.L. Li, C.Y. Tian, C.X. Tang, and Z. Rengel (2002). Uptake of nitrogen from indigenous soil pool by cotton plant inoculated with arbuscular mycorrhizal fungi. *Communications in Soil Science and Plant Analysis* 33:3825-3836.
Frey, B. and H. Schüepp (1993). Acquisition of nitrogen by external hyphae of arbuscular mycorrhizal fungi associated with *Zea mays* L. *New Phytologist* 124:221-230.
Gardiner, D.T. and N.W. Christensen (1991). Pear seedling responses to phosphorus, fumigation and mycorrhizal inoculation. *Journal of Horticultural Science* 66:775-780.
Gavito, M.E. and P.A. Olsson (2003). Allocation of plant carbon to foraging and storage in arbuscular mycorrhizal fungi. *FEMS Microbiology Ecology* 45:181-187.

George, E., H. Marschner, and I. Jakobsen (1995). Role of arbuscular mycorrhizal fungi in uptake of phosphorus and nitrogen from soil. *Critical Reviews in Biotechnology* 15:257-270.

Gerdemann, J.W. (1964). The effect of mycorrhiza on the growth of maize. *Mycologia* 56:342-349.

Govindarajulu, M., P.E. Pfeffer, J. Hairu, J. Abubaker, D.D. Douds, J.W. Allen, H. Bücking, P.J. Lammers, and Y. Shachar-Hill. (2005). Nitrogen transfer in the arbuscular mycorrhizal symbiosis. *Nature* 435:819-823.

Hamel, C. and D.L. Smith (1991). Interspecific N-transfer and plant development in a mycorrhizal field-grown mixture. *Soil Biology and Biochemistry* 23:661-665.

Hardie, K. (1985). The effect of removal of extraradical hyphae on water uptake by vesicular-arbuscular mycorrhizal plants. *New Phytologist* 101:677-684.

Harrison, M.J. and M.I. van Buuren (1995). A phosphate transporter from the mycorrhizal fungus *Glomus versiforme*. *Nature* 378:626-629.

Haselwandter, K. (1995). Mycorrhizal fungi: Siderophore production. *Critical Review in Biotechnology* 15:287-291.

Hawkins, H.J. and E. George (2001). Reduced ^{15}N-nitrogen transport through arbuscular mycorrhizal hyphae to *Triticum aestivum* L. supplied with ammonium *vs.* nitrate nutrition. *Annals of Botany* 87:303-311.

Hawkins, H.J., A. Johansen, and E. George (2000). Uptake and transport of organic and inorganic nitrogen by arbuscular mycorrhizal fungi. *Plant and Soil* 226: 275-285.

Heggo, A., J.S. Angle, and R.L. Chaney (1990). Effects of vesicular-arbuscular mycorrhizal fungi on heavy metal uptake by soybeans. *Soil Biology and Biochemistry* 22:267-277.

Hodge, A. (2001). Arbuscular mycorrhizal fungi influence decomposition of, but not plant nutrient capture from, glycine patches in soil. *New Phytologist* 151: 725-734.

Hodge, A. (2003a). N capture by *Plantago lanceolata* and *Brassica napus* from organic material: The influence of spatial dispersion, plant competition and an arbuscular mycorrhizal fungus. *Journal of Experimental Botany* 54:2331-2342.

Hodge, A. (2003b). Plant nitrogen capture from organic matter as affected by spatial dispersion, interspecific competition and mycorrhizal colonization. *New Phytologist* 157:303-314.

Hodge, A., C.D. Campbell, and A.H. Fitter (2001). An arbuscular mycorrhizal fungus accelerates decomposition and acquires nitrogen directly from organic material. *Nature (London)* 413:297-299.

Jansson, S.L. and J. Persson (1982). Mineralization and immobilization of soil nitrogen. In *Nitrogen in Agricultural Soils*. Madison, WI: American Society of Agronomy, pp. 229-252.

Johansen, A., I. Jakobsen, and E.S. Jensen (1993). Hyphal transport by a vesicular-arbuscular mycorrhizal fungus of N applied to the soil as ammonium or nitrate. *Biology and Fertility of Soils* 16:66-70.

Johansen, A., I. Jakobsen, and E.S. Jensen (1994). Hyphal N transport by a vesicular-arbuscular mycorrhizal fungus associated with cucumber grown at three nitrogen levels. *Plant and Soil* 160:1-9.

Jungk, A.O. (1996). Dynamics of nutrient movement at the soil-root interface. In *Plant Roots: The Hidden Half,* Y. Waisel, A. Eshel, and U. Kafkafi (eds.). New York: Marcel Dekker, pp. 529-556.

Kabir, Z., I.P. O'Halloran, P. Widden, and C. Hamel (1998). Vertical distribution of arbuscular mycorrhizal fungi under corn (*Zea mays* L.) in no-till and conventional tillage systems. *Mycorrhiza* 8:53-55.

Koide, R.T. and Z. Kabir (2000). Extraradical hyphae of the mycorrhizal fungus *Glomus intraradices* can hydrolyse organic phosphate. *New Phytologist* 148: 511-517.

Kothari, S.K., H. Marschner, and V. Römheld (1990). Direct and indirect effects of VA mycorrhizal fungi and rhizosphere microorganisms on acquisition of mineral nutrients by maize (*Zea mays* L.) in a calcareous soil. *New Phytologist* 116:637-645.

Kothari, S.K., H. Marschner, and V. Römheld (1991). Effect of a vesicular-arbuscular mycorrhizal fungi and rhizosphere microorganisms on manganese reduction in the rhizosphere and manganese concentrations in maize (*Zea mays* L.). *New Phytologist* 117:649-655.

Lambert, D.H., D.E. Baker, and H. Cole Jr. (1979). The role of mycorrhizae in the interactions of phosphorus with zinc, copper, and other elements. *Soil Science Society of America Journal* 43:976-980.

Lewis, D.J. and J.P. Quirk (1967). Phosphate diffusion in soil and uptake by plants. III. ^{31}P movements and uptake by plants as indicated by ^{32}P autoradiography. *Plant and Soil* 27:446-453.

Li, X-L., H. Marschner, and V. Römheld (1991). Acquisition of phosphorus and copper by VA-mycorrhizal hyphae and root-to-shoot transport in white clover. *Plant and Soil* 136:49-57.

Liu, A., C. Hamel, R.I. Hamilton, and D.L. Smith (2000). Mycorrhizae formation and nutrient uptake of new corn (*Zea mays* L.) hybrids with extreme canopy and leaf architecture as influenced by soil N and P levels. *Plant and Soil* 221:157-166.

Liu, A., C. Hamel, R.I. Hamilton, B.L. Ma, and D.L. Smith (2000). Acquisition of Cu, Zn, Mn, and Fe by mycorrhizal maize (*Zea mays* L.) grown in soil at different P and micronutrient levels. *Mycorrhiza* 9:331-336.

Liu, A., C. Hamel, A. Elmi, C. Costa, B. Ma, and D.L. Smith (2002). Concentrations of K, Ca, and Mg in maize colonized by arbuscular mycorrhizal fungi under field conditions. *Canadian Journal of Soil Science* 82:271-278.

Marschner, H. (1988). Mechanisms of manganese acquisition by roots from soils. In *Manganese in Soils and Plants,* D. Graham, R.J. Hannam, and N.D. Uren (eds.). London: Kluwer, pp. 191-204.

Marschner, H. and B. Dell (1994). Nutrient uptake in mycorrhizal symbiosis. *Plant and Soil* 159:89-102.

Marulanda, A., R. Azcón, and J.M. Ruiz-Lozano (2003). Contribution of six arbuscular mycorrhizal fungal isolates to water uptake by *Lactuca sativa* plants under drought stress. *Physiologia Plantarum* 119:526-533.

Medeiros, C.A.B., R.B. Clark, and J.R. Ellis (1994). Effects of excess manganese on mineral uptake in mycorrhizal sorghum. *Journal of Plant Nutrition* 17:2203-2219.

Morel, C. (1988). Analyse par traçage isotopique du comportement du phosphore dans les systèmes sol-engrais-plante: conséquences en matière de fertilisation. Thèse de l'Université de Droit, d'Economie et des Sciences d'Aix-Marseille. Spécialité: Science de la terre.

Morel, C. and C. Plenchette (1994). Is the isotopically exchangeable phosphate of a loamy soil the plant available P? *Plant and Soil* 158:287-297.

Mosse, B., D.S. Hayman, and D.J. Arnold (1973). Advances in the study of vesicular-arbuscular mycorrhiza. V. Phosphate uptake by three plant species from P-deficient soils labelled with ^{32}P. *New Phytologist* 72:802-805.

Nielsen, J.D. and A. Jensen (1983). Influence of vesicular-arbuscular mycorrhiza fungi on growth and uptake of various nutrients as well as uptake ratio of fertilizer P for lucerne (*Medicago sativa*). *Plant and Soil* 70:165-172.

Nye, P. and P.B. Tinker (1977). *Solute movement in the soil—root system.* Oxford, UK: Blackwell Scientific.

Pacovsky, R.S., E.A. Paul, and G.J. Bethlenfalvay (1986). Response of mycorrhizal and P-fertilized soybeans to nodulation by *Bradyrhizobium* or ammonium nitrate. *Crop Science* 26:145-150.

Pearson, J.N. and I. Jakobsen (1993). The relative contribution of hyphaea and roots to phosphorus uptake by arbuscular mycorrhizal plants, measured by dual labelling with ^{32}P and ^{33}P. *New Phytologist* 124:489-494.

Persad-Cinnery, S.B. and L.E. Chinnery (1996). Vesicular-arbuscular mycorrhizae and micronutrient nutrition. In *Advancements in Micronutrient Research*, A. Hemantaranjan (ed.). Jodhpur, India: Scientific Publishers, pp. 376-382.

Pinochet, J., C. Fernandez, M. de C. Jaizme, and P. Tenoury (1997). Micropropagated banana infected with *Meloidogyne javanica* responds to *Glomus intraradices* and phosphorus. *Horticultural Science* 32:101-103.

Pirazzi, R., E. Rea, and M. Bragaloni (1999). Improvement of micronutrient uptake of valuable broadleaves in interaction with *Glomus mosseae. Geomicrobiological Journal* 16:79-84.

Plenchette, C. and J.C. Fardeau (1988). Effet du pouvoir fixateur du sol sur le prélèvement de phosphore par les racines et les mycorrhizes. *Compte-rendus de l'Académie des Sciences* 306, III:201-206.

Plenchette, C. and C. Morel (1994). External and internal P-requirements of mycorrhizal and non mycorrhizal oats and soybean plants. *Biology and Fertility of Soils* 21:303-308.

Posta, K., H. Marschner, and V. Römheld (1994). Manganese reduction in the rhizosphere of mycorrhizal and nonmycorrhizal maize. *Mycorrhiza* 5:119-124.

Raju, P.S., R.B. Clark, J.R. Ellis, and J.W. Maranville (1990). Mineral uptake and growth of sorghum colonized with VA mycorrhiza at varied soil phosphorus levels. *Journal of Plant Nutrition* 13:843-859.

Rhodes, L.H. and J.W. Gerdemann (1978). Translocation of calcium and phosphate by external hyphae of vesicular-arbuscular mycorrhizae. *Soil Science* 126: 125-126.

Rigou, L. and E. Mignard (1994). Factors of acidification of the rhizosphere of mycorrhizal plants. Measurement of pCO_2 in the rhizosphere. *Acta Botanica Gallica* 141:533-539.

Ruiz-Lozano, J.M. (2003). Arbuscular mycorrhizal symbiosis and alleviation of osmotic stress. New perspectives for molecular studies. *Mycorrhiza* 13:309-317.

Schubert, A., C. Marzachi, M. Mazzitelli, M.C. Cravero, and P. Bonfante-Fasolo (1987). Development of total and viable extraradical mycelium in the vesicular-arbuscular mycorrhizal fungus *Glomus clarum* Nicol. & Schenck. *New Phytologist* 107:183-190.

Schweiger, P.F. and I. Jakobsen (1999). Direct measurement of arbuscular mycorrhizal phosphorus uptake into field-grown winter wheat. *Agronomy Journal* 91:998-1002.

Siqueira, J.O. and O.J. Saggin (1995). The importance of mycorrrhizae association in natural low-fertility soils. In *International Symposium on Environmental Stress: Maize in Perspective,* A.T. Machado, R. Magnavaca, S. Panday, S. Panday, and A.F. da Silva (eds.). Lagoas: Brazilian Department Center for Maize and Sorghum Research (CNPMS), Sete Lagoas, pp. 239-280.

Smith, S.E., S. Dickson, and F.A. Smith (2001). Nutrient transfer in arbuscular mycorrhizas: how are fungal and plant processes integrated? *Australian Journal of Plant Physiology* 28:683-694.

Smith, S.E. and D.J. Read (1997). *Mycorrhizal symbiosis.* 2nd edition. San Diego, London: Academic Press. 605 pp.

Smith, S.E., F.A. Smith, and I. Jakobsen (2004). Functional diversity in arbuscular mycorrhizal (AM) symbioses: the contribution of the mycorrhizal P uptake pathway is not correlated with mycorrhizal responses in growth or total P uptake. *New Phytologist* 162:511-524.

Smith, F.A., S.E. Smith, B.J. St. John, and D.J.D. Nicholas (1986). Inflow of N and P into roots of mycorrhizal and non mycorrhizal onions. In *Physiological and Gentical Aspects of Mycorrhizae,* V. Gianinazzi-Perason and S. Gianinazzi (eds.). Paris, France: INRA, pp. 371-375.

Smith, S.E., B.J. St. John, F.A. Smith, and D.J.D. Nicholas (1985). Activity of glutamine synthetase and glutamate dehydrogenase in *Trifolium subterraneum* L. and *Allium cepa* L.: Effects of mycorrhizal infection and phosphate nutrition. *New Phytologist* 99:211-227.

Staddon, P.L., C.B. Ramsey, N. Ostle, P. Ineson, and A.H. Fitter (2003): Rapid turnover of hyphae of mycorrhizal fungi determined by AMS microanalysis of C-14. *Science* 300:1138-1140.

Stevenson, F.J. and M.A. Cole (1986). The phosphorus cycle. In *Cycles of Soil: Carbon, Nitrogen, Phosphorus, Sulfur, Micronutrients.* New York: John Wiley and Sons, pp. 231-284.

Subramanian, K.S. and C. Charest (1998). Arbuscular mycorrhizae and nitrogen assimilation in maize after drought and recovery. *Physiologia Plantarum* 102: 285-296.

Tarafdar, J.C. and H. Marschner (1994). Efficiency of VAM hyphae in utilisation of organic phosphorus by wheat plants. *Soil Science and Plant Nutrition* 40: 593-600.

Tilak, K.V.B.R. and A. Dwivedi (1990). Nitrate reductase activity of vesicular-arbuscular mycorrhizal fungi. In *Current Trends in Mycorrhizal Research.*

Proceedings of the National Conference on Mycorrhiza. Hisar, India: Haryana Agricultural University, Feb. 14-16, 1990, pp. 59-60.
Tisdale, S.L., W.L. Nelson, and J.D. Beaton (1985). *Soil fertility and fertilizers.* 4th edition. New York, NY: Macmillan.
Tisdall, J.M. (1991). Fungal hyphae and structural stability of soil. *Australian Journal of Soil Research* 29:729-743.
Toussaint, J.P., M. St-Arnaud, and C. Charest (2004). Nitrogen transfer and assimilation between the arbuscular mycorrhizal fungus *Glomus intraradices* Schenck & Smith and Ri T-DNA roots of *Daucus carota* L. in an *in vitro* compartmented system. *Canadian Journal of Microbiology* 50:251-260.
Treseder, K. and M.F. Allen (2002). Direct nitrogen and phosphorus limitation of arbuscular mycorrhizal fungi: a model and field test. *New Phytologist* 155: 507-515.
Uetake, Y., T. Kojima, T. Ezawa, and M. Saito (2002). Extensive tubular vacuole system in an arbuscular mycorrhizal fungus, *Gigaspora margarita. New Phytologist* 154:761-768.
Viereck, N., P.E. Hansen, and I. Jakobsen (2004). Phosphate pool dynamics in the arbuscular mycorrhizal fungus *Glomus intraradices* studied by in vivo P-31 NMR spectroscopy. *New Phytologist* 162:783-794.
Weissenhorn, I., C. Leyval, G. Belgy, and J. Berthelin (1995). Arbuscular mycorrhizal contribution to heavy metal uptake by maize (*Zea mays* L.) in pot culture with contaminated soil. *Mycorrhiza* 5:245-251.
Yao, Q., X. L. Li, G. Feng, and P. Christie (2001). Influence of extramatrical hyphae on mycorrhizal dependency of wheat genotypes. *Communications in Soil Science and Plant Analysis* 32:3307-3317.

Chapter 3

Effects of the Arbuscular Mycorrhizal Symbiosis on Plant Diseases and Pests

Marc St-Arnaud
Vladimir Vujanovic

Today, plant disease control measures are highly significant due to the intimate relationship between plant health and the well-being of people, animals, and the environment (Harrington, 1995). To safely meet the food needs of the growing human population, we will have to continue to develop better management practices and sustainable plant disease management strategies. Plant diseases can significantly reduce crop yield and quality, and can also affect human and animal health through the accumulation of toxic residues in consumable products. Highly efficient, environmentally sound control strategies are therefore critical to the future of agriculture, with a strong emphasis on the need for biological-based approaches of disease management. Apart from being effective in controlling plant pathogens, biocontrol products should not negatively impact non-target organisms (Brimner and Boland, 2003), or create tolerance development in pest organisms (Gossen and Rimmer, 2001). We should find adequate and efficient alternatives to chemical pesticides with minimal environmental impact (Ekelund, Westergaard, and Soe, 2000). Of various potential pathways, arbuscular mycorrhizal (AM) fungi management has raised considerable interest in the sustainable agriculture community (Bethlenfalvay and Linderman, 1992; Schreiner and

Mycorrhizae in Crop Production
© 2007 by The Haworth Press, Inc. All rights reserved.
doi:10.1300/5425_03

Bethlenfalvay, 1995). Mycorrhizas constitute the most important mutualistic symbioses on earth (Strack et al., 2003). The most common type, arbuscular mycorrhiza, develops in the roots of up to 80 percent of all terrestrial plant species and involves more than 150 species of the Glomeromycota (Schüßler, 2002). It is now well established that AM fungi exhibit no host specificity, but rather host preference. These soil organisms play key roles in the diversity and productivity of natural ecosystems' (Hart and Klironomos, 2002; Hart, Reader, and Klironomos, 2003; van der Heijden et al., 1998), and are associated with most agricultural crops (Smith and Read, 1997). AM fungi are major components of the rhizosphere that can affect both the incidence and severity of plant diseases in agriculture and horticulture. Several reviews and books have been published over the years, generally reporting on specific aspects of interactions between AM fungi and plant pathogens (Dehne, 1982; Garcia-Garrido and Ocampo, 2002; Graham, 1986; Linderman, 1992, 1994, 2000, 2001; Miller, Rajapakse, and Garber, 1986; Perrin, 1990, 1991; Schönbeck, 1979; Smith, 1988; St-Arnaud and Elsen, 2005; St-Arnaud, Hamel, Caron, and Fortin, 1995). The purpose of the present chapter is to summarize the research on various agricultural crops, with particular emphasis on the modes of action through which AM fungi could reduce plant disease damage in agriculture.

EFFECT OF AM FUNGI ON ROOT DISEASES

Soilborne diseases are by far the most difficult to manage, and no efficient control measures are generally available on a long-term basis. The main soilborne pathogens belong to the fungi or nematodes groups, and most plant disease studies dealing with AM fungi were focused on root diseases. The interactions between mycorrhizal fungi and root diseases are well documented in scientific literature (see Table 3.1). Overall, most of the crop-pathogen combinations examined benefited from mycorrhizal inoculation and showed lower disease symptoms or higher yields when plants were mycorrhizal. The effect of AM fungi on the pathogen population or on disease development processes at infection sites is, however, less clear. Of the reported interactions with fungal pathogens, very few were unfavorable

TABLE 3.1. Interaction between arbuscular mycorrhizal fungi and soilborne fungal pathogens.

Crop species	Pathogen[a]	Mycorrhizal species[a]	References
Albizia procera	*Fusarium oxysporum*	*Glomus fasciculatus* *G. tenuis*	Chakravarty and Mishra, 1986b
Allium cepa	*Fusarium oxysporum* f.sp.*cepa* *Phoma terrestris* *Pyrenochaeta terrestris* *Sclerotium cepivorum* *Thielaviopsis basicola*	*Glomus* sp. Unidentified species	Becker, 1976; Cole and Lim in Dehne, 1982; Cole and Mokhtar in Dehne, 1982; Safir, 1968; Torres-Barragán et al., 1996
Agrostis stolonifera	*Microdochium nivale*	Wild population	Gange and Case, 2003
Ananas comosus	*Phytophthora cinnamomi*	*Glomus* sp.	Guillemin et al., 1994
Arachis hypogaea	*Sclerotium rolsfii* *Fusarium solani* *Rhizoctonia solani*	*G. fasciculatum* *G. mosseae*	Abdalla and Abdel-Fattah, 2000; Krishna and Bagyaraj, 1983
Arachis sp.	*Sclerotium rolsfii*	Unidentified species	Jalali and Hisar, 1991
Asparagus officinalis	*Fusarium oxysporum* *Fusarium* sp. *Helicobasidium mompa*	*G. fasciculatum* *Glomus* sp. *Gigaspora margarita*	Hamel et al., 2004; Matsubara et al., 2000, 2001, 2002; Wacker et al., 1990; Yergeau, 2004
Brassica napus	*Rhizoctonia solani*	Unidentified species	Iqbal and Mahmood, 1986; Mahmood and Iqbal, 1982
B. oleracea	*Fusarium monoliforme* *Rhizoctonia solani*	*G. mosseae* Unidentified species	Iqbal et al., 1987, 1988a, 1988b
Cajanus cajan	*Fusarium udum* *Phytophthora dreschleri*	*Glomus* sp. *Gi. Calospora*	Jalali and Hisar, 1991
Carica papaya	*Phytophthora palmivora*	Unidentified species	Ramirez, 1974
Cassia tora	*Fusarium oxysporum*	*G. fasciculatus G tenuis*	Chakravarty and Mishra, 1986a

TABLE 3.1 *(continued)*

Crop species	Pathogen[a]	Mycorrhizal species[a]	References
Chamaecyparis lawsoniana	*Phytophthora cinnamomi*	Wild population *G. mosseae*	Bartschi et al., 1981
Cicer arietinum	*Fusarium oxysporum*	*G. fasciculatum*	Jalali and Hisar, 1991
Citrus sp.	*Fusarium oxysporum* *Phytophthora parasitica* *Thielaviopsis basicola*	*G. etunicatum* *G. intraradices* *G. fasciculatus* Unidentified species	Davis, 1980; Davis and Menge, 1980; Davis et al., 1978, 1979; Graham, 1988; Nemec, 1979; Schenck et al., 1977
C. sinensis	*Phytophthor. Parasitica* *Thielaviopsis basicola*	*Gi. margarita* *G. constrictus* *G. fasciculatum* *G. intraradices* *G. mosseae* *Sclerocystis sinuosa*	Davis, 1980; Davis and Menge, 1981; Graham and Egel, 1988; Nemec et al., 1996
Corchorus ollitorius	*Fusarium solani*	*G. macrocarpum*	Bali and Mukerji, 1988
Cucumis sativum	*Fusarium oxysporum f.sp. cucumemerinum*	Unidentified species	Dehne, 1977
Cymbopogon winterianus	*Pythium aphanidermatum*	*G. aggregatum*	Ratti et al., 1998
Dalbergia sissoo	*Fusarium solani*	*G. fasciculatus* *G tenuis*	Chakravarty and Mishra, 1986b
Daucus carota	*Fusarium oxysporum* f.sp. *chrysanthemi*	*G. intraradices*	Benhamou et al., 1994; St-Arnaud, Hamel, Vimard, et al., 1995

Dianthus caryophyllus	*Fusarium oxysporum* f.sp. *dianthi*	*G. intraradices*	St-Arnaud et al., 1997
Elettaria cardomomum	*Fusarium moniliformae*	*G. fasciculatum*	Thomas et al., 1994
Euphorbia pulcherrima	*Pythium ultimum* *Rhizoctonia solani*	*G. fasciculatum* Unidentified species	Idczak et al., 1991; Kaye et al., 1984; Stewart and Pfleger, 1977
Fragaria vesca	*Phytophthora fragariae*	*G. caledonium* *G. mosseae* *G. fistulosum*	Bååh and Hayman, 1983; Mark and Cassells, 1996
Fragaria x *ananassa*	*Phytophthora fragariae*	*G. etunicatum* *G. fasciculatum*	Norman et al., 1996
Glycine max	*Fusarium solani* *Macrophomina phaseolina* *Phytophthora megasperma* *P. megasperma* f.sp. *glycinea* *Pythium ultimum* *Rhizoctonia solani*	*G. mosseae* *G. etunicatus* *Endogone* sp *E. mosseae* Unidentified species	Chou and Schmitthener, 1974; Ross, 1972; Stewart and Pfleger, 1977; Whatley and Gerdemann, 1981; Zambolim and Schenck, 1981, 1983
Gossypium hirsutum	*Fusarium oxysporum* f.sp. *vasifectum* *Sclerocystis sinuosa* *Thielaviopsis basicola* *Verticillium dahliae*	*G. macrocarpum* *G. versiforme* *G. mosseae* *G. fasciculatus* *S. sinuosa*	Bali and Mukerji, 1988; Davis et al., 1979; Liu, 1995; Schönbeck and Dehne, 1977
Gossypium barbadense	*Verticillium dahliae*	*G. mosseae* *G. versiforme* *S. sinuosa*	Liu, 1995
Hordeum sp.	*Cochliobolus sativus*	*G. etunicatum*	Dehne, 1987
Hordeum vulgare	*Bipolaris sorokiniana*	*G. dimorphicum* *G. intraradices* *G. mosseae* *Glomus* sp.	Boyetchko and Tewari, 1988, 1990

TABLE 3.1 *(continued)*

Crop species	Pathogen[a]	Mycorrhizal species[a]	References
Lactuca sativa	*Olpidium brassicae*	*G. mosseae* Unidentified species	Schönbeck and Dehne, 1979, 1981
Liriodendron tulipifera	*Cylindrocladium scoparium*	Unidentified species	Barnard, 1977
Lycopersicon esculentum	*Fusarium oxysporum* f.sp. *lycopersici* *F. oxysporum* f. sp. *radicis-lycopersici* *Phytophthora nicotinae* f.sp. *parasitica* *Pyrenochaeta lycopersici* *Pythium aphanidermatum* *Verticillium albo-atrum* *Verticillium dahliae*	*G. caledonium* *G. etunicatum* *G. intraradices* *G. mosseae* Unidentified species	Bååth and Hayman, 1983; Bochow and Abou-Shaar, 1990; Caron et al., 1985, 1986a, 1986b, 1986c, Caron et al., 1986; Dehne and Schönbeck, 1975, 1979a; Jalali and Hisar, 1991; Karagiannidis et al., 2002; Lioussanne et al., 2003; McGraw, 1983; McGraw and Schenck, 1981; Nemec et al., 1996; Pozo et al., 1996; Ramaraj et al., 1988
Linum usitatissimum	*Fusarium oxysporum* f.sp. *lini*	*G. intraradices*	Dugassa et al., 1996
Medicago sativa	*Fusarium oxysporum* f sp. *Medicaginis* *Phytophthora megasperma* *Pythium paraecandrum* *Verticillium albo-atrum*	*G. fasciculatus* *G. mosseae* *Glomus* sp.	Davis et al., 1978; Hwang, 1988; Hwang et al., 1992
Musa acuminata	*Cylindrocladium spathiphylli*	*Glomus* sp.	Declerck et al., 2002
Nicotiana tabacum	*Olpidium brassicae* *Pythium* sp. *Thielavioipsis basicola*	*G. microcarpum* *G. monosporum* *G. mosseae* Unidentified species	Baltruschat and Schönbeck, 1972; 1975; Giovannetti et al., 1991; Jalali and Hisar, 1991; Schönbeck and Dehne, 1979; Tosi et al., 1988

Onobrychis viciifolia	*Pythium* spp.	*G. fasciculatus* *G. intraradices*	Hwang et al., 1993
Persea Americana	*Phytophthora cinnamomi*	*G. fasciculatus*	Davis et al., 1978; Mataré and Hattingh, 1978
Pisum sativum	*Rhizoctonia solani*	*G. mossae*	Morandi et al., 2002
Phaseolus vulgaris	*Fusarium solani* *Fusarium solani* f.sp. *phaseoli* *Rhizoctonia solani*	*G. intraradices* *G. macrocarpum*	Filion et al., 2003b; Gonçalves et al., 1991a, 1991b; Guillon et al., 2002
Pisum sativum	*Aphanomyces euteiches*	*G. fasciculatum* *G. intraradices* *G. mosseae* Wild inoculum	Bødker et al., 2002; Kjøller and Rosendahl, 1996; Larsen and Bødker, 2001; Rosendahl, 1985
Poncirus trifoliata X *Citrus sinensis*	*Phytophthora parasitica*	*Gi. Margarita* *G. constrictus* *G. fasciculatus* *G. mosseae* *S. sinuosa*	Davis and Menge, 1981
Prunus persica	*Cylindrocarpon destructans*	*G. aggregatum* *G. intraradices*	Traquair and Pohlman, 1990; Traquair, 1995
Solanum melongena	*Verticillium dahliae*	*Gi. Etunicatum* *G mosseae*	Karagiannidis et al., 2002; Matsubara et al., 1995
Tagetes patula	*Pythium ultimum*	*G. intraradices*	St-Arnaud, 1998; St-Arnaud et al., 1994
Tagetes erecta	*Pythium ultimum*	*G. mosseae*	Calvet et al., 1993
Theobroma cacao	*Ganoderma pseudoferreum*	Unidentified species	Chulan et al., 1990
Trifolium sp.	*Fusarium avenaceum*	Unidentified species	Cole and Lim in Dehne, 1982

TABLE 3.1 *(continued)*

Crop species	Pathogen[a]	Mycorrhizal species[a]	References
Triticum aestivum	*Cochliobolus sativus* (=*Bipolaris sorokiana*) *Gaeumannomyces graminis tritici*	*G. fasciculatus* *G. intraradices*	Graham and Menge, 1982; Rempel, 1989
Triticum sp.	Root infecting fungi	Unidentified species	Mahmood and Khurshid, 1988
Vigna radiata	*Macrophomina phaseolina*	*G. coronatum* *G. mosseae*	Jalali and Hisar, 1991; Kasiamdari et al., 2002
Vigna unguilata	*Macrophomina* sp. *Phytophthora vignae* *Rhizoctonia solani*	*G. clarum* *G. etunicatum* *G. intraradices*	Abdel-Fattah and Shabana, 2002; Ramaraj et al., 1988; Fernando and Linderman, 1997
Vulpia ciliata ssp. *Ambigua*	*Embellisia chlamydospora* *Fusarium oxysporum*	*Glomus* sp.	Newsham et al., 1994, 1995
Zea mays	*Rhizoctonia solani*	*G. mosseae*	Khadge et al., 1990
37 pasture species	*Bipolaris sorokiniana*	Wild inoculum	Thompson and Wildermuth, 1989

[a]Species names are those used in the cited references.

to the host plant. In most cases, mycorrhizal plants showed less significant disease impacts, while in many cases, the pathogen inoculum or the number of infection sites were not necessarily reduced.

The need for significant mycorrhizal colonization prior to pathogenic infection is not a clearly established condition for disease inhibition. Several studies have found that a high level of mycorrhizal root colonization prior to infection by the pathogen was essential for effective biocontrol. For example, Bärtschi, Gianinazzi-Pearson, and Vegh (1981) observed that mycorrhizal inoculation did not protect *Chamaecyparis lawsoniana* roots against *Phytophthora cinnamomi* unless the mycorrhizal infection was well established prior to pathogen inoculation. This observation was corroborated by the results of Rosendahl (1985) on mycorrhizal peas infected with *Aphanomyces euteiches*. More recently, Slezack et al. (2000), working with the same organisms, showed that the bioprotection effect coincided with an induction of mycorrrhiza-related chitinolytic enzymes, and appeared to depend on a fully established mycorrhizal symbiosis. Many studies have also shown that a minimum colonization level is not always required for mycorrhizal-induced disease reduction. For example, Caron, Fortin, and Richard (1986b) infected tomato plants with *Fusarium oxysporum* f. sp. *radicis-lycopersici* four weeks before, simultaneously with, and four weeks after inoculation with the AM fungus *Glomus intraradices*. They noted a reduction in *Fusarium* root necroses in all cases, even when *G. intraradices* was inoculated four weeks after infection with the pathogen. Similarly, St-Arnaud et al. (1994) studied *Tagetes patula* plants inoculated with *G. intraradices* and infected with the root pathogen *Pythium ultimum*. Mycorrhizal colonization significantly lowered pathogen development in roots independently of the extent of mycorrhizal colonization. In a later study, *Dianthus caryophyllus,* a non-mycorrhizal species, was shown to be protected against the root pathogen *F. oxysporum* f. sp. *dianthi* when the plant was grown in coculture with a mycorrhizal species colonized by *G. intraradices*. Disease symptoms and plant mortality were significantly reduced even if *D. caryophyllus* plants were sown in a soil preinfested with both AM fungus and pathogen, and then exposed to the two fungi simultaneously (St-Arnaud et al., 1997). Current data also suggest the importance of cultivar-related AM efficiency. In plantlets of two micropropagated potato cultivars infected with

Rhizoctonia solani, inoculation with *G. etunicatum* or *G. intraradices* significantly reduced the mortality rate by 77 and 26 percent respectively in the cv Goldrush, whereas mycorrhizal colonization did not change the mortality rate in the cv LP89221 (Yao, Tweddell, and Desilets, 2002).

The literature also contains many studies on AM fungi interaction with parasitic nematodes (see Table 3.2). As for fungal root pathogens, most studies showed that being mycorrhizal was beneficial to the host plant, and to our knowledge, no one interaction showed an increase in the extent of nematode root infection. In roughly one quarter of the cases, nematode population was increased, but the plant biomass was also higher than in non-mycorrhizal control plants. Management of the antagonistic potential of AM symbiosis to suppress nematodes in agricultural ecosystems was proposed more than a decade ago (Sikora, 1992). The biocontrol effects observed varied from very low to very high, depending on the plant species, nematode species, and the AM fungal species involved (Roncadori, 1997). Of the AM fungi, some species, such as *G. mosseae*, were more consistently effective than others in suppressing plant parasitic nematodes. Under controlled conditions, this species reduced damage to papaya by the root-knot nematode *Meloidogyne javanica* by 37 percent. Similarly, the same fungus reduced *M. incognita* damage to white clover by 68 percent.

Compensation for nematode damage in mycorrhizal plants is frequently observed. For example, Little and Maun (1996) measured a significant improvement in foliar growth rate and root dry weight of mycorrhizal *Ammophila breviligulata* plants infected with nematodes. More recently, Talavera, Itou, and Mizukubo (2001) reported that mycorrhizas did not protect tomato plants from *M. incognita* when inoculated simultaneously with AM fungi. However, if tomato seedlings were inoculated with nematodes three weeks after inoculation with AM fungi, there was compensation for the reduction of plant growth caused by nematode infection. Similar results were obtained with carrot roots infected with *Pratylenchus penetrans*, where mycorrhizas compensated for the damage caused by nematodes. Moreover, nematode counts in soil were reduced by 49 percent in pots inoculated with the AM fungus. Furthermore, various authors have recently provided evidence supporting an effect of AM symbiosis on nematode biology. Ryan et al. (2000) observed that hatching

TABLE 3.2. Interaction between arbuscular mycorrhizal fungi and nematodes.

Crop species	Pathogen[a]	Mycorrhizal species[a]	References
Arachis hypogaea	*Meloidogyne arenaria*	*Gigaspora margarita* *Glomus etunicatus*	Hussey and Roncadori, 1982
Avena sativa	*Meloidogyne incognita*	*G. mosseae* Unidentified species	Hussey and Roncadori, 1982; Sikora and Schönbeck, 1975
Citrus jambhiri	*Tylenchulus semipenetrans*	*G. mosseae*	Baghel et al., 1990
Citrus limon	*Radopholus similis* *Tylenchulus semipenetrans*	*G. etunicatum* *G. mosseae*	O'Bannon and Nemec, 1979
Citrus sp.	*Radopholus similis* *Tylenchulus semipenetrans*	*G. etunicatus* *G. fasciculatus* *G. mosseae* Unidentified species	Hussey and Roncadori, 1982; O'Bannon and Nemec, 1979
Cucumis sativus	*Meloidogyne incognita* *Pratylenchus penetrans*	Unidentified species	Priestel, 1980
Cyphomandra betacea	*Meloidogyne incognita*	*Gi. margarita* *G. fasciculatum* *G. mosseae* *G. tenue* Unidentified species	Cooper and Grandison, 1987
Daucus carota	*Meloidogyne hapla* *Pratylenchus coffeae* *Radopholus similis*	*G. mosseae* Unidentified species	Elsen et al., 2001, 2003; Hussey and Roncadori, 1982; Sikora and Schönbeck, 1975

TABLE 3.2 *(continued)*

Crop species	Pathogen[a]	Mycorrhizal species[a]	References
Glycine max	*Heterodera glycines* *Meloidogyne incognita*	*Endogone* sp *E. calospora* *Gi. heterogama* *Gi. margarita* *G. etunicatus* *G. fasciculatum* *G. macrocarpum* *Scutellospora heterogama*	Francl and Dropkin, 1985; Hussey and Roncadori, 1982; Kellam and Schenck, 1980; Schenck et al., 1975; Schenck and Kinlock, 1974
Gossypium hirsutum	*Meloidogyne incognita* *Pratylenchus brachyurus*	*Gi. margarita* *G. intraradices*	Hussey and Roncadori, 1977, 1978; Roncadori and Hussey, 1977; Smith et al., 1986
Gossypium sp.	*Meloidogyne incognita* *Pratylenchus brachyurus*	*Gi. margarita* *G. etunicatus* *G. mosseae* Unidentified species	Hussey and Roncadori, 1977, 1978, 1982; Roncadori and Hussey, 1977
Lycopersicon esculentum	*Meloidogyne incognita* *Meloidogyne javanica* *Rotylenchulus reniformis*	*Gi. margarita* *G. fasciculatum* *G. mosseae* Unidentified species	Bagyaraj et al., 1979; Hussey and Roncadori, 1982; Jalali and Hisar, 1991; Orolfo, 1990; Sikora, 1979; Sikora and Schönbeck, 1975; Singh et al., 1990; Sitaramaiah and Sikora, 1982
Musa cv. Dwarf Cavendish	*Radopholus similis*	*G. fasciculatum*	Umesh et al., 1988

Nicotiana tabacum	*Heterodera solanacearum* *Meloidogyne incognita* Root-knot nematode	*Gi. gigantea* *G. mosseae* Unidentified species	Fox and Spasoff, 1972; Hussey and Roncadori, 1982; Sikora and Schönbeck, 1975; Subhashini, 1990
Phaseolus vulgaris	*Rotylenchulus reniformis*	Unidentified species	Sitaramaiah and Sikora, 1982
Piper nigrum	*Meloidogyne incognita*	*G. etunicatum*	Sivaprasad et al., 1990
Prunus persica	*Meloidogyne hapla*	*Gi. margarita* *G. etunicatus*	Hussey and Roncadori, 1982
Vicia faba	*Meloidogyne javanica*	Vesicular-arbuscular mycorrhizal fungi[b]	Salem et al., 1984
Vigna unguiculata	*Heterodera cajani*	*G. epigaeus* *G. fasciculatum*	Jain and Sethi, 1987
Vitis sp	*Meloidogyne arenaria*	*G. fasciculatus*	Hussey and Roncadori, 1982
Vitis vinifera	*Meloidogyne arenaria*	*G. fasciculatum*	Atilano et al., 1976

[a]Species names are those used in the cited references.

[b]The authors considered *Fusarium equiseti* and *Pythium butleri*[2] as vesicular-arbuscular mycorrhizal fungi.

activity of the cyst nematode *Globodera pallida* was increased in the presence of root leachates from mycorrhizal potato plants. Mycorrhizal inoculation of potato microplants with a mix of three AM fungal species (*Glomus* spp.) resulted in an increased hatch of *G. pallida,* but not *G. rostochiensis* (Ryan et al., 2003). Number of feeding nematodes and plant biomass were similarly increased in mycorrhizal plants, suggesting an increased tolerance to potato cyst nematodes of the mycorrhizal plants. Elsen, Declerck, and De Waele (2001) also stressed the importance of *G. intraradices* in decreasing the reproduction of the burrowing nematode *(Radopholus similis).* In carrot roots grown in vitro, the AM fungus suppressed the *R. similis* population by almost 50 percent, while there was no correlation between nematode population density and either AM fungus internal root colonization or extraradical mycelium development. The same approach was used to study the interaction between *G. intraradices* and the lesion nematode *Pratylenchus coffeae* (Elsen, Declerck, and De Waele, 2003). In the presence of mycorrhizas, the nematode population was similarly reduced as in the previous experiment.

Arbuscular mycorrhizas have also been shown to control plant parasitic nematodes in combination with non-pathogenic strains of soilborne pathogens. Because AM fungi and non-pathogenic fungi improve plant health by different mechanisms, the combination of two such microbes with complementary mechanisms might increase the overall control efficiency, and therefore, provide an environmentally safe alternative to nematicide application. Experiments conducted by Diedhiou et al. (2003) showed a positive interaction between the AM fungus *G. coronatum* and the non-pathogenic *F. oxysporum* strain Fo162 in the control of *Meloidogyne incognita* on tomato.

EFFECT OF AM FUNGI ON DISEASES CAUSED BY FOLIAR OR SYSTEMIC PATHOGENS

Very few reports are available in the literature on these topics (Tables 3.3 and 3.4), as compared with root diseases. Plant colonization by AM fungi usually causes an increase in the activity of biotrophic leaf pathogens (Gernns, von Alten, and Poehling, 2001) and of disease symptoms. Schönbeck and Dehne (1979) compared the effect of

TABLE 3.3. Interaction between arbuscular mycorrhizal fungi and foliar fungal pathogens.

Crop species	Pathogen[a]	Mycorrhizal species[a]	References
Cucumis sativus	*Erysiphe cichoracearum*	*Glomus mosseae*	Schönbeck and Dehne, 1979
Hevea brasiliensis	*Microcyclus ulei*	*G. etunicatum*	Feldmann et al., 1989
Hordeum vulgare	*Erysiphe graminis* *Helminthosporium sativum*	*G. mosseae*	Schönbeck and Dehne, 1979
Lactuca sativa	*Botrytis cinerea* *Bremia lactucae*	*G. mosseae* *G. etunicatum*	Meyer and Dehne, 1986; Schönbeck and Dehne, 1979
Linum usitatissimum	*Oidium lini*	*G. intraradices*	Dugassa et al., 1996
Lycopersicon esculentum	Stolbur group Phytoplasma	*G. mosseae*	Lingua et al., 2002
Nicotinia tabacum	*Erysiphe graminis* f.sp. *hordei* (=*Blumeria graminis* f.sp. *hordei*) *Botrytis cinerea*	*G. etunicatum* *G. intraradices*	Gernns et al., 2001; Shaul et al., 1999
Phaseolus vulgare	*Uromyces phaseoli*	*G. mosseae* *G. constrictum* *G. etunicatum*	Meyer and Dehne, 1986; Schönbeck and Dehne, 1979
Triticum aestivum	*Puccinia graminis* f.sp. *tritici*	*G. intraradices*	Rempel, 1989

[a]Species names are those used in the cited references.

TABLE 3.4. Interaction between arbuscular mycorrhizal fungi and bacteria, virus or phytoplasma causing systemic diseases.

Crop species	Pathogen[a]	Mycorrhizal species[a]	References
Bacteria			
Lycopersicon esculentum	*Erwinia carotovora pv carotovora* *Pseudomonas solanacearum* *Pseudomonas syringae* *Clavibacter michiganensis* subsp. *Michiganensis*	*Glomus mosseae* *G. intraradices* Unidentified species	Filion, 1998; Garcia-Garrido and Ocampo, 1988a, 1988b, 1989; Halos and Zorilla, 1979
Virus and Phytoplasma			
Citrus aurentium	CLRV-2 Tristeza virus T3	*G. etunicatum*	Nemec and Myhre, 1984
Citrus macrophylla	Tristeza virus T1	*G. etunicatum*	Nemec and Myhre, 1984
Fragaria chiloensis ananassa	Arabis mosaic virus	*Endogone macrocarpa* var. *geospora*	Daft and Okusanya, 1973
Lycopersicon esculentum	Tobacco mosaic virus Potato virus X Phytoplasma of the Stolbur group	*E. macrocarpa* var. *geospora* *G. mosseae* *Glomus sp.* Unidentified species	Daft and Okusanya, 1973; Jabaji-Hare and Stobbs, 1984; Lingua et al., 2002; Schönbeck and Spengler, 1979
Nicotiana tabacum	Tobacco mosaic virus	*G. mosseae* *G. intraradices*	Schönbeck and Dehne, 1979; Schönbeck and Schinzer, 1972; Shaul et al., 1999
Petunia sp.	Arabis mosaic virus	*E. macrocarpa* var. *geospora*	Daft and Okusanya, 1973
Solanum tuberosum	virus	Unidentified species	Jalali and Hisar, 1991

[a]Species or virus names are those used in the cited references.

G. mosseae on several foliar diseases in various plant hosts and noted that, in all cases, mycorrhizal inoculation increased the disease symptoms. When *Phaseolus vulgaris* was infected with *Uromyces phaseoli,* the production of spores was also increased on mycorrhizal plants, a result also noted by Meyer and Dehne (1986). Observing a similar effect on the symptoms of *Bremia lactucae* inoculated on *Lactuca sativa,* these authors suggested that the effect was related to a symbiosis-related hormonal modification that delayed senescence of the leaves. Rempel (1989) reported a similar effect on *Triticum aestivum* infected by *Puccinia graminis* f. sp. *tritici* and under the influence of *G. intraradices.* Contrary to the previously cited studies, Feldmann, Junquiera, and Lieberer (1989) observed a reduction in the diameter of the lesions and sporulation of *Microcyclus ulei,* the causal agent of South American leaf blight on *Hevea brasiliensis,* even if the number of lesions was not influenced. They suggested that mycorrhizal colonization had caused a physiological change enhancing plant defense mechanisms.

Only a handful of studies have been conducted on diseases caused by bacteria. Garcia-Garrido and Ocampo (1988a, 1988b, 1989) studied the impact of *G. mosseae* colonization on two species of pathogenic bacteria in tomato plants, *Erwinia carotovora* pv *carotovora* and *Pseudomonas syringae.* In both cases, mycorrhizal colonization significantly decreased the pathogenic bacterial population in the rhizosphere. Interestingly, no relationship was found between phosphorus concentration in plant tissue and the above results. Moreover, the time lapse between inoculation with the pathogen and inoculation with the AM fungus was found to have no effect. A compensation of the plant dry weight loss induced by *P. syringae* or *E. carotovora* was also observed when pathogenic bacteria and AM fungus were inoculated simultaneously. Filion (1998), using a polyclonal anti-*Clavibacter michiganensis* subsp. *michiganensis* antibody in DAS-ELISA, observed that *C. m. michiganensis* population was significantly reduced in the mycorrhizosphere of mycorrhizal tomato plants as compared to AM hyphosphere and bulk soil compartments. In contrast, the population of *C. m. michiganensis* was not significantly different between rhizosphere, hyphosphere, and bulk soil compartments in the non-mycorrhizal systems.

With viral infections, results are comparable to those obtained with foliar pathogenic fungi, though more variable. Viral diseases are often more prominent in vigorously growing plants. The available reports suggest that mycorrhizal plants become more susceptible to viral infections, possibly due to the improvement in plant nutrient status resulting from the symbiosis. Daft and Okusanya (1973) noted an increase in the number of particles of three different viruses in the leaves and roots of three different host plants, when the plants were colonized by *Endogone macrocarpa geospora* (=*Glomus*). They obtained similar results by increasing the amount of phosphorus in the nutrient solution and thus suggested that the increase in viral particles was a consequence of the increased P absorption in AM plants. Schönbeck and Spengler (1979) used immunofluorescence to observe Tobacco Mosaic Virus (TMV) particles, and noted that they were mainly present in the arbuscule-containing cells. Jabaji-Hare and Stobbs (1984) tested the possible transfer of virus between plants via the mycorrhizal fungal hypha, but were unable to show any viral particles in arbuscules or transfer between mycorrhizal plants. Schönbeck and Dehne (1981) reported an increase in local lesions typical of TMV infection in mycorrhizal plants. Similarly, Nemec and Myhre (1984) observed that mycorrhizal infection was unable to minimize the extent of viral infections in *Citrus,* as expected. As in the case of nematodes, viruses seem therefore to benefit from the mycorrhizal host-fungus association, even though plant yield is also often increased.

A protective effect induced by AM fungi against a phytoplasma of the Stolbur group has recently been observed in tomato (Lingua et al., 2002). Symptoms induced by the phytoplasma were less severe when the plants also harbored AM fungi. Morphological parameters such as shoot and root fresh weights, shoot height, internode length, leaf number, and adventitious root diameter were closer to those of healthy plants when arbuscular mycorrhiza was present. Reduced nuclear senescence was also observed in AM plants infected with the phytoplasma. The percentage of nuclei with different ploidy levels was intermediate between AM and phytoplasma-infected plants. While the mechanisms underlying these interactions are not clear, it appears that arbuscular mycorrhizal colonization delays nucleus senescence in colonized roots (Lingua et al., 1999).

EFFECT OF AM FUNGI ON INSECTS AND OTHER INVERTEBRATES

AM fungi impact on plant diseases cannot be fully understood without considering all multitrophic interactions between host plants and their environment. Over the past decade, many papers have been published on mycorrhiza-herbivore interactions, covering a large spectrum of interactions between AM fungi and phytophagous or soil insects and other fauna. More details can be found in various review papers (Gange, 2000; Gange and Bower, 1997; Gange and Brown, 2002a; Gehring and Whitham, 2002). Different studies examined whether insect herbivory affects arbuscular mycorrhizal symbiosis. For example, Gange, Bower, and Brown (2002) reported that foliar-feeding herbivores reduced AM root colonization level. The authors observed that a mycotrophic plant had little damage by herbivory, whereas a non-mycotrophic plant was affected more. In another study, *Plantago lanceolata,* a plant species that is naturally highly mycorrhizal and suffers low continual insect damage over a growing season, was compared with *Senecio jacobaea,* a weakly mycorrhizal species that is frequently subject to rapid and total defoliation by moth larvae. Herbivory was found to reduce AM colonization in *P. lanceolata,* but had no effect in *S. jacobaea* (Gange, Bower, and Brown, 2002). Interestingly, AM colonization reduced the level of leaf damage in *P. lanceolata* but not in *S. jacobaea.*

Other studies were focused mainly on the effect of mycorrhizal colonization on the plant response to herbivory. Gange, Brown, and Sinclair (1994) reported a reduction in black vine weevil *(Otiorhynchus sulcatus)* larval growth by AM infection in *Taraxacum officinale.* In strawberry (*Fragaria* x *ananassa*) plants, mycorrhizal colonization by *G. mosseae* and *G. fasciculatum* reduced larval survival and biomass when inoculated singly, but not together (Gange, 2001). In *P. lanceolata,* infection by AM fungi lowered damage caused by foliar-feeding insects (*Arctia caja* and *Myzus persicae*). The authors suggested that AM infection can alter the carbon/nutrient balance of plants, leading to an increased allocation to carbon-based defenses (Gange and West, 1994).

In *Cirsium arvense* infested by the thistle gall fly *(Urophora cardui),* mycorrhizal colonization reduced fly performance presumably by modifying the nutritive value of gall tissues (Gange and Nice, 1997). In contrast, in another experiment where two aphid species (*Myzus ascalonicus* and *M. persicae*) were reared on *P. lanceolata* plants, AM colonization increased aphid weight and fecundity (Gange, Bower, and Brown, 1999). The relation between a leaf-mining insect (*Chromatomyia syngenesiae*) and a parasitoid (*Diglyphus isaea*) was studied with *Leucanthemum vulgare* inoculated with three species of AM fungi (Gange, Brown, and Aplin, 2003). Results showed that the parasitism rate was mycorrhizal fungal species-dependent. Some AM fungal-host combinations increased parasitism, some decreased it, while others had no effect. It was concluded that the cause of these differences was most likely plant size, with parasitoid-searching efficiency being reduced on the larger plants, resulting from certain AM fungal species combinations. However, a mycorrhizal effect on herbivore-produced plant volatiles cannot be ruled out. A similar result was obtained by Goverde et al. (2000) showing that AM fungi influenced life history traits of a lepidopteran herbivore. This study examined the interaction between larvae of common blue butterfly (*Polyommatus icarus*) and AM fungi. Survival of larvae fed with AM plants was greater than those fed with non-mycorrhizal plants.

Wamberg, Christensen, and Jakobsen (2003) suggested complex multitrophic interactions between foliar-feeding insects, mycorrhizas, and rhizosphere protozoa on pea plants. Feeding of adult weevils (*Sitona lineatus*) on pea plants (*Pisum sativum*) stimulated *G. intraradices* pea root colonization at the first harvest and stimulated rhizosphere protozoa in the absence of AM fungus. At the second harvest, herbivory decreased AM fungus colonization and had no effect on rhizosphere protozoa. AM colonization had no effect on herbivory at the first harvest but decreased herbivory at the second harvest. Belowground respiration was stimulated by herbivory and this effect was most pronounced during vegetative growth. The results therefore suggest that herbivory stimulated belowground carbon transfer in young plants in the nutrient acquisition phase as opposed to the reproductive phase, where herbivory had no such effect. This herbivory-induced carbon transfer will benefit the AM fungus symbiont when present and the free-living rhizosphere microorganisms—here represented by a bacterivorous protozoa—in the absence of the AM fungus. Based on these studies, it was recently proposed that AM fungi can affect the composition of insect assemblages on plants, and should therefore be considered as a factor in the evolution of insect specialism (Gange, Stagg, and Ward, 2002). Strong links exist between above- and belowground food webs, and would also affect the structure and development of successional plant communities (Gange and Brown, 2002b).

EFFECT OF PATHOGENS ON AM FUNGI

Bååth and Hayman (1983), Davis and Menge (1980), Krishna and Bagayraj (1983), and Rosendahl (1985), observed a reduction of AM colonization by various plant pathogens. These authors suggested that this inhibition would be related to the reduction of the photosynthetic effectiveness of the plant leaves and thus of root exudation. Hwang, Chang, and Chakravarty (1992) noted a reduction of AM colonization and number of vesicles produced in the presence of *F. oxysporum* f. sp. *medicaginis* and of *Verticillium albo-atrum* in the roots of *Medicago sativa*. Afek et al. (1991) also observed a reduction of root infection by *G. intraradices* in the presence of *P. ultimum* in the field. However, St-Arnaud et al. (1994) observed no inhibition of colonization by *G. intraradices* on *T. patula* roots in the presence of *P. ultimum* in containers. Similarly, Garcia-Garrido and Ocampo (1988a, 1988b, 1989) observed no inhibition of *G. mosseae* when tomato plants were infected with *Pseudomonas syringae* or *E. carotovora*. Ravnskov, Larsen, and Jakobsen (2002) reported no impact of the biocontrol bacterium *Burkholderia cepacia* on either mycorrhiza formation or the functioning of the AM fungus *G. intraradices* in terms of P transport, whereas the results suggested that mycorrhiza might have adverse effects on *B. cepacia,* with the presence of mycelium of *G. intraradices* reducing the biomass of three out of five *B. cepacia* strains. Nemec and Myhre (1984) observed a reduction in root colonization by *G. etunicatum* in *Citrus macrophylla* and *C. aurantium* infected by the Tristeza virus but not in *C. paradisii* infected by the Clrv-2 virus; they suggested that the reduction in AM colonization might have resulted from the deterioration of the roots caused by the viral infection. On the contrary, Kaye, Pfleger, and Stewart (1984) reported greater colonization of roots of poinsettia by *G. fasciculatum* in the presence of *P. ultimum* than in the absence of the pathogen. Moreover, Caron Fortin, and Richard et al. (1986b) found an increase in AM colonization of tomato roots when plants were infected by *F. oxysporum* f. sp. *radicis-lycopersici* simultaneously or four weeks before inoculation with *G. intraradices,* but not when the pathogen was inoculated four weeks after inoculation with the AM fungus. While the authors cannot explain this effect, they indicated that it clearly demonstrated the capacity of the AM fungus to colonize roots extensively despite the presence of the pathogen.

Muller (2003) recently studied the infection of *Lolium perenne* by endophytes of the *Clavicipitaceae* that live in aboveground parts of many grasses of temperate regions. The two fungal endophytes, *Epichloe typhina* and *Neotyphodium lolii,* significantly decreased root mycorrhizal colonization. This decrease was in some cases correlated with alterations of growth. Depending on the endophyte strain and on the mycorrhizal status, shoot-root biomass ratios were also significantly affected. The authors concluded that effects of endophytes may be enhanced or counterbalanced in the presence of AM fungi.

AM fungi could also be the target of mycoparasitic microorganisms used to biologically control plant diseases (Brimner and Boland, 2003). Biological control agents could interact with pathogens through different mechanisms, including mycoparasitism, production of antibiotics or enzymes, competition for nutrients, and induction of plant host defenses. While useful for controlling pathogens, these organisms may pose risks to non-target species, such as AM fungi. For example, the interaction between the biocontrol organism *T. harzianum* and *G. intraradices* growing on pea roots was studied in vitro (Rousseau et al., 1996) using a compartmented root-organ culture system (Fortin et al., 2002). A marked antagonism of *T. harzianum* to *G. intraradices* spores and hyphae was described. This study shows that the extramatrical phase of AM fungi may be adversely affected by biocontrol microorganisms, which should be considered when developing biocontrol strategies. However, positive (Datnoff, Nemec, and Pernezny, 1995), neutral (Fracchia et al., 1998), and negative (McGovern, Datnoff, and Tripp, 1992) interactions between *Trichoderma* species and AM fungi root colonization or biocontrol potential have also been reported. Ravnskov, Larsen, and Jakobsen (2002) studied the impact of the biocontrol bacterium *B. cepacia* on *G. intraradices*. This bacterium is known to suppress a broad range of root pathogenic fungi. In root-free soil compartments, the AM fungus reduced the biomass of three out of five *B. cepacia* strains as determined using specific phospholipid fatty acids (PLFA) and neutral lipid fatty acids (NLFA) profiles. However, *B. cepacia* had no impact on AM fungal biomass and energy reserves. Further, Larsen, Ravnskov, and Jakobsen (2003) examined the combined effect of the same AM fungus and biocontrol bacterium against the root pathogen *P. ultimum*. Both the bacterium and AM fungus reduced the population density of the pathogen, while *B. cepacia* biomass was also reduced by *G. intraradices,* but not the AM fungus which was not significantly affected by the biocontrol bacterium.

ROLES AND MECHANISMS OF AM FUNGI IN BIOCONTROL

More than 30 years ago, knowledge of the role of AM symbiosis in biocontrol of plant diseases could be summarized by the following sentence by Gerdemann (1968, p. 414) in a paper published in the *Annual Review of Phytopathology:* "Vesicular-arbuscular mycorrhizae may have no significant effect on the susceptibility of plants to disease, but without evidence we must not assume that this is true." Some years later, the same author (Gerdemann, 1975) tried to define the mechanisms involved in the observed reduction of disease symptoms in AM colonized plants, on the basis of a few pioneering studies. He suggested, on a purely speculative basis, that many mechanisms proposed by Zak (1964) to explain the increased resistance to diseases of ectomycorrhizal plants might also apply to the AM symbiosis. These mechanisms included (1) production of antibiotics by the AM fungus, (2) production of inhibiting molecules by the host plant in response to AM infection, (3) change in root exudation reducing attraction for zoospores, and (4) modification of rhizosphere microbial populations, that is, more non-pathogenic microorganisms and fewer pathogens.

The role of AM fungi in disease resistance then became the subject of more studies, and some years later, Dehne (1982) published a synthesis in *Phytopathology*. In that paper, he stated that (1) AM fungi could delay pathogen development in roots but the effect would be limited to the colonized tissues, (2) AM fungi could also increase the disease symptoms systemically, especially in the non-colonized portions of the plant. He concluded by proposing that the main, if not the only, mechanism involved in the response of AM colonized plants to pathogens would be the increased assimilation of nutrients, as a result of increased plant growth and physiological activity. He also indicated that other mechanisms might be involved, but are minor and strictly limited to the area colonized by the AM fungus. Since that time, there has been an explosion of research interest in this subject, and our view of the mechanisms regulating the interaction between AM plants and pathogens has evolved considerably.

The impact of AM fungi was examined under various ecological conditions and in a large number of host-pathogen associations (Tables 3.1-3.4). The literature indicates that plant diseases may be strongly influenced by AM fungi and generally by the combined action of one or more mode of actions (Linderman, 1994, 1996; St-Arnaud et al., 1995). These could be described as direct actions through direct competition or inhibition of the pathogen; and indirect actions including (1) alleviation of abiotic stress including enhanced nutrition, (2) biochemical induced changes, and (3) interactions with microbiota. The many studies published in the past few decades tend to largely confirm reduced disease incidence as a result of AM colonization. Exceptions in this regard, however, also exist. Such contrasting reports of mycorrhiza-pathogen interactions appear to be the result of the various mechanisms involved. It appears that more than one mechanism may be operative in a single AM fungus-plant-pathogen-microbiota combination, and any such interaction should be considered to have a continuum of possible modes of action. In the following sections, we will attempt to synthesize the current knowledge and to describe the main pathways involved.

Enhanced Assimilation of Nutrients

Enhancement of the ability of an AM plant to access nutrients, especially phosphorus, is well-known (Smith and Read, 1997) and was often postulated to be the main mechanism to explain the lower level of disease development in AM plants. More than 30 years ago, Daft and Okusanya (1973) observed an increase in the number of three different viruses in leaves and roots of different crops when inoculated with a mycorrhizal fungus. As similar results were obtained by increasing phosphorus fertilization, it was suggested that the increase in viruses was the result of the higher P level in AM plants. In contrast, a mycorrhiza-mediated increase in plant tolerance to fungal root pathogens has often been attributed to a concurrent increase in root P concentration (Davis and Menge, 1980; Graham and Menge, 1982). According to Graham and Menge (1982), a reduction in root exudation induced by increasing P content in AM roots (Graham, Leonard, and Menge, 1981) was the key mechanism by which the activity of *Gaeumannomyces graminis* var. *tritici* could be reduced. Schwab, Menge, and Leonard (1983) showed that, as P fertilization increased, AM colonization may lower the concentration of carbohydrates, carboxylic acids, and amino acids in root exudates. Comparisons between exudation rates and AM fungal growth also suggested that the AM fungus is able to change membrane permeability and accelerate exudation. Through increased P nutrition, AM fungi also enhance root growth, increase absorptive capacity of the root system for nutrients and water, and affect cellular processes in roots (Smith and Gianinazzi-Pearson, 1988). Such mycorrhiza-induced processes may change both the quantity and quality of exudates released from roots and involved in the changes of pathogen attraction observed in AM colonized plants (Norman and Hooker, 2000; Lioussanne, Jolicoeur, and St-Arnaud, 2003).

Rosendahl (1985) observed no differences in the production of oospores by *A. euteiches* between AM-colonized and non-colonized parts of pea roots in a split-root system, while P concentration was higher in the portion colonized by the AM fungi. Garcia-Garrido and Ocampo (1988b) also reported that the protection of tomato plants against *Erwinia carotovora* and the reduction of the bacterium population in the rhizosphere of the AM plants were not associated with P concentration in the host plant. Caron, Fortin, and Richard (1986c) were the first to demonstrate, with a *F. oxysporum* f. sp. *radicis-lycopersici*—tomato system inoculated or not with *G. intraradices*—that P nutrition is not the only factor involved in disease reduction and consequently that other mechanisms should necessarily be involved. It has been further shown, using two distinct plant-pathogen associations, that increased P assimilation is not essential to AM-mediated disease protection. The growth of *P. ultimum* in the roots and soil of *T. patula* plants was shown to be inhibited by *G. intraradices* inoculation, unrelated to P availability (St-Arnaud et al., 1994). More significant is the demonstration that a wilt disease caused by *F. oxysporum* f. sp. *dianthi* in the non-mycorrhizal species *D. caryophyllus* could be controlled by growing a companion plant colonized by *G. intraradices* in the same pot (St-Arnaud et al., 1997). Because carnation is a non-mycorrhizal species that did not develop functional AM colonization, the possibility that disease suppression was related to a nutritional effect was virtually nil. The absence of arbuscule development in carnation roots and the similar nutrient content (N, P, K, Ca, Mg, B, Zn, Fe, Mn, and Cu) among plants inoculated or not with the AM fungus, but growing in close contact with the mycorrhizal host *(T. patula),* strongly supported the hypothesis that the biocontrol effect was not related to symbiosis-mediated changes in host plant metabolism and was clearly not related to nutrition. To our knowledge, this work was the first and only demonstration that AM fungi could lower disease development in a non-AM plant species. These results strongly supported those of Caron, Fortin, and Richard (1986c) with tomato-*Fusarium,* reducing the possibility that their results were an exception, as suggested by Smith (1988). Trotta et al. (1996) also observed a reduction in disease caused by *Phytophthora nicotianae* in tomato plants colonized by *G. mosseae,* unrelated to P availability. Niemira, Hammerschhnidt, and Safir (1996) similarly reported a suppression of postharvest damages caused by *Fusarium sambucinum* on potato minitubers, without a corresponding increase in P nutrition in AM inoculated plants. Marschner and Baumann (2003) also reported that the observed effect of mycorrhizal colonization on the bacterial community structure changes in the rhizosphere of maize may be, at least in part, plant-mediated, since two P levels in the non-mycorrhizal treatments had no significant effect on the bacterial communities.

Morphological and Biochemical Induced Changes and Defense Mechanisms

Triggering of defense responses have been repeatedly proposed to explain the lower level of disease development in AM hosts. AM fungi may induce a slight and transient activation of the defense-related metabolic pathways (Harrison and Dixon, 1993; Volpin et al.,

1994; Garcia-Garrido and Ocampo, 2002) taking place during symbiotic recognition, but suppression of defense pathways has also been shown (Dassi, Dumas-Gaudot, and Gianinazzi, 1998; Guenoune et al., 2001). Research on this topic has therefore produced contradictory results over the years (Shaul et al., 2000). Earlier studies proposed that the increased resistance in AM plants was determined by the induced synthesis of inhibiting compounds by the host (Baltruschat and Schönbeck, 1975; Krishna and Bagyaraj, 1983). Rosendahl (1985) had postulated a systemic effect of *G. fasciculatum* on *P. sativum* infected by *A. euteiches*. The AM fungus lowered root necrosis both on colonized and non-colonized roots when inoculated with the pathogen in the same pot, but only on the colonized root parts in a split-root system. However, zoospore production by *A. euteiches* was reduced in mycorrhizal plants even if the AM fungus and pathogen were inoculated in different compartments. Caron, Richard, and Fortin (1986) reported that the roots of tomato plants inoculated with *F. oxysporum* f. sp. *radicis-lycopersici* showed less root necroses when transplanted into a substrate in which an AM fungus was preestablished. The authors suggested that disease inhibition occurred following a stimulation of defense mechanisms. Feldmann, Juquiera, and Lieberei (1989) reported the accumulation of cyanoglucosides to a level comparable to that obtained by the infection of *Thanatephorus cucumeris* in the roots of *Hevea brasiliensis* colonized by a strain of *Glomus etunicatum*. However, the root content in scopoletin (a phytoalexin), increased by the pathogen, was significantly decreased by one AM fungus strain, but not by another strain. Other defense responses were also reported in AM plants subjected to pathogen infection, such as formation of papillae (Nehemiah, 1977 in Lieberei and Feldmann, 1989), synthesis of chitinases (Dehne, Schönbeck, and Baltruschat, 1978), and lignification and increase in peroxidase activity (Dehne and Schönbeck, 1979b). Other reports indicated little or no synthesis of phytoalexins or hydrolytic enzymes (Morandi, Bailey, and Gianinazzi-Pearson, 1984; Dumas, Gianinazzi-Pearson, and Gianinazzi, 1989; Wyss, Boller, and Wiemken, 1989; Dumas et al., 1990) or deposition of callose or phenolic compounds around AM fungal hyphae (Maffei et al., 1986; Spanu and Bonfante-Fasolo, 1988). In *Glycine max, Rhizoctonia solani* induced an increase in glyceollin 20 days after inoculation, while no accumulation was noted in similar plants colonized by *G. mosseae* only. However, when the AM fungus was inoculated along with *Trichoderma harzianum,* the level of phytoalexin accumulation was comparable to that obtained with *R. solani* or *T. harzianum* only (Wyss, Boller, and Wiemken, 1992). Spanu and Bonfante-Fasolo (1988) monitored the activity of peroxidases from the beginning of the AM infection to the maturity of the roots in *Allium porrum.* They showed an increase in peroxidase activity at the beginning of interaction, in response to the penetration and intercellular growth of the fungal hyphae. Peroxidase activity later quickly fell to a level comparable to that observed in non-colonized roots. They concluded that the cellular walls were not modified by the symbiotic interaction and were not acting as barriers to fungal penetration.

Some authors have hypothesized that AM fungal colonization may impact soilborne plant pathogens through alteration in root architecture, which may contribute to protection against fungal pathogenic infections (Fusconi et al., 1999; Matsubara, Ohba, and Fukui, 2001) and nematodes (Fassuliotis, 1979). Localized morphological effects have been shown to occur in AM roots. For example, morphological and physical changes, such as cell wall lignification and production of other polysaccharides, have been reported to prevent tomato, onion, and cucumber cell penetration by *F. oxysporum* or *Phoma terrestris* (Sharma and Johri, 2002).

Direct involvement of defense responses in the disease resistance of AM plants has rarely been suggested (Benhamou et al., 1994; Slezack et al., 2000), but induction of PR protein by AM fungi has been extensively demonstrated (Blilou, Bueno et al., 2000; Guillon et al., 2002; Pozo et al., 1998; Pozo, Cordier et al., 2002; Pozo, Slezack-Deschaumes et al., 2002; Shaul et al., 1999; Slezack et al., 1999, 2000, 2001). Kjøller and Rosendahl (1996) observed that although pea root infection by *A. euteiches* was not modified by inoculation of *G. intraradices,* the plants preinoculated with *G. intraradices* showed no symptoms of severe root rot. Tests with split-root systems, where AM fungi and AM infected roots are separated from the site of pathogen challenge, have clearly indicated that systemic induced resistance can occur (Pozo, Cordier et al., 2002).

In the past decade, a great deal of research has been conducted in the field of molecular genetics and biochemical technologies, allowing identification of changes in gene expression and detection of specific molecules associated with disease biocontrol in field conditions. The rapid recognition between plants and AM fungi include signal perception, signal transduction, and defense gene activation (Garcia-Garrido and Ocampo, 2002). Salzer and Boller (2000) pointed out that AM fungi, as with ectomycorrhizal fungi, secrete chitin elicitors, which could induce a defense response. For instance, the elicitor derived from an extract of extraradical mycelium of *G. intraradices* was able to induce phytoalexin synthesis in soybean cotyledons (Lambais, 2000). Moreover, in *Medicago truncatula,* the same fungus was able to induce the expression of a chalcone synthase, the first enzyme in the metabolism of flavonoid compounds, such as phytoalexin (Bonanomi et al., 2001). Furthermore, a hypersensitive-like response, an oxidative burst, could be observed in compatible AM associations at sites where hyphal tips of *G. intraradices* attempted to penetrate cortical root cells (Salzer, Corbiere, and Boller, 1999). During early stages of AM formation, both an increase in catalase and peroxidase activity in roots of tobacco and onion and a salicylic acid accumulation were observed and correlated with an increase in the expression of genes encoding lipid transfer protein and phenylalanine ammonia-lyase, indicating that the first reaction of the root cells to the invasion of AM fungi is a defense response (Blilou, Ocampo, and Garcia-Garrido, 2000). In this context, the induction of defense gene expression could be considered to be a result of fungal elicitor recognition and signal transduction pathway activation. Also, according to Strack et al. (2003), colonization of root cells by AM fungi induces many dramatic changes in cytoplasmic organization: vacuole fragmentation, transformation of the plasma membrane to a peri-arbuscular membrane covering the arbuscule, increase in cytoplasm volume and in number of cell organelles, as well as movement of the nucleus into central position. In some of these changes, microtubules are most likely involved in changes of host cell morphology and cytoplasmic architecture (Balestrini et al., 1994). With regard to the molecular crosstalk between the two organisms (Garcia-Garrido and Ocampo, 2002), a number of phytohormones (cytokinins, abscisic acid, jasmonate) as well as various secondary metabolites, such as isopropanoids, have also been observed characterizing specific symbiotic pathways. The triggering of defense-related metabolic pathways by AM fungi remains therefore one of the main hypotheses to explain disease inhibition in AM hosts, and may constitute an important contributing factor after a pathogen has contacted and begun penetrating plant tissues.

Interaction with Pathogens and Other Soil Microbiota

The rhizosphere is the soil zone surrounding the roots of plants in which complex relations exist among the plant, the soil, and microorganisms (Curl and Truelove, 1986). The roots provide nutrients that influence the activity of a broad range of microbiota (Marschner and Baumann, 2003). Altered exudation induces changes in the composition of microbial populations (Linderman, 1988). Physiological changes of AM roots are major, but have not yet been completely characterized. Recently, Labour, Jolicoeur, and St-Arnaud (2003) conducted a rigorous comparison of nutrient metabolism and AM receptiveness between transformed and non-transformed tomato root lines. Many root lines, produced from different *Agrobacterium* strains and tomato cultivars, were characterized for their growth kinetics, nutrient use, and responsiveness to AM colonization. The results showed a large variability in growth responses among lines from similar origin. More importantly, mycorrhizal responsiveness was unrelated to root line origin or nutrient-uptake efficiency, but depended instead on initial P availability. This pointed to the role of intracellular storage and use efficiency in regulating symbiosis. The nitrogen metabolism was also analyzed and results showed a significant alteration in the N-key enzyme activities, N transfer, and assimilation between the symbiotic partners, and different glutamine synthetase isoforms in roots and AM mycelium (Toussaint, St-Arnaud, and Charest, 2004).

These changes in root physiology certainly have significant impacts on the rhizosphere microflora through alteration of root exudation and other nutrition-related mechanisms. The changes are so significant that the term *mycorrhizosphere* was created to describe the soil zone influenced by the AM mycorrhizas (Linderman, 1988; Rambelli, 1973). Linderman and Paulitz (1990) further suggested that space distribution of microbes residing on or close to the AM hyphae might be strongly influenced. In a study of the impact of three AM species on *Sorghum bicolor* rhizosphere, Andrade et al. (1997)

observed that AM mycelium development had little influence on the composition of the microflora in the hyphosphere, but AM root colonization was positively correlated with bacterial numbers in the hyphosphere and with the presence of *Pseudomonas* in the rhizosphere. On the basis of these results, they suggested that root exudates would be more important to rhizobacteria population than the development of AM mycelia. Further, they showed that both the root and fungal components of mycorrhizae enhance water-stable soil aggregates stability, which affect microorganism numbers indirectly by providing a favorable habitat (Andrade et al., 1998). Caron, Fortin, and Richard (1985), based on the interaction between *F. oxysporum* f. sp. *radicis-lycopersici* and tomato colonized by *G. intraradices,* suggested that there is probably a direct interaction between the AM fungi and the pathogens near the mycorrhizal mycelium in the soil. Other authors had made a similar suggestion after having noted that *G. mosseae* reduces the population of *Pseudomonas* and *Erwinia* bacteria in tomato, that the AM fungi precedes the pathogenic bacterium or that the inoculation is simultaneous therefore under conditions where it was not very probable that the effect of AM fungi was mediated via nutritional effects on the host plant (Garcia-Garrido and Ocampo, 1988b, 1989). Caron, Fortin, and Richard (1986c) underlined the need for research on the possible production of antibiotics or inhibiting substances by the AM fungi or associated plant host. A direct stimulation or inhibition of microbial activity by AM fungi had often been postulated (Caron, Fortin, and Richard, 1985; Garcia-Garrido and Ocampo, 1988b, 1989; Linderman, 1988, 1992, 1994; Linderman and Paulitz, 1990) but the lack of an adequate experimental system has long prevented any direct experimental evidence. Based on the technology of root-organ culture of AM fungi in vitro (Bécard and Fortin, 1988), an appropriate axenic system has been designed (St-Arnaud et al., 1996) and used to demonstrate a direct interaction between *G. intraradices* and *F. oxysporum* (St-Arnaud, Hamel, Vimard, et al., 1995). Differential effects of exudates from AM mycelium have been further shown on various soil fungal and bacterial species (Filion, St-Arnaud, and Fortin, 1999), supporting the existence of mechanisms independent of the root colonization in the AM-mediated biocontrol. The in vitro culture of AM has been proven to be a tool of choice that has greatly influenced our understanding of this symbiosis (Fortin et al., 2002), and its use to analyze the interaction between AM fungi and soilborne pathogens or non-pathogenic rhizosphere microorganisms has also been recently reviewed (St-Arnaud and Elsen, 2005).

As pointed out by Mathesius (2003), plant root development and performance are largely affected by a multitude of soil microorganisms to which they are exposed. Their interactions can be of a symbiotic, neutral, or pathogenic nature, and occur with the plant, as well as between microbial species. These rhizospheric interactions involve a variety of bacteria, fungi, and nematodes and can very well have an effect on pathogenic interactions via various mechanisms such as antibiosis, competition for space or nutrients, or parasitism. This highly complex situation stresses the need to examine the AM fungi-pathogen interactions within multispecific and multitrophic plant microbial systems, in non-sterile soil conditions, and in various field conditions. Highly specific and sensitive biochemical or molecular tools to quantitatively analyze microbial interactions are increasingly available. For example, PCR primers specific to *G. intraradices* or to *F. solani* f. sp. *phaseoli* were recently used for direct quantification of fungi from soil and roots, using real-time PCR technology (Filion, St-Arnaud, and Jabaji-Hare, 2003a). This approach, combined with the use of axenically produced AM fungus spores (Vimard et al., 1999) and a compartmentalized soil microcosm simulating the AM mycorrhizosphere (Filion et al., 2001) was used to examine the AM fungus/*F. solani* interaction in the bean rhizosphere. Results showed a reduction in root rot symptoms and pathogen biomass in roots, as well as a differential effect on the pathogen in the AM rhizosphere compared to the hyphosphere (Filion, St-Arnaud, and Jabaji-Hare, 2003b). Using the same host, a concurrent study showed that the AM fungus may also induce a systemic alteration of defense-related gene transcripts (Guillon et al., 2002). However, the lack of any clear relation with root rot supports the involvement of mechanisms other than host defense responses in the AM fungus/*F. solani* interaction.

Mansfeld-Giese, Larsen, and Bødker (2002) studied a bacterial population associated with the mycelium of the AM fungus *G. intraradices*. They found that *G. intraradices* influences the culturable aerobic-heterotrophic bacterial communities in the rhizosphere and hyphosphere of cucumber plants *(Cucumis sativus)*. In this study, 1,400 bacterial colonies were isolated and identified by fatty acid methyl ester (FAME) analysis, 87 species within 48 genera being identified with a similarity index >0.30. *Pseudomonas, Arthrobacter,* and *Burkholderia* were the genera most frequently encountered. Large differences in bacterial community structure were observed between rhizosphere soil, root-free soil/sand, and washed sand extract, whereas major differences between mycorrhizal and nonmycorrhizal treatments were observed for a few bacterial species only. Isolates identified as *Paenibacillus* spp. were more frequently found in the mycorrhizal treatment with mycelium of *G. intraradices,* indicating that bacteria of this genus may live in close association with mycelium of AM fungi. Similarly, Andrade, Linderman, and Bethlenfalvay (1998) studied the persistence of *Alcaligenes eutrophus* and *Arthrobacter globiformis,* two bacteria isolated from AM hyphosphere or mycorrhizosphere soils, in the rhizosphere of *S. bicolor.* Ten weeks after inoculation, populations of two species were found at a high level in the AM hyphosphere and mycorrhizosphere soil, while they were barely detectable near non-AM roots or in noncolonized soil compartments. Furthermore, Sood (2003) found a chemotactic response of plant-growth-promoting bacteria toward roots of arbuscular mycorrhizal *(G. fasciculatum)* tomato plants. A significantly higher number of bacterial cells of wild strains were attracted toward mycorrhizal tomato roots compared to non-mycorrhizal roots. Substances exuded by roots served as chemo-attractants for these bacteria. In particular, *P. fluorescens* showed strong chemotactic responses toward citric and malic acids, which were predominant constituents in root exudates of tomato plants. The attraction of zoospores of *P. nicotianae* by exudates from mycorrhizal tomato roots was recently found to differ significantly from non-colonized roots (Lioussanne, Jolicoeur, and St-Arnaud, 2003, 2004). In *P. lanceolata* rhizosphere soil, van Aarle, Soderstrom, and Olsson (2003) observed an increased microbial biomass in the presence of AM fungal hyphae, as measured from increased amount of NLFA. They concluded that AM fungal hyphae can thus stimulate microorganisms. A competition between AM fungi and root pathogens for carbon compounds derived from photosynthesis and stored in roots was also recently proposed as a potential mechanism of pathogen inhibition in AM plants (Graham, 2001).

A large-scale study of commercial asparagus fields was used to study the impact of mutual feedback interactions between plants and soil microorganisms on a crown and root rot disease of asparagus (Hamel et al., 2004) associated with large pathogenic and non-pathogenic *Fusarium* communities (Vujanovic et al., 2006). Results showed an association between disease incidence and reorganization of the soil microbial community. Microbial biodiversity as well as bacterial and fungal abundance decreased sharply with the onset of disease, followed by a similar decrease in the AM fungi population. Biodiversity and microbial population size then increased to finally reach a new equilibrium. The authors hypothesized that higher disease incidence is a consequence of the soil microbial community reorganization. They also suggested that successful disease management strategies may be based on cultural practices that would dilute the negative influence of asparagus on the soil microbial community. Using molecular primers specific to *Fusarium* (Yergeau et al., 2005) and AM fungi species (Kowalchuk, de Souza, and Van Veen, 2002), Yergeau (2004) studied the relation between these two groups in asparagus fields. *Fusarium* and AM fungi species were assessed using PCR-DGGE (denaturing gradient gel electrophoresis) in both short- and long-term experiments. Results showed that sampling locality and plant age significantly influenced both AM fungi and *Fusarium* communities. Some species are clearly associated with plants of a specific age and almost absent on others. Particular *Fusarium* taxa are also reported as being inversely linked to various AM fungi taxa.

Various levels of synergism between biocontrol agents or plant-growth-promoting-rhizomicroorganisms (PGPRs) and AM fungi have been reported with various crops. For example, Srinath, Bagyaraj, and Satyanarayana (2003) studied the influence of the AM fungus *G. mosseae* inoculated alone or with the PGPRs *Bacillus coagulans* and *Trichoderma harzianum* on the growth of *Ficus benjamina*. The plants showed maximum plant height, biomass, P content, mycorrhizal root colonization, AM fungal spore numbers, and populations of *T. harzianum* and *B. coagulans* in root zone soil when all three organisms were inoculated together. On the other hand, Diedhiou et al. (2003) studied the interaction between *G. coronatum* and a non-pathogenic isolate of *F. oxysporum* strain Fo162 in the biocontrol of *M. incognita* on tomato. While preinoculation of tomato plants with

G. coronatum or Fo162 stimulated plant growth and reduced *M. incognita* infestation, combined application of the fungi enhanced mycorrhization of tomato roots but did not increase overall nematode control or plant growth.

Multitrophic interactions are highly complex in soil and other types of microbiota, mesobiota, and macrobiota have also been shown to interact with AM fungi. Fracchia et al. (2003) reported a beneficial effect of the soil yeast *Rhodotorula mucilaginosa* on root colonization by *G. mosseae* in soybean *(Glycine max)* and *Gigaspora rosea* in red clover *(Trifolium pratense).* The percentage of root length with AM colonization was increased only when the soil yeast was inoculated before the AM fungus was introduced, but yeast exudates applied to soil similarly increased AM root colonization. Ronn et al. (2002) observed that free-living soil bacterial-feeding protozoa are significantly reduced in the rhizosphere of pea *(P. sativum* L. Cv. *Solara)* colonized with *G. caledonium,* as compared to non-mycorrhizal treatments. They attributed the lower protozoa population to a depressing effect on bacterial production by AM colonization. Collembolan grazing on AM fungi and its impact on plant growth was studied in pea rhizosphere utilizing a mix of AM fungi and the dominant collembolan species (*Isotoma* sp.) indigenous to the experimental soil (Schreiner and Bethlenfalvay, 2003). The results showed that collembolan grazing on mycorrhizae can be detrimental to plant growth when other fungal food sources are limited, but grazing on mycorrhizal fungi does not occur in these experimental conditions when ample organic matter and associated saprophytic fungi are present in soils. Diaye et al. (2003) found that a soil-feeding termite *(Cubitermes nikoloensis)* influenced the development of symbiotic microflora (rhizobia, arbuscular mycorrhizas) associated with a fallow leguminous plant, *Crotalaria ochroleuca.* The results indicated higher contents of available P and mineral-N in the mound wall of termite nests, promoting AM spore production and the number of rhizobia nodules per root system.

CONCLUSIONS

Although disease management using chemicals has met with good success in the past, it is now obvious that a more environmentally sound approach should be favored. In this context, biological control of plant pathogens through inoculation of antagonistic organisms, or by favoring the growth of useful indigenous species already present, is one of the most promising ways. Naturally occurring rhizosphere microorganisms are choice candidates for biological control because they are already part of the equilibrium between plants, pathogens, and the soil environment. Moreover, AM fungi play a central role in soil biology and contribute to increasing growth, improving nutritional status, and stress tolerance by modifying the host plant metabolism, and inducing defense mechanisms against pathogens (Smith and Read, 1997). Today, AM fungal inoculum is increasingly available and is produced by over 30 companies worldwide (Gianinazzi and Vosátka, 2004). Despite that and the extensive evidence of the beneficial impact of AM fungi in plant health, none of the biocontrol products currently on the market contains AM fungi (Whipps, 2004). AM fungi are generally registered as biostimulants or biofertilizers to avoid the more complex and expensive procedure required for the registration of biocontrol products. This situation may have slowed down the introduction of mycorrhizal inocula in agriculture. Nevertheless, these ubiquitous fungi exist in almost any soil and play a central role in ecosystem functioning (Klironomos, 2002; van der Heijden et al., 1998). The understanding of the role played by AM fungi in soil microbial population equilibrium and biodiversity is therefore a key factor in strategies to modify agricultural practices and develop biological control approaches in the context of sustainable development (Jeffries et al., 2003; Johansson, Paul, and Finlay, 2004). AM fungi are major components of soil microbiota and obviously interact with other microorganisms in the rhizosphere, that is, the zone of influence of plant roots on microbial populations and other soil constituents. Mycorrhiza formation changes several aspects of plant physiology and some nutritional and physical properties of the rhizospheric soil. These effects modify the colonization patterns of the roots or mycorrhizosphere by soil microorganisms. The rhizosphere of mycorrhizal plants, in practice a mycorrhizosphere, supports a wide range of microbial activities responsible for several key ecosystem processes. Microbial interactions in the rhizosphere of mycorrhizal plants improve plant fitness, soil quality, and mycorrhizosphere activity of non-pathogenic organisms against pathogens, which are critical issues for a sustainable agricultural development and ecosystem functioning. In the context of biological control of plant diseases, AM fungal agents must be seriously considered and examined as alternatives to chemical pesticides due to their perceived increased level of safety and minimal environmental impacts.

REFERENCES

Abdalla, M., and G. Abdel-Fattah (2000). Influence of the endomycorrhizal fungus *Glomus mosseae* on the development of peanut pod rot disease in Egypt. *Mycorrhiza* 10:29-35.

Abdel-Fattah, G.M., and Y.M. Shabana (2002). Efficacy of the arbuscular mycorrhizal fungus *Glomus clarum* in protection of cowpea plants against root rot pathogen *Rhizoctonia solani. Zeitschrift Fur Pflanzenkrankheiten und Pflanzenschutz* 109:207-215.

Afek, U., J.A. Menge, and E.L.V. Johnson (1991). Interaction among mycorrhizae, soil solarization, metalaxyl, and plants in the field. *Plant Disease* 75:665-671.

Andrade, G., R.G. Linderman, and G.J. Bethlenfalvay (1998). Bacterial associations with the mycorrhizosphere and hyphosphere of the arbuscular mycorrhizal fungus *Glomus mosseae. Plant and Soil* 202:79-87.

Andrade, G., K.L. Mihara, R.G. Linderman, and G.J. Bethlenfalvay (1997). Bacteria from rhizosphere and hyphosphere soils of different arbuscular-mycorrhizal fungi. *Plant and Soil* 192:71-79.

Andrade, G., K.L. Mihara, R.G. Linderman, and G.J. Bethlenfalvay (1998). Soil aggregation status and rhizobacteria in the mycorrhizosphere. *Plant and Soil* 202:89-96.

Atilano, R.A., J.R. Rich, H. Ferris, and J.A. Menge (1976). *Meloidogyne arenaria* on endomycorrhizal grape *(Vitis vinifera)* rootings. *Journal of Nematology* 8:278-279.

Bååth, E., and D.S. Hayman (1983). Plant growth responses to vesicular-arbuscular mycorrhizae XIV. Interactions with Verticillium wilt on tomato plants. *New Phytologist* 95:419-426.

Baghel, P.P.S., D.S. Bhatti, and B.L. Jalali (1990). Interaction of VA mycorrhizal fungus and *Tylenchulus semipenetrans* on Citrus. In *Current Trends in Mycorrhizal Research. Proceedings of the National Conference on Mycorrhiza,* B.L. Jalali, and H. Chand (eds.). February 14-16, 1990, Hisar, India, pp. 118-119.

Bagyaraj, D.J., A. Manjunath, and D.D.R. Reddy (1979). Interaction of vesicular-arbuscular mycorrhiza with root knot nematodes in tomato. *Plant and Soil* 51:397-403.

Balestrini, R., C. Romera, P. Puigdomenech, and P. Bonfante (1994). Location of a cell-wall hydroxyproline-rich glycoprotein, cellulose and beta-1,3-glucans in apical and differentiated regions of maize mycorrhizal roots. *Planta* 195:201-209.

Bali, M., and K.G. Mukerji (1988). Effect of VAM fungi on Fusarium wilt of cotton and jute. In *Proceedings of the Asiatic Conference on Mycorrhizae,* A. Mahadevan, N. Raman, and K. Natarajan (eds.). January 29-31, 1988, Madras, India, pp. 233-234.

Baltruschat, H., and F. Schönbeck (1972). Untersuchungen über den Einfluß der endotrophen Mycorrhiza auf die Chalmydosporen Bildung von *Thielaviopsis basicola* in Tabakwurzeln. *Phytopathologische Zeitschrift* 74:358-361.

Baltruschat, H., and F. Schönbeck (1975). Untersuchungen über den Einfluß der endotrophen Mycorrhiza auf den Befall von Tabak mit *Thielaviopsis basicola. Phytopathologische Zeitschrift* 84:172-188.

Barnard, E.L. (1977). *The Mycorrhizal Biology of Liriodendron Tulipifera and its Relationship to Cylindrocladium Root Rot.* PhD thesis, Duke University, North Carolina, 147 pp.

Bartschi, H., V. Gianinazzi-Pearson, and I. Vegh (1981). Vesicular-arbuscular mycorrhiza formation and root rot disease *(Phytophthora cinnamomi)* development in *Chamaecyparis lawsoniana. Phytopathologische Zeitschrift* 102:213-218.

Bécard, G., and J.A. Fortin (1988). Early events of vesicular-arbuscular mycorrhiza formation on Ri T-DNA transformed roots. *New Phytologist* 108:211-218.

Becker, W.N. (1976). *Quantification of Onion Vesicular-Arbuscular Mycorrhizae andt heir Resistance to Pyrenochaeta Terrestris.* PhD thesis, University of Illinois, 72 pp.

Benhamou, N., J.A. Fortin, C. Hamel, M. St-Arnaud, and A. Shatilla (1994). Resistance responses of mycorrhizal Ri T-DNA-transformed carrot roots to infection by *Fusarium oxysporum* f. sp. *chrysanthemi. Phytopathology* 84:958-968.

Bethlenfalvay, G., and R. Linderman (1992). *Mycorrhizae in Sustainable Agriculture.* ASA special publication number 54, Madison, WI, American Society of Agronomy, 124 pp.

Blilou, I., P. Bueno, J.A. Ocampo, and J. Garcia-Garrido (2000). Induction of catalase and ascorbate peroxidase activities in tobacco roots inoculated with the arbuscular mycorrhizal *Glomus mosseae. Mycological Research* 104:722-725.

Blilou, I., J.A. Ocampo, and J.M. Garcia-Garrido (2000). Induction of Ltp (Lipid transfer protein) and Pal (Phenylalanine ammonia-lyase) gene expression in rice roots colonized by the arbuscular mycorrhizal fungus *Glomus mosseae. Journal of Experimental Botany* 51:1969-1977.

Bochow, H., and M. Abou-Shaar (1990). On the phytosanitary effect of VA-mycorrhiza in tomatoes to the corky-root disease. *Zentralblatt fur Mikrobiologie* 145:171-176.

Bødker, L., R. Kjøller, K. Kristensen, and S. Rosendahl (2002). Interactions between indigenous arbuscular mycorrhizal fungi and *Aphanomyces euteiches* in field-grown pea. *Mycorrhiza* 12:7-12.

Bonanomi, A., J.H. Oetiker, R. Guggenheim, T. Boller, A. Wiemken, and R. Vögeli-Lange (2001). Arbuscular mycorrhiza in mini-mycorrhizotrons: first contact of *Medicago truncatula* roots with *Glomus intraradices* induces chalcone synthase. *New Phytologist* 150:573-582.

Boyetchko, S.M., and J.P. Tewari (1988). The effect of VA mycorrhizal fungi on infection by *Bipolaris sorokiniana* in barley. *Canadian Journal of Plant Pathology* 10:361.

Boyetchko, S.M., and J.P. Tewari (1990). Effect of phosphorus and VA mycorrhizal fungi on common root rot of barley. In *Innovation and Integration. Proceedings of the 8th North American Conference on Mycorrhizae*, September 5-8, 1990, Jackson, Wyoming, p. 33.

Brimner, T.A., and G.J. Boland (2003). A review of the non-target effects of fungi used to biologically control plant diseases. *Agriculture Ecosystems and Environment* 100:3-16.

Calvet, C., J.M. Barea, and J. Pera (1993). Growth response of marigold (*Tagetes erecta* L.) to inoculation with *Glomus mosseae, Trichoderma aureoviride* and *Pythium ultimum* in a peat-perlite mixture. *Plant and Soil* 148:1-6.

Caron, M., J.A. Fortin, and C. Richard (1985). Influence of substrate on the interaction of *Glomus intraradices* and *Fusarium oxysporum* f. sp. *radicis-lycopersici* on tomatoes. *Plant and Soil* 87:233-239.

Caron, M., J.A. Fortin, and C. Richard (1986a). Effect of *Glomus intraradices* on infection by *Fusarium oxysporum* f. sp. *radicis-lycopersici* on tomatoes over a twelve-week period. *Canadian Journal of Botany* 64:552-556.

Caron, M., J.A. Fortin, and C. Richard (1986b). Effect of inoculation sequence on the interaction between *Glomus intraradices* and *Fusarium oxysporum* f. sp. *radicis-lycopersici* in tomatoes. *Canadian Journal of Plant Pathology* 8:12-16.

Caron, M., J.A. Fortin, and C. Richard (1986c). Effect of phosphorus concentration and *Glomus intraradices* on Fusarium root rot of tomatoes. *Phytopathology* 76:942-946.

Caron, M., C. Richard, and J.A. Fortin (1986). Effect of preinfestation of the soil by a vesicular-arbuscular mycorrhizal fungus, *Glomus intraradices*, on Fusarium crown and root rot of tomatoes. *Phytoprotection* 67:15-19.

Chakravarty, P., and R.R. Mishra (1986a). Influence of endotrophic mycorrhizae on the fusarial wilt of *Cassia tora* L. *Journal of Phytopathology* 115:130-133.

Chakravarty, P., and R.R. Mishra (1986b). The influence of VA mycorrhizae on the wilting of *Albizia procera* and *Dalbergia sissoo. European Journal of Forest Pathology* 16:91-97.

Chou, L.C., and A.F. Schmitthener (1974). Effect of *Rhizobium japonicum* and *Endogone mosseae* on soybean root rot caused by *Pythium ultimum* and *Phytophthora megasperma* var. *sojae. Plant Disease Reporter* 58:221-225.

Chulan, A.H., G.T. Shaji, and Z.A. Christine (1990). Interactions of mycorrhizal fungi with root pathogen of cocoa. In *Current Trends in Mycorrhizal Research. Proceedings of the National Conference on Mycorrhiza,* B.L. Jalali, and H. Chand (eds.). February 14-16, 1990, Hisar, India, pp. 78-79.

Cooper, K.M., and G.S. Grandison (1987). Effects of vesicular-arbuscular mycorrhizal fungi on infection of tamarillo *(Cyphomandra betacea)* by *Meloidogyne incognita* in fumigated soil. *Plant Disease* 71:1101-1107.

Curl, E.A., and B. Truelove (1986). *The Rhizosphere.* Advanced series in agriculture sciences, Vol. 15. New York: Springer-Verlag, 288 pp.

Daft, M.J., and B.O. Okusanya (1973). Effect of Endogone mycorrhizae on plant growth. V. Influence of infection on the multiplication of virus in tomato, petunia and strawberry. *New Phytologist* 72:975-983.

Dassi, B., E. Dumas-Gaudot, and S. Gianinazzi (1998). Do pathogenesis-related (PR) proteins play a role in bioprotection of mycorrhizal tomato roots towards *Phytophthora parasitica*? *Physiological and Molecular Plant Pathology* 52: 167-183.

Datnoff, L.E., S. Nemec, and K. Pernezny (1995). Biological control of Fusarium crown and root rot of tomato in Florida using *Trichoderma harzianum* and *Glomus intraradices*. *Biological Control* 5:127-131.

Davis, R.M. (1980). Influence of *Glomus fasciculatus* on *Thielaviopsis basicola* root rot of Citrus. *Plant Disease* 64:839-840.

Davis, R.M., and J.A. Menge (1980). Influence of *Glomus fasciculatus* and soil phosphorus on Phytophthora root rot of Citrus. *Phytopathology* 70:447-452.

Davis, R.M., and J.A. Menge (1981). *Phytophthora parasitica* inoculation and intensity of vesicular-arbuscular mycorrhizae in Citrus. *New Phytologist* 87: 705-715.

Davis, R.M., J.A. Menge, and D.C. Erwin (1979). Influence of *Glomus fasciculatus* and soil phosphorus on Verticillium wilt of cotton. *Phytopathology* 69:453-456.

Davis, R.M., J.A. Menge, and G.A. Zentmyer (1978). Influence of vesicular-arbuscular mycorrhizae on Phytophthora root rot of three crop plants. *Phytopathology* 68:1614-1617.

Declerck, S., J. Risede, G. Rufyikiri, and B. Delvaux (2002). Effects of arbuscular mycorrhizal fungi on severity of root rot of bananas caused by *Cylindrocladium spathiphylii*. *Plant Pathology* 51:109-115.

Dehne, H.-W. (1977). *Untersuchungen über den Einfluss der Endotrophen Mycorrhiza auf die Fusarium-Welke an Tomate und Gurke*. PhD thesis, University of Bonn, Germany, 150 pp.

Dehne, H.-W. (1982). Interaction between vesicular-arbuscular mycorrhizal fungi and plant pathogens. *Phytopathology* 72:1115-1119.

Dehne, H.-W. (1987). VA mycorrhizae and plant health. In *Mycorrhizae in the Next Decade. Practical Applications and Research Priorities. Proceedings of the 7th North American Conference on Mycorrhizae*, D.M. Sylvia, L.L. Hung, and J.H. Graham (eds.). Gainesville, FL, p. 192.

Dehne, H.-W., and F. Schönbeck (1975). The influence of endotrophic mycorrhizae on the Fusarium wilt of tomato. *Zeitschrift für Pflanzenkrankheiten und Pflanzenschutz* 82:630-632.

Dehne, H.-W., and F. Schönbeck (1979a). Untersuchungen zum Einfluß der endotrophen Mycorrhiza auf Pflanzenkrankheiten I. Ausbreitung von *Fusarium oxysporum* f. sp. *lycopersici* in Tomaten. *Phytopathologische Zeitschrift* 95:105-110.

Dehne, H.-W., and F. Schönbeck (1979b). Untersuchungen zum Einfluß der endotrophen Mycorrhiza auf Pflanzenkrankheiten II. Phenolstoffwechsel und Lignifizierung. *Phytopathologische Zeitschrift* 95:210-216.

Dehne, H.-W., F. Schönbeck, and H. Baltruschat (1978). Untersuchungen zum Einfluß der endotrophen Mykorrhiza auf Pflanzenkrankheiten 3. Chitinaseaktivität und Ornithinzyklus. *Zeitschrift für Pflanzenkrankheiten und Pflanzenschutz* 85:666-678.

Diaye, D.N., R. Duponnois, A. Brauman, and M. Lepage (2003). Impact of a soil feeding termite, *Cubitermes nikoloensis*, on the symbiotic microflora associated with a fallow leguminous plant *Crotalaria ochroleuca*. *Biology and Fertility of Soils* 37:313-318.

Diedhiou, P.M., J. Hallmann, E.C. Oerke, and H.W. Dehne (2003). Effects of arbuscular mycorrhizal fungi and a non-pathogenic *Fusarium oxysporum* on *Meloidogyne incognita* infestation of tomato. *Mycorrhiza* 13:199-204.

Dugassa, G.D., H. von Alten, and F. Schönbeck (1996). Effects of arbuscular mycorrhiza (AM) on health of *Linum usitatissimum* L. infected by fungal pathogens. *Plant and Soil* 185:173-182.

Dumas, E., V. Gianinazzi-Pearson, and S. Gianinazzi (1989). Production of new soluble proteins during VA endomycorrhiza formation. *Agriculture Ecosystems and Environment* 29:111-114.

Dumas, E., A. Tahiri-Alaoui, S. Gianinazzi, and V. Gianinazzi-Pearson (1990). Observations on modifications in gene expression with VA endomycorrhiza development in tobacco: qualitative and quantitative changes in protein profiles. In *Endocytobiology*, P. Nardon, V. Gianinazzi-Pearson, A.M. Grenier, L. Margulis, and D.C. Smith (eds.). Paris: INRA Press, pp. 153-157.

Ekelund, F., K. Westergaard, and D. Soe (2000). The toxicity of the fungicide propiconazole to soil flagellates. *Biology and Fertility of Soils* 31:70-77.

Elsen, A., S. Declerck, and D. De Waele (2001). Effects of *Glomus intraradices* on the reproduction of the burrowing nematode *(Radopholus similis)* in dixenic culture. *Mycorrhiza* 11:49-51.

Elsen, A., S. Declerck, and D. De Waele (2003). Use of root organ cultures to investigate the interaction between *Glomus intraradices* and *Pratylenchus coffeae*. *Applied and Environmental Microbiology* 69:4308-4311.

Fassuliotis, G. (1979). Plant breeding for root-knot nematode resistance. In *Root-knot Nematodes (Meloidogyne sp.): Systematics, Biology, and Control*, F. Laberti, and C.E. Taylor (eds.). New York: Academic Press, pp. 425-453.

Feldmann, F., N.T.V. Junquiera, and R. Lieberei (1989). Utilization of VA-mycorrhiza as a factor in integrated plant protection. *Agriculture Ecosystems and Environment* 29:131-135.

Fernando, W., and R. Linderman (1997). The effect of mycorrhizal (*Glomus intraradices*) colonization on the development of root and stem rot (*Phytophthora vignae*) on cowpea. *Journal of the National Science Council of Sri Lanka* 25:39-47.

Filion, M. (1998). *Interactions Mycorhizosphériques entre un Champignon Endomycorhizien Arbusculaire (Glomus intraradices) et Différents Micro-organismes Edaphiques*. M.Sc. thesis, Université de Montréal, 103 pp.

Filion, M., M. St-Arnaud, and J.A. Fortin (1999). Direct interaction between the arbuscular mycorrhizal fungus *Glomus intraradices* and different rhizosphere microorganisms. *New Phytologist* 141:525-533.

Filion, M., M. St-Arnaud, C. Guillon, C. Hamel, and S.H. Jabaji-Hare (2001). Suitability of *Glomus intraradices in vitro* produced spores and root segment inoculum for the establishment of a mycorrhizosphere in an experimental microcosm. *Canadian Journal of Botany* 79:879-885.

Filion, M., M. St-Arnaud, and S.H. Jabaji-Hare (2003a). Direct quantification of fungal DNA from soil substrate using real-time PCR. *Journal of Microbiological Methods* 53:67-76.

Filion, M., M. St-Arnaud, and S.H. Jabaji-Hare (2003b). Quantification of *Fusarium solani* f. sp. *phaseoli* in mycorrhizal bean plants and surrounding mycorrhizosphere soil using real-time polymerase chain reaction and direct isolations on selective media. *Phytopathology* 93:229-235.

Filtenborg, O., J. Frisvad, and U. Thrane (1996). Moulds in food spoilage. *International Journal of Food Microbiology* 33:85-102.

Fortin, J.A., G. Bécard, S. Declerck, Y. Dalpé, M. St-Arnaud, A.P. Coughlan, and Y. Piché (2002). Arbuscular mycorrhiza on root-organ cultures. *Canadian Journal of Botany* 80:1-20.

Fox, J.A., and L. Spasoff (1972). Interaction of *Heterodera solanacearum* and *Endogone gigantea* on tobacco. *Journal of Nematology* 4:224-225.

Fracchia, S., A. Godeas, J.M. Scervino, I. Sampedro, J.A. Ocampo, and I. Garcia-Romera (2003). Interaction between the soil yeast *Rhodotorula mucilaginosa* and the arbuscular mycorrhizal fungi *Glomus mosseae* and *Gigaspora rosea*. *Soil Biology and Biochemistry* 35:701-707.

Fracchia, S., M.T. Mujica, I. Garcia-Romera, J.M. Garcia-Garrido, J. Martin, J.A. Ocampo, and A. Godeas (1998). Interactions between *Glomus mosseae* and arbuscular mycorrhizal sporocarp-associated saprophytic fungi. *Plant and Soil* 200:131-137.

Francl, L.J., and V.H. Dropkin (1985). *Glomus fasciculatum*, a weak pathogen of *Heterodera glycines*. *Journal of Nematology* 17:470-475.

Fusconi, A., E. Gnavi, A. Trotta, and G. Berta (1999). Apical meristems of tomato roots and their modifications induced by arbuscular mycorrhizal and soilborne pathogenic fungi. *New Phytologist* 142:505-516.

Gange, A.C. (2000). Arbuscular mycorrhizal fungi, Collembola and plant growth. *Trends in Ecology & Evolution* 15:369-372.

Gange, A.C. (2001). Species-specific responses of a root- and shoot-feeding insect to arbuscular mycorrhizal colonization of its host plant. *New Phytologist* 150: 611-618.

Gange, A.C., and E. Bower (1997). Interactions between insects and mycorrhizal fungi. In *Multitrophic Interactions in Terrestrial Systems*, A.C. Gange, and V.K. Brown (eds.). Oxford: Blackwell Science, pp. 115-132.

Gange, A.C., and V.K. Brown (2002a). Actions and interactions of soil invertebrates and arbuscular mycorrhizal fungi in affecting the structure of plant communities. In *Mycorrhizal Ecology*, M.G.A. VanderHeijden, and I.R. Sanders (eds.). Berlin: Springer-Verlag, pp. 321-344.

Gange, A.C., and V.K. Brown (2002b). Soil food web components affect plant community structure during early succession. *Ecological Research* 17:217-227.

Gange, A.C., and S.J. Case (2003). Incidence of microdochium patch disease in golf putting greens and a relationship with arbuscular mycorrhizal fungi. *Grass and Forage Science* 58:58-62.

Gange, A.C., and H.E. Nice (1997). Performance of the thistle gall fly, *Urophora cardui*, in relation to host plant nitrogen and mycorrhizal colonization. *New Phytologist* 137:335-343.
Gange, A.C., and H.M. West (1994). Interactions between arbuscular mycorrhizal fungi and foliar-feeding insects in *Plantago lanceolata* L. *New Phytologist* 128:79-87.
Gange, A.C., E. Bower, and V.K. Brown (1999). Positive effects of an arbuscular mycorrhizal fungus on aphid life history traits. *Oecologia* 120:123-131.
Gange, A.C., E. Bower, and V.K. Brown (2002). Differential effects of insect herbivory on arbuscular mycorrhizal colonization. *Oecologia* 131:103-112.
Gange, A.C., V.K. Brown, and D.M. Aplin (2003). Multitrophic links between arbuscular mycorrhizal fungi and insect parasitoids. *Ecology Letters* 6:1051-1055.
Gange, A.C., V.K. Brown, and G.S. Sinclair (1994). Reduction of black vine weevil larval growth by vesicular-arbuscular mycorrhizal infection. *Entomologia Experimentalis et Applicata* 70:115-119.
Gange, A.C., P.G. Stagg, and L.K. Ward (2002). Arbuscular mycorrhizal fungi affect phytophagous insect specialism. *Ecology Letters* 5:11-15.
Garcia-Garrido, J.M., and J.A. Ocampo (1988a). Interaccion entre *G. mosseae* y *Pseudomonas syringae* en la rizosfera de plantas de tomate. *Anales de Edafologia y Agrobiologia* 47:1679-1686.
Garcia-Garrido, J.M., and J.A. Ocampo (1988b). Interaction between *Glomus mosseae* and *Erwinia carotovora* and its effects on the growth of tomato plants. *New Phytologist* 110:551-555.
Garcia-Garrido, J.M., and J.A. Ocampo (1989). Effect of VA mycorrhizal infection of tomato on damage caused by *Pseudomonas syringae*. *Soil Biology and Biochemistry* 21:165-167.
Garcia-Garrido, J.M., and J.A. Ocampo (2002). Regulation of the plant defence response in arbuscular mycorrhizal symbiosis. *Journal of Experimental Botany* 53:1377-1386.
Gehring, C.A., and T.G. Whitham (2002). Mycorrhizae-herbivore interactions: Population and community consequences. In *Mycorrhizal Ecology,* M.G.A. van der Heijden, and I.R. Sanders (eds.). Berlin: Springer-Verlag, pp. 295-320.
Gerdemann, J. (1968). Vesicular-arbuscular mycorrhiza and plant growth. *Annual Review of Phytopathology* 6:397-418.
Gerdemann, J.W. (1975). Vesicular-arbuscular mycorrhizae. In *The development and Function of Roots,* J.G. Torrey, and D.T. Clarkson (eds.). New York: Academic Press, pp. 575-591.
Gernns, H., H. von Alten, and H.M. Poehling (2001). Arbuscular mycorrhiza increased the activity of a biotrophic leaf pathogen—Is a compensation possible? *Mycorrhiza* 11:237-243.
Gianinazzi, S., and M. Vosátka (2004). Inoculum of arbuscular mycorrhizal fungi for production systems: science meets business. *Canadian Journal of Botany* 82:1264-1271.
Giovannetti, M., L. Tosi, G. Delatorre, and A. Zazzerini (1991). Histological, physiological and biochemical interactions between vesicular-arbuscular mycorrhizae

and *Thielaviopsis basicola* in tobacco plants. *Phytopathologische Zeitschrift* 131:265-274.

Gonçalves, E.J., J.J. Muchovej, and R.M.C. Muchovej (1991a). Effect of kind and method of fungicidal treatment of bean seed on infections by the VA-mycorrhizal fungus *Glomus macrocarpum* and by the pathogenic fungus *Fusarium solani*. I. Fungal and plants parameters. *Plant and Soil* 132:41-46.

Gonçalves, E.J., J.J. Muchovej, and R.M.C. Muchovej (1991b). Effect of kind and method of fungicidal treatment of bean seed on infections by the VA-mycorrhizal fungus *Glomus macrocarpum* and by the pathogenic fungus *Fusarium solani*. 2. Temporal-spatial relationships. *Plant and Soil* 132:47-51.

Gossen, B., and S. Rimmer (2001). First report of resistance to benomyl fungicide in *Sclerotinia sclerotiorum*. *Plant Disease* 48:121-125.

Goverde, M., M.G.A. van der Heijden, A. Wiemken, I.R. Sanders, and A. Erhardt (2000). Arbuscular mycorrhizal fungi influence life history traits of a lepidopteran herbivore. *Oecologia* 125:362-369.

Graham, J.H. (1986). Citrus mycorrhizae: Potential benefits and interactions with pathogens. *Hortscience* 21:1302-1306.

Graham, J.H. (1988). Interaction of mycorrhizal fungi with soilborne plant pathogens and other organisms: An introduction. *Phytopathology* 78:365-366.

Graham, J.H. (2001). What do root pathogens see in mycorrhizas? *New Phytologist* 149:357-359.

Graham, J.H., and D.S. Egel (1988). Phytophthora root rot development on mycorrhizal and phosphorus-fertilized nonmycorrhizal sweet orange seedlings. *Plant Disease* 72:611-614.

Graham, J.H., and J.A. Menge (1982). Influence of vesicular-arbuscular mycorrhizae and soil phosphorus on take-all disease of wheat. *Phytopathology* 72:95-98.

Graham, J.H., R.T. Leonard, and J.A. Menge (1981). Membrane-mediated decrease in root exudation responsible for phosphorus inhibition of vesicular-arbuscular mycorrhiza formation. *Plant Physiology* 68:548-552.

Guenoune, D., S. Galili, D.A. Phillips, H. Volpin, I. Chet, Y. Okon, and Y. Kapulnik (2001). The defense response elicited by the pathogen *Rhizoctonia solani* is suppressed by colonization of the AM-fungus *Glomus intraradices*. *Plant Science* 160:925-932.

Guillemin, J.P., S. Gianinazzi, V. Gianinazzi-Pearson, and J. Marchal (1994). Contribution of arbuscular mycorrhizas to biological protection of micropropagated pineapple *Ananas comosus* (L) Merr) against *Phytophthora cinnamomi* Rands. *Agricultural Science in Finland* 3:241-251.

Guillon, C., M. St-Arnaud, C. Hamel, and S.H. Jabaji-Hare (2002). Differential and systemic alteration of defence-related gene transcript levels in mycorrhizal bean plants infected with *Rhizoctonia solani*. *Canadian Journal of Botany* 80:305-315.

Halos, P.M., and R.A. Zorilla (1979). Vesicular-arbuscular mycorrhizae increase growth and yield of tomatoes and reduce infection by *Pseudomonas solanacearum*. *Philippine Agricultural Scientist* 62:309-315.

Hamel, C., V. Vujanovic, R. Jeannotte, A. Nakano-Hylander, and M. St-Arnaud (2004). Negative feedback on a perennial crop: Fusarium crown and root rot of asparagus is related to changes in soil microbial community structure. *Plant and Soil* 268:75-87.

Harrington, L. (1995). Sustainability in perspective: strengths and limitations of farming systems research in contributing to sustainable agriculture. *Journal of Sustainable Agriculture* 5:41-59.

Harrison, M.J., and R.A. Dixon (1993). Isoflavonoid accumulation and expression of defense gene transcripts during the establishment of vesicular-arbuscular mycorrhizal associations in roots of *Medicago truncatula. Molecular Plant Microbe interactions* 6:643-654.

Hart, M.M., and J.N. Klironomos (2002). Diversity of arbuscular mycorrhizal fungi and ecosystem functioning. In *Mycorrhizal Ecology,* M.G.A. van der Heijden, and I.R. Sanders (eds.). Berlin: Springer-Verlag, pp. 225-242.

Hart, M.M., R.J. Reader, and J.N. Klironomos (2003). Plant coexistence mediated by arbuscular mycorrhizal fungi. *Trends in Ecology & Evolution* 18:418-423.

Hussey, R.S., and R.W. Roncadori (1977). Interaction of *Pratylenchus brachyurus* and an endomycorrhizal fungus on cotton. *Journal of Nematology* 9:270-271.

Hussey, R.S., and R.W. Roncadori (1978). Interaction of *Pratylenchus brachyurus* and *Gigaspora margarita* on cotton. *Journal of Nematology* 10:16-20.

Hussey, R.S., and R.W. Roncadori (1982). Vesicular-arbuscular mycorrhizae may limit nematode activity and improve plant growth. *Plant Disease* 66:9-14.

Hwang, S. (1988). Effects of VA mycorrhizae and metalaxyl on growth of alfalfa seedlings in soils from fields with alfalfa sickness in Alberta. *Plant Disease* 72:448-452.

Hwang, S.F., P. Chakravarty, and D. Prévost (1993). Effects of rhizobia, metalaxyl, and VA mycorrhizal fungi on growth, nitrogen fixation, and development of *Pythium* root rot of sainfoin. *Plant Disease* 77:1093-1098.

Hwang, S.F., K.F. Chang, and P. Chakravarty (1992). Effects of vesicular-arbuscular mycorrhizal fungi on the development of Verticillium and Fusarium wilts of alfalfa. *Plant Disease* 76:239-243.

Idczak, E., F. Feldmann, and R. Lieberei (1991). Specific response of two Poinsettia varieties to *Pythium ultimum* and VAM In *Mycorrhizas in Ecosystems—Structure and Function. Proceedings of the 3rd European Symposium on Mycorrhiza,* August 19-23, 1991, Sheffield.

Iqbal, S.H., and T. Mahmood (1986). Vesicular-arbuscular mycorrhiza as a deterrent to damping-off caused by *Rhizoctonia solani* in *Brassica napus. Biologia (Lahore)* 32:193-200.

Iqbal, S.H., G. Nasim, and M. Niaz (1987). I. Role of VA mycorrhiza as a deterrent against pathogenic infections caused by *Fusarium moniliforme* in *Brassica oleracea. Biologia (Lahore)* 33:271-278.

Iqbal, S.H., G. Nasim, and M. Niaz (1988a). II. Role of vesicular-arbuscular mycorrhiza as a deterrent to damping-off caused by *Rhizoctonia solani* in *Brassica oleracea. Biologia (Lahore)* 34:79-84.

Iqbal, S.H., G. Nasim, and M. Niaz (1988b). IV. VA mycorrhiza as a deterrent to damping-off caused by *Rhizoctonia solani* at different temperatures regimes. *Biologia (Lahore)* 34:215-222.

Jabaji-Hare, S.H., and L.W. Stobbs (1984). Electron microscopic examination of tomato roots infected with *Glomus* sp. and tobacco mosaic virus. *Phytopathology* 74:277-279.

Jain, R.K., and C.L. Sethi (1987). Pathogenecity of *Heterodera cajani* on cowpea as influenced by the presence of VAM fungi, *Glomus fasciculatum* or *G. epigaeus. Indian Journal of Nematology* 17:165-170.

Jalali, B.L., and H.A.U. Hisar (1991). Mycorrhizal systems in management of plant diseases. *Mycorrhiza news (Asia)* 3:3.

Jeffries, P., S. Gianinazzi, S. Perotto, K. Turnau, and J.M. Barea (2003). The contribution of arbuscular mycorrhizal fungi in sustainable maintenance of plant health and soil fertility. *Biology and Fertility of Soils* 37:1-16.

Johansson, J.F., L.R. Paul, and R.D. Finlay (2004). Microbial interactions in the mycorrhizosphere and their significance for sustainable agriculture. *FEMS Microbiology Ecology* 48:1-13.

Karagiannidis, N., F. Bletsos, and N. Stavropoulos (2002). Effect of Verticillium wilt (*Verticillium dahliae* Kleb.) and mycorrhiza *(Glomus mosseae)* on root colonization, growth and nutrient uptake in tomato and eggplant seedlings. *Scientia Horticulturae* 94:145-156.

Kasiamdari, R.S., S.E. Smith, F.A. Smith, and E.S. Scott (2002). Influence of the mycorrhizal fungus, *Glomus coronatum*, and soil phosphorus on infection and disease caused by binucleate Rhizoctonia and *Rhizoctonia solani* on mung bean *(Vigna radiata). Plant and Soil* 238:235-244.

Kaye, J.W., F.L. Pfleger, and E.L. Stewart (1984). Interaction of *Glomus fasciculatum* and *Pythium ultimum* on greenhouse-grown poinsettia. *Canadian Journal of Botany* 62:1575-1579.

Kellam, M.K., and N.C. Schenck (1980). Interaction between a vesicular-arbuscular mycorrhizal fungus and root-knot nematode on soybean. *Phytopathology* 70: 293-296.

Khadge, B.R., L.L. Ilag, and T.W. Mew (1990). Interaction study of *Glomus mosseae* and *Rhizoctonia solani*. In *Current Trends in Mycorrhizal Research. Proceedings of the National Conference on Mycorrhiza,* B.L. Jalali, and H. Chand (eds.). February 14-16, 1990, Hisar, India, pp. 94-95.

Kjøller, R., and S. Rosendahl (1996). The presence of the arbuscular mycorrhizal fungus *Glomus intraradices* influences enzymatic activities of the root pathogen *Aphanomyces euteiches* in pea roots. *Mycorrhiza* 6:487-491.

Klironomos, J.N. (2002). Feedback with soil biota contributes to plant rarity and invasiveness in communities. *Nature* 417:67-70.

Kowalchuk, G.A., F.A. De Souza, and J.A. Van Veen (2002). Community analysis of arbuscular mycorrhizal fungi associated with *Ammophila arenaria* in Dutch coastal sand dunes. *Molecular Ecology* 11:571-581.

Krishna, K.R., and D.J. Bagyaraj (1983). Interaction between *Glomus fasciculatum* and *Sclerotium rolfsii* in peanut. *Canadian Journal of Botany* 61:2349-2351.

Labour, K., M. Jolicoeur, and M. St-Arnaud (2003). Arbuscular mycorrhizal responsiveness of *in vitro* tomato root lines is not related to growth and nutrient uptake rates. *Canadian Journal of Botany* 81:645-656.

Lambais, M.R. (2000). Regulation of plant defense-related genes in arbuscular mycorrhizae. In *Current Advances in Mycorrhizae Research,* G.K. Podila, and D.D. Douds (eds.). St Paul, Minnesota: American Phytopathological Society Press, pp. 45-59.

Larsen, J., and L. Bødker (2001). Interactions between pea root-inhabiting fungi examined using signature fatty acids. *New Phytologist* 149:487-493.

Larsen, J., S. Ravnskov, and I. Jakobsen (2003). Combined effect of an arbuscular mycorrhizal fungus and a biocontrol bacterium against *Pythium ultimum* in soil. *Folia Geobotanica* 38:145-154.

Lieberei, R., and F. Feldmann (1989). Physiological changes in roots colonized by vesicular arbuscular mycorrhizal fungi—reactions in mutualistic and parasitic interactions. *Agriculture Ecosystems and Environment* 29:251-255.

Linderman, R.G. (1988). Mycorrhizal interactions with the rhizosphere microflora: the mycorrhizosphere effect. *Phytopathology* 78:366-371.

Linderman, R.G. (1992). Vesicular-arbuscular mycorrhizae and soil microbial interactions. In *Mycorrhizae in Sustainable Agriculture,* G.J. Bethlenfalvay, and R.G. Linderman (eds.). Madison, Wisconsin: American Society of Agronomy, pp. 45-70.

Linderman, R.G. (1994). Role of VAM fungi in biocontrol. In *Mycorrhizae and Plant Health,* F.L. Pfleger, and R.G. Linderman (eds.). St. Paul, Minnesota: American Phytopathological Society Press, pp. 1-25.

Linderman, R.G. (1996). Managing soilborne diseases: The microbial connection. In *Management of Soil-Borne Diseases,* V. Gupta, and R. Utkhede (eds.). New Delhi, India: M/S Narosa Publ. House, pp. 3-20.

Linderman, R.G. (2000). Effects of mycorrhizas on plant tolerance to diseases. In *Arbuscular Mycorrhizas: Physiology and Functions,* Y. Kapulnik, and D. Douds (eds.). Netherlands: Kluwer Academic Publishers, pp. 345-365.

Linderman, R.G. (2001). Mycorrhizae and their effects on diseases. In *Diseases of Woody Ornamentals and Trees in Nurseries,* R.K. Jones, and D.M. Benson (eds.). St Paul, Minnesota: American Phytopathological Society, pp. 433-434.

Linderman, R.G., and T.C. Paulitz (1990). Mycorrhizal-rhizobacterial interactions. In *Biological Control of Soil-Borne Plant Pathogens,* D. Hornby, R.J. Cook, Y. Henis, W.H. Ko, A.D. Rovira, B. Schippers, and P.R. Scott (eds.). Wallingford, UK: CAB International, pp. 261-283.

Lingua, G., G. D'Agostino, N. Massa, M. Antosiano, and G. Berta (2002). Mycorrhiza-induced differential response to a yellows disease in tomato. *Mycorrhiza* 12:191-198.

Lingua, G., S. Sgorbati, A. Citterio, A. Fusconi, A. Trotta, E. Gnavi, and G. Berta (1999). Arbuscular mycorrhizal colonization delays nucleus senescence in leek root cortical cells. *New Phytologist* 141:161-169.

Lioussanne, L., M. Jolicoeur, and M. St-Arnaud (2003). Effects of the alteration of tomato root exudation by *Glomus intraradices* colonization on *Phytophthora parasitica* var. *nicotianae* zoospores. In *Proceedings of the 4th International Conference on Mycorrhizae,* August 10-15, 2003, Montréal, p. 291.

Lioussanne, L., M. Jolicoeur, and M. St-Arnaud (2004). Transformed tomato root exudates are less attractive to zoospores of *Phytophthora nicotianae* after colonization by the arbuscular mycorrhizal fungus *Glomus intraradices*. In *Proceedings of the 75th Canadian Phytopathogical Society Annual Congress,* June 13-16, 2004, Ottawa, ON, p. 39.

Little, L.R., and M.A. Maun (1996). The "Ammophila problem" revisited: a role for mycorrhizal fungi. *Journal of Ecology* 84:1-7.

Liu, R.J. (1995). Effect of vesicular-arbuscular mycorrhizal fungi on verticillium wilt of cotton. *Mycorrhiza* 5:293-297.

Maffei, M., A. Codignola, P. Spanu, S. Scannerini, and P. Bonfante-Fasolo (1986). Costituenti fenolici in pareti cellulari di radici in piante axeniche e mycorrizate. *Giornale Botanico Italiano* 120(S 2):22.

Mahmood, T., and S.H. Iqbal (1982). Influence of soil moisture contents on VA mycorrhizal and pathogenic infection by *Rhizoctonia solani* in *Brassica napus*. *Pakistan Journal of Agricultural Research* 3:45-49.

Mahmood, T., and T. Khurshid (1988). Comparative study of VA mycorrhizae and root infecting fungi in wheat roots. *Biologia (Lahore)* 34:303-308.

Mansfeld-Giese, K., J. Larsen, and L. Bødker (2002). Bacterial populations associated with mycelium of the arbuscular mycorrhizal fungus *Glomus intraradices*. *FEMS Microbiology Ecology* 41:133-140.

Mark, G.L., and A.C. Cassells (1996). Genotype-dependence in the interaction between *Glomus fistulosum, Phytophthora fragariae* and the wild strawberry *(Fragaria vesca). Plant and Soil* 185:233-239.

Marschner, P., and K. Baumann (2003). Changes in bacterial community structure induced by mycorrhizal colonisation in split-root maize. *Plant and Soil* 251:279-289.

Mataré, R., and M.J. Hattingh (1978). Effect of mycorrhizal status of avocado seedlings on root rot caused by *Phytophthora cinnamomi. Plant and Soil* 49:433-435.

Mathesius, U. (2003). Conservation and divergence of signalling pathways between roots and soil microbes—the Rhizobium-legume symbiosis compared to the development of lateral roots, mycorrhizal interactions and nematode-induced galls. *Plant and Soil* 255:105-119.

Matsubara, Y., N. Hasegawa, and H. Fukui (2002). Incidence of Fusarium root rot in asparagus seedlings infected with arbuscular mycorrhizal fungus as affected by several soil amendments. *Journal of the Japanese Society for Horticultural Science* 71:370-374.

Matsubara, Y., N. Ohba, and H. Fukui (2001). Effect of arbuscular mycorrhizal fungus infection on the incidence of fusarium root rot in asparagus seedlings. *Journal of the Japanese Society for Horticultural Science* 70:202-206.

Matsubara, Y., H. Tamura, and T. Harada (1995). Growth enhancement and verticillium wilt control by vesicular-arbuscular mycorrhizal fungus inoculation in eggplant. *Journal of the Japanese Society for Horticultural Science* 64:555-561.

Matsubara, Y., Y. Kayukawa, M. Yano, and H. Fukui (2000). Tolerance of asparagus seedlings infected with arbuscular mycorrhizal fungus to violet root rot caused by *Helicobasidium mompa. Journal of the Japanese Society for Horticultural Science* 69:552-556.

McGovern, R.J., L.E. Datnoff, and L. Tripp (1992). Effect of mixed infection and irrigation method on colonization of tomato roots by *Trichoderma harzianum* and *Glomus intraradices*. *Proceedings of the Florida State Horticultural Society* 105:361-363.

McGraw, A.C. (1983). *The Influence of Inoculum Density of Vesicular-Arbuscular Mycorrhizal Fungi on their Development and on Fusarium Wilt of Tomato*. PhD thesis, University of Florida, 131 pp.

McGraw, A.C., and N.C. Schenck (1981). Effects of two species of vesicular-arbuscular mycorrhizal fungi on the development of Fusarium wilt of tomato. *Phytopathology* 71:894.

Meyer, J., and H.-W. Dehne (1986). The influence of VA mycorrhizae on biotrophic leaf pathogens. In Physiological and genetical aspects of mycorrhizae. *Proceedings of the 1st European Symposium on Mycorrhizae,* Dijon July 1-5, 1985. Institut national de la recherche agronomique, pp. 781-786.

Miller, J.C., S. Rajapakse, and R.K. Garber (1986). Vesicular-arbuscular mycorrhizae in vegetables crops. *Hortscience* 21:974-984.

Morandi, D., J.A. Bailey, and V. Gianinazzi-Pearson (1984). Isoflavonoid accumulation in soybean roots infected with vesicular-arbuscular mycorrhizal fungi. *Physiological Plant Pathology* 24:357-364.

Morandi, D., A. Gollotte, and P. Camporota (2002). Influence of an arbuscular mycorrhizal fungus on the interaction of a binucleate *Rhizoctonia* species with Myc(+) and Myc(−) pea roots. *Mycorrhiza* 12:97-102.

Muller, J. (2003). Artificial infection by endophytes affects growth and mycorrhizal colonisation of *Lolium perenne*. *Functional Plant Biology* 30:419-424.

Nemec, S. (1979). *Fusarium oxysporum* wilt disease development in key lime infected with *Glomus etunicatus*. In *Proceedings of the 4th North American Conference on Mycorrhizae,* Colorado University, Fort Collins, p. 72

Nemec, S., and D. Myhre (1984). Virus-*Glomus etunicatum* interactions in citrus rootstocks. *Plant Disease* 68:311-314.

Nemec, S., L.E. Datnoff, and J. Strandberg (1996). Efficacy of biocontrol agents in planting mixes to colonize plant roots and control root diseases of vegetables and citrus. *Crop Protection* 15:735-742.

Newsham, K.K., A.H. Fitter, and A.R. Watkinson (1994). Root pathogenic and arbuscular mycorrhizal fungi determine fecundity of asymptomatic plants in the field. *Journal of Ecology* 82:805-814.

Newsham, K.K., A.H. Fitter, and A.R. Watkinson (1995). Arbuscular mycorrhiza protect an annual grass from root pathogenic fungi in the field. *Journal of Ecology* 83:991-1000.

Niemira, B.A., R. Hammerschmidt, and G.R. Safir (1996). Postharvest suppression of potato dry rot (*Fusarium sambucinum*) in prenuclear minitubers by arbuscular mycorrhizal fungal inoculum. *American Potato Journal* 73:509-515.

Norman, J.R., and J.E. Hooker (2000). Sporulation of *Phytophthora fragariae* shows greater stimulation by exudates of non-mycorrhizal than by mycorrhizal strawberry roots. *Mycological Research* 104:1069-1073.

Norman, J.R., D. Atkinson, and J.E. Hooker (1996). Arbuscular mycorrhizal fungal-induced alteration to root architecture in strawberry and induced resistance to the root pathogen *Phytophthora fragariae*. *Plant and Soil* 185:191-198.

O'Bannon, J.H., and S. Nemec (1979). The response of *Citrus limon* seedlings to a symbiont, *Glomus etunicatus*, and a pathogen, *Radopholus similis*. *Journal of Nematology* 11:270-275.

Orolfo, E.B. (1990). Effect of the biological interactions between mycorrhiza and nematophagous fungi for the control of plant parasitic nematodes. In *Innovation and Integration. Proceedings of the 8th North American Conference on Mycorrhizae,* September 5-8, 1990, Jackson, WY, p. 229.

Perrin, R. (1990). Interactions between mycorrhizae and diseases caused by soil-borne fungi. *Soil Use Management* 6:189-195.

Perrin, R. (1991). Mycorhizes et protection phytosanitaire. In *Les mycorhizes des arbres et plantes cultivées,* D.G. Strullu (eds.). Paris: Technique et documentation-Lavoisier, pp. 93-130.

Pozo, M., C. Azcón-Aguilar, and J. Barea (1996). Defence-related enzyme activation in tomato roots infected by *Phytophthora*. In *Biological Control of Root Pathogens by VA Mycorrhizas: Research into the Mechanisms Involved,* (eds.). Brussels: Official Publications of the European Communities, pp. 47-58.

Pozo, M.J., C. Azcon-Aguilar, E. Dumas-Gaudot, and J.M. Barea (1998). Chitosanase and chitinase activities in tomato roots during interactions with arbuscular mycorrhizal fungi or *Phytophthora parasitica*. *Journal of Experimental Botany* 49:1729-1739.

Pozo, M.J., C. Cordier, E. Dumas-Gaudot, S. Gianinazzi, J.M. Barea, and C. Azcon-Aguilar (2002). Localized versus systemic effect of arbuscular mycorrhizal fungi on defence responses to *Phytophthora* infection in tomato plants. *Journal of Experimental Botany* 53:525-534.

Pozo, M.J., S. Slezack-Deschaumes, E. Dumas-Gaudot, S. Gianinazzi, and C. Azcon-Aguilar (2002). Plant defense responses induced by arbuscular mycorrhizal fungi. In *Mycorrhizal Technology in Agriculture: From Genes to Bioproducts,* S. Gianinazzi, H. Schuepp, J.M. Barea, and K. Haselwandter (eds.). Basel, Switzerland: Birkhauser Verlag Ag, pp. 103-111.

Priestel, G. (1980). *Wechselbeziehung zwischen der endotrophen Mycorrhiza und dem Wurzelgallennematoden Meloidogyne incognita (Kofoid & White, 1919) Chitwood, 1949 an Gurke.* PhD thesis, University of Hannover, 103 pp.

Ramaraj, B., N. Shanmugam, and A. Dwarakanath Reddy (1988). Biocontrol of Macrophomina root rot of cowpea and Fusarium wilt of tomato by using VAM fungi. In *Proceedings of the 1st Asiatic Conference on Mycorrhizae,* A. Mahadevan, N. Raman, and K. Natarajan (eds.). January 29-31, 1988, Madras, India, pp. 250-251.

Rambelli, A. (1973). The rhizosphere of mycorrhizae. In *Ectomycorrhizae,* G.L. Marks, and T.T. Koslowski (eds.). New York: Academic Press, pp. 299-343.

Ramirez, B.N. (1974). *Influence of Endomycorrhizae on the Relationship of Inoculum Density of Phytophthora Palmivora in Soil to Infection of Papaya Roots.* M.Sc thesis, University of Florida, 103 pp.

Ratti, N., M. Alam, S. Sharma, and K. Janardhanan (1998). Effects of *Glomus agregatum* on lethal yellowing disease of Java citronella by *Pythium aphanidermatum*. *Symbiosis* 24:115-126.

Ravnskov, S., J. Larsen, and I. Jakobsen (2002). Phosphorus uptake of an arbuscular mycorrhizal fungus is not effected by the biocontrol bacterium *Burkholderia cepacia*. *Soil Biology and Biochemistry* 34:1875-1881.

Rempel, C.B. (1989). *Interactions between Vesicular-Arbuscular Mycorrhizae (VAM) and Fungal Pathogens in Wheat.* M.Sc thesis, University of Manitoba, 134 pp.

Roncadori, R.W. (1997). Interactions between arbuscular mycorrhizas and plant parasitic nematodes in agro-ecosystems. In *Multitrophic Interactions in Terrestrial Systems,* A.C. Gange, and V.K. Brown (eds.). Oxford: Blackwell Science, pp. 101-113.

Roncadori, R.W., and R.S. Hussey (1977). Interaction of the endomycorrhizal fungus *Gigaspora margarita* and root-knot nematode on cotton. *Phytopathology* 67:1507-1511.

Ronn, R., M. Gavito, J. Larsen, I. Jakobsen, H. Frederiksen, and S. Christensen (2002). Response of free-living soil protozoa and microorganisms to elevated atmospheric CO_2 and presence of mycorrhiza. *Soil Biology and Biochemistry* 34:923-932.

Rosendahl, S. (1985). Interactions between the vesicular-arbuscular mycorrhizal fungus *Glomus fasciculatum* and *Aphanomyces euteiches* root rot of peas. *Phytopathologische Zeitschrift* 114:31-40.

Ross, J.P. (1972). Influence of Endogone mycorrhiza on Phytophthora rot of soybean. *Phytopathology* 62:896-897.

Rousseau, A., N. Benhamou, I. Chet, and Y. Piché (1996). Mycoparasitism of the extramatrical phase of *Glomus intraradices* by *Trichoderma harzianum. Phytopathology* 86:434-443.

Ryan, N.A., T. Deliopoulos, P. Jones, and P.P.J. Haydock (2003). Effects of a mixed-isolate mycorrhizal inoculum on the potato—potato cyst nematode interaction. *Annals of Applied Biology* 143:111-119.

Ryan, N.A., E.M. Duffy, A.C. Cassells, and P.W. Jones (2000). The effect of mycorrhizal fungi on the hatch of potato cyst nematodes. *Applied Soil Ecology* 15:233-240.

Safir, G. (1968). *The Influence of Vesicular-Arbuscular Mycorrhiza on the Resistance of Onion to Pyrenochaeta terrestris.* M.Sc thesis, University of Illinois, 36 pp.

Salem, F.M., M.A. Salem, K. Fawaz, and S.H. Michail (1984). Studies on the interaction between certain mycorrhizal fungi and *Meloidogyne javanica* (Treub) (Nematoda) on root-knot severity and growth on broad bean-plants. *Anzeiger fur Schaedlingskunde Pflanzenschutz Umweltschutz* 57:72-74.

Salzer, P., and T. Boller (2000). Elicitor-induced reactions in mycorrhizae and their suppression. In *Current Advances in Mycorrhizae Research,* G.K. Podila, and D.D. Douds (eds.). St Paul, Minesotta: American Phytopathological Society Press, pp. 1-10.

Salzer, P., H. Corbiere, and T. Boller (1999). Hydrogen peroxide accumulation in *Medicago truncatula* roots colonized by the arbuscular mycorrhiza-forming fungus *Glomus intraradices. Planta* 208:319-325.

Schenck, N.C., and R.A. Kinlock (1974). Pathogenic fungi, parasitic nematodes, and endomycorrhizal fungi associated with soybean roots in Florida. *Plant Disease Reporter* 58:169-173.

Schenck, N.C., R.A. Kinloch, and D.W. Dickson (1975). Interaction of endomycorrhizal fungi and root-knot nematode on soybean. In *Endomycorrhizas*, F.E. Sanders, B. Mosse, and P.B. Tinker (eds.). London: Academic Press, pp. 607-617.

Schenck, N.C., W.H. Ridings, and J.A. Cornell (1977). Interaction of two vesicular-arbuscular mycorrhizal fungi and *Phytophthora parasitica* on two citrus root stocks. In *Proceedings of the 3rd North American Conference on Mycorrhizae*, Corvallis, Oregon, p. 9.

Schönbeck, F. (1979). Endomycorrhiza in relation to plant diseases. In *Soil-Borne Plant Pathogens*, B. Schippers, and W. Gams (eds.). London: Academic Press, pp. 271-280.

Schönbeck, F., and H.-W. Dehne (1977). Damage to mycorrhizal and non mycorrhizal cotton seedlings by *Thielaviopsis basicola*. *Plant Disease Reporter* 61: 266-267.

Schönbeck, F., and H.-W. Dehne (1979). Untersuchungen zum Einfluß der endotrophen Mykorrhiza auf Pflanzenkrankheiten 4. Pilzliche Sproßparasiten, *Olpidium brassicae*, TMV. *Zeitschrift für Pflanzenkrankheiten und Pflanzenschutz* 86:103-112.

Schönbeck, F., and H.-W. Dehne (1981). Mycorrhiza and plant health. *Gesunde Pflanzen* 33:186-190.

Schönbeck, F., and U. Schinzer (1972). Untersuchungen über den Einfluß der endotrophen Mycorrhiza auf die TMV-Läsionenbildung in *Nicotiana tabacum* L. var. *Xanthi*-nc. *Phytopathologische Zeitschrift* 73:78-80.

Schönbeck, F., and G. Spengler (1979). Nachweis von TMV in Mycorrhiza-haltigen Zellen der Tomate mit Hilfe der Immunofluoreszenz. *Phytopathologische Zeitschrift* 94:84-86.

Schreiner, R.P., and G.J. Bethlenfalvay (1995). Mycorrhizal interactions in sustainable agriculture. *Critical Reviews in Biotechnology* 15:271-285.

Schreiner, R.P., and G.J. Bethlenfalvay (2003). Crop residue and Collembola interact to determine the growth of mycorrhizal pea plants. *Biology and Fertility of Soils* 39:1-8.

Schüßler, A. (2002). Molecular phylogeny, taxonomy, and evolution of *Geosiphon pyriformis* and arbuscular mycorrhizal fungi. *Plant and Soil* 244:75-83.

Schwab, S.M., J.A. Menge, and R.T. Leonard (1983). Quantitative and qualitative effects of phosphorus on extracts and exudates of sudangrass roots in relation to vesicular-arbuscular mycorrhiza formation. *Plant Physiology* 73:761-765.

Seitz, L., D. Sauer, H. Mohr, and D. Aldis (1982). Fungal growth and dry matter loss during bin storage of high-moisture corn. *Cereal Chemistry* 59:9-14.

Sharma, A.K. and B.N. Johri (2002). Arbuscular-mycorrhiza and plant disease. In *Arbuscular Mycorrhizae: Interactions in Plants, Rhizosphere and Soils*, A.K. Shaxena, and B.N. Johri (eds.). Enfield, New Hampshire: Science Publishers Inc, pp. 69-96.

Shaul, O., R. David, G. Sinvani, I. Ginzberg, D. Ganon, S. Wininger, B. BenDor, H. Badani, N. Ovdat, and Y. Kapulnik (2000). Plant defense responses during arbuscular mycorrhiza symbiosis. In *Current Advances in Mycorrhizae Research,* G.K. Podila, and D.D. Douds (eds.). St Paul, Minesotta: Americal Phytopathological Society Press, pp. 61-68.

Shaul, O., S. Galili, H. Volpin, I. Ginzberg, Y. Elad, I. Chet, and Y. Kapulnik (1999). Mycorrhiza-induced changes in disease severity and PR protein expression in tobacco leaves. *Molecular Plant Microbe Interactions* 12:1000-1007.

Sikora, R.A. (1979). Predisposition to *Meloidogyne* infection by the endotrophic mycorrhizal fungus *Glomus mosseae.* In *Root-knot Nematodes (Meloidogyne species). Systematics, Biology and Control,* F. Lamberti, and C.E. Taylor (eds.). New York: Academic Press, pp. 399-404.

Sikora, R.A. (1992). Management of the antagonistic potential in agricultural ecosystems for the biological control of plant parasitic nematodes. *Annual Review of Phytopathology* 30:245-270.

Sikora, R.A., and F. Schönbeck (1975). Effect of vesicular arbuscular mycorrhiza *(Endogone mosseae)* on the population dynamics of the root-knot nematodes *Meloidogyne incognita* and *M. hapla.* In *Proceedings of the VIII International Congress of Plant Protection,* Moscow, pp. 158-166.

Singh, Y.P., R.S. Singh, and K. Sitaramaiah (1990). Mechanism of resistance of mycorrhizal tomato against root-knot nematode. In *Current Trends in Mycorrhizal Research. Proceedings of the National Conference on Mycorrhiza,* B.L. Jalali, and H. Chand (eds.). February 14-16, 1990, Hisar, India, pp. 96-97.

Sitaramaiah, K., and R.A. Sikora (1982). Effect of the mycorrhizal fungus, *Glomus fasciculatum* on the host parasite relationship of *Rotylenchus reniformis* in tomato. *Nematologica* 28:412-419.

Sivaprasad, P., A. Jacob, S.K. Nair, and B. George (1990). Influence of VA mycorrhizal colonisation of root-knot nematode infestation in *Piper nigrum* L. In *Current Trends in Mycorrhizal Research. Proceedings of the National Conference on Mycorrhiza,* B.L. Jalali, and H. Chand (eds.). February 14-16, 1990, Hisar, India, pp. 100-101.

Slezack, S., E. Dumas-Gaudot, M. Paynot, and S. Gianinazzi (2000). Is a fully established arbuscular mycorrhizal symbiosis required for bioprotection of *Pisum sativum* roots against *Aphanomyces euteiches*? *Molecular Plant Microbe Interactions* 13:238-241.

Slezack, S., E. Dumas-Gaudot, S. Rosendahl, R. Kjøller, M. Paynot, J. Negrel, and S. Gianinazzi (1999). Endoproteolytic activities in pea roots inoculated with the arbuscular mycorrhizal fungus *Glomus mosseae* and/or *Aphanomyces euteiches* in relation to bioprotection. *New Phytologist* 142:517-529.

Slezack, S., J. Negrel, G. Bestel-Corre, E. Dumas-Gaudot, and S. Gianinazzi (2001). Purification and partial amino acid sequencing of a mycorrhiza-related chitinase isoform from *Glomas mosseae*-inoculated roots of *Pisum sativum* L. *Planta* 213:781-787.

Smith, G.S. (1988). The role of phosphorus nutrition in interactions of vesicular-arbuscular mycorrhizal fungi with soilborne nematodes and fungi. *Phytopathology* 78:371-374.

Smith, S.E., and V. Gianinazzi-Pearson (1988). Physiological interactions between symbionts in vesicular-arbuscular mycorrhizal plants. *Annual Review of Plant Physiology and Plant Molecular Biology* 39:221-244.

Smith, S.E., and D.J. Read (1997). *Mycorrhizal Symbiosis.* 2nd edition. San Diego, London: Academic Press, 605 pp.

Smith, G.S., R.S. Hussey, and R.W. Roncadori (1986). Penetration and postinfection development of *Meloidogyne incognita* on cotton as affected by *Glomus intraradices* and phosphorus. *Journal of Nematology* 18:429-435.

Sood, S.G. (2003). Chemotactic response of plant-growth-promoting bacteria towards roots of vesicular-arbuscular mycorrhizal tomato plants. *FEMS Microbiology Ecology* 45:219-227.

Spanu, P., and P. Bonfante-Fasolo (1988). Cell-wall bound peroxydase activity in roots of mycorrhizal *Allium porrum. New Phytologist* 109:119-124.

Srinath, J., D. Bagyaraj, and B. Satyanarayana (2003). Enhanced growth and nutrition of micropropagated *Ficus benjamina* to *Glomus mosseae* co-inoculated with *Trichoderma harzianum* and *Bacillus coagulans. World Journal of Microbiology & Biotechnology* 19:69-72.

St-Arnaud, M. (1998). *Effet de la Symbiose Endomycorhizienne à Vésicules et Arbuscules sur le Développement de Mycoses Racinaires: Identification des Mécanismes dAaction.* PhD thesis, Université de Montréal, 202 pp.

St-Arnaud, M., and A. Elsen (2005). Interaction with soil borne pathogens and non-pathogenic rhizosphere micro-organisms. In *Root-Organ Culture of Mycorrhizal Fungi,* S. Declerck, D.-G. Strullu, and J.A. Fortin (eds.). New York: Springer-Verlag, pp. 217-231.

St-Arnaud, M., C. Hamel, M. Caron, and J.A. Fortin (1994). Inhibition of *Pythium ultimum* in roots and growth substrate of mycorrhizal *Tagetes patula* colonized with *Glomus intraradices. Canadian Journal of Plant Pathology* 16:187-194.

St-Arnaud, M., C. Hamel, M. Caron, and J.A. Fortin (1995). Endomycorhizes VA et sensibilité aux maladies: synthèse de la littérature et mécanismes d'interaction potentiels. In *La Symbiose Mycorhizienne—État des Connaissances,* J.A. Fortin, C. Charest, and Y. Piché (eds.). Frelighsburg, Québec: Orbis Publishing, pp. 51-87.

St-Arnaud, M., C. Hamel, B. Vimard, M. Caron, and J.A. Fortin (1995). Altered growth of *Fusarium oxysporum* f. sp. *chrysanthemi* in an *in vitro* dual culture system with the vesicular arbuscular mycorrhizal fungus *Glomus intraradices* growing on *Daucus carota* transformed roots. *Mycorrhiza* 5:431-438.

St-Arnaud, M., C. Hamel, B. Vimard, M. Caron, and J.A. Fortin (1996). Enhanced hyphal and spore production of the arbuscular mycorrhizal fungus *Glomus intraradices* in an *in vitro* system in the absence of host roots. *Mycological Research* 100:328-332.

St-Arnaud, M., C. Hamel, B. Vimard, M. Caron, and J.A. Fortin (1997). Inhibition of *Fusarium oxysporum* f.sp. *dianthi* in the non-VAM species *Dianthus caryophyllus* by co-culture with *Tagetes patula* companion plants colonized by *Glomus intraradices. Canadian Journal of Botany* 75:998-1005.

Stewart, E.L., and F.L. Pfleger (1977). Development of poinsettia as influenced by endomycorrhizae, fertilizer and root rot pathogens *Pythium ultimum* and *Rhizoctonia solani. Florist's Review* 159:37-79.

Strack, D., T. Fester, B. Hause, W. Schliemann, and M.H. Walter (2003). Arbuscular mycorrhiza: biological, chemical, and molecular aspects. *Journal of Chemical Ecology* 29:1955-1979.

Subhashini, D.V. (1990). The role of VA mycorrhiza in controlling certain root diseases of tobacco. In *Current Trends in Mycorrhizal Research. Proceedings of the National Conference on Mycorrhiza,* B.L. Jalali, and H. Chand (eds.). February 14-16, 1990, Hisar, India, p. 102.

Talavera, M., K. Itou, and T. Mizukubo (2001). Reduction of nematode damage by root colonization with arbuscular mycorrhiza (*Glomus* spp.) in tomato-*Meloidogyne incognita* (Tylenchida: Meloidognidae) and carrot-*Pratylenchus penetrans* (Tylenchida: Pratylenchidae) pathosystems. *Applied Entomology and Zoology* 36:387-392.

Thomas, L., B.C. Mallesha, and D.J. Bagyaraj (1994). Biological control of damping-off of cardamom by the VA mycorrhizal fungus, *Glomus fasciculatum. Microbiological Research* 149:413:417.

Thompson, J.P., and G.B. Wildermuth (1989). Colonization of crop and pasture species with vesicular-arbuscular mycorrhizal fungi and a negative correlation with root infection by *Bipolaris sorokiniana. Canadian Journal of Botany* 69:687-693.

Torres-Barragán, A., E. Zavaleta-Mejía, C. González-Chávez, and R. Ferrera-Cerrato (1996). The use of arbuscular mycorrhizae to control onion white rot (*Sclerotium cepivorum* Berk) under field conditions. *Mycorrhiza* 6:253-257.

Tosi, L., M. Giovannetti, A. Zazzerini, and G. Della Torre (1988). Influence of mycorrhizal tobacco roots, incorporated into the soil, on the development of *Thielaviopsis basicola. Phytopathology* 122:186-189.

Toussaint, J.-P., M. St-Arnaud, and C. Charest (2004). Nitrogen transfer and assimilation between the arbuscular mycorrhizal fungus *Glomus intraradices* Schenck & Smith and Ri T-DNA roots of *Daucus carota* L. in an *in vitro* compartmented system. *Canadian Journal of Microbiology* 50:251-260.

Traquair, J.A., and D.L. Pohlman (1990). Endomycorrhizal biocontrol of Cylindrocarpon root rot of peach trees in field soil. In *Innovation and Integration. Proceedings of the 8th North American Conference on Mycorrhizae,* September 5-8, 1990, Jackson, WY, p. 288.

Trotta, A., G.C. Varese, E. Gnavi, A. Fusconi, S. Sampo, and G. Berta (1996). Interactions between the soilborne root pathogen *Phytophthora nicotianae* var *parasitica* and the arbuscular mycorrhizal fungus *Glomus mosseae* in tomato plants. *Plant and Soil* 185:199-209.

Umesh, K.C., K. Krishnappa, and D.J. Bagyaraj (1988). Interaction of burrowing nematode, *Radopholus similis* (Cobb, 1893) Thorne 1949, and VA mycorrhiza, *Glomus fasciculatum* (Thaxt). Gerd and Trappe in banana (*Musa acuminata* Colla.). *Indian Journal of Nematology* 18:6-11.

van Aarle, I.M., B. Soderstrom, and P.A. Olsson (2003). Growth and interactions of arbuscular mycorrhizal fungi in soils from limestone and acid rock habitats. *Soil Biology and Biochemistry* 35:1557-1564.

van der Heijden, M.G.A., J.N. Klironomos, M. Ursic, P. Moutoglis, R. Streitwolf-Engel, T. Boller, A. Wiemken, and I.R. Sanders (1998). Mycorrhizal fungal diversity determines plant biodiversity, ecosystem variability and productivity. *Nature* 396:69-72.

Vimard, B., M. St-Arnaud, V. Furlan, and J.A. Fortin (1999). Colonization potential of *in vitro*-produced arbuscular mycorrhizal fungus spores compared with a root-segment inoculum from open pot culture. *Mycorrhiza* 8:335-338.

Volpin, H., Y. Elkind, Y. Okon, and Y. Kapulnik (1994). A vesicular arbuscular mycorrhizal fungus *(Glomus intraradix)* induces a defense response in alfalfa roots. *Plant Physiology* 104:683-689.

Vujanovic, V., C. Hamel, E. Yergeau, and M. St-Arnaud (2006). Biodiversity and biogeography of *Fusarium* species from northeastern North American asparagus fields based on microbiological and molecular approaches. *Microbial Ecology*. In press.

Wacker, T.L., G.R. Safir, and C.T. Stephens (1990). Effect of *Glomus fasciculatum* on the growth of asparagus and the incidence of Fusarium root rot. *Journal of the American Society of Horticultural Sciences* 115:550-554.

Wamberg, C., S. Christensen, and I. Jakobsen (2003). Interaction between foliar-feeding insects, mycorrhizal fungi, and rhizosphere protozoa on pea plants. *Pedobiologia* 47:281-287.

Whatley, T.L., and J.W. Gerdemann (1981). The effect of *Glomus etunicatus* and soil phosphorus on Phytophthora root rot of soybean. *Phytopathology* 71:912.

Whipps, J.M. (2004). Prospects and limitations for mycorrhizas in biocontrol of root pathogens. *Canadian Journal of Botany* 82:1198-1227.

Wyss, P., T. Boller, and A. Wiemken (1989). Glyceollin production in soybean during the process of infection by *Glomus mosseae* and *Rhizoctonia solani*. *Agriculture Ecosystems and Environment* 29:451-456.

Wyss, P., T. Boller, and A. Wiemken (1992). Testing the effect of biological control agents on the formation of vesicular arbuscular mycorrhiza. *Plant and Soil* 147: 159-162.

Yao, M.K., R.J. Tweddell, and H. Desilets (2002). Effect of two vesicular-arbuscular mycorrhizal fungi on the growth of micropropagated potato plantlets and on the extent of disease caused by *Rhizoctonia solani*. *Mycorrhiza* 12:235-242.

Yergeau, E. (2004). *Caractérisation moléculaire de la biodiversité des Fusarium associés à la fusariose de l'asperge (*Asparagus officinalis *L.) au Québec*. M.Sc. thesis, Université de Montréal, 97 p.

Yergeau, E., M. Filion, V. Vujanovic, and M. St-Arnaud (2005). A PCR-denaturing gradient gel electrophoresis (DGGE) approach to assess *Fusarium* diversity in asparagus. *Journal of Microbiological Methods* 60:143-154.

Zak, B. (1964). Role of mycorrhizae in root disease. *Annual Review of Phytopathology* 2:377-392.

Zambolim, L., and N.C. Schenck (1981). Interactions between a vesicular-arbuscular mycorrhiza and root-rot infecting fungi on soybean. *Phytopathology* 71:267.

Zambolim, L., and N.C. Schenck (1983). Reduction of the effects of pathogenic root-infecting fungi on soybean by mycorrhizal fungus, *Glomus mosseae*. *Phytopathology* 73:1402-1405.

Chapter 4

Capturing the Benefits of Arbuscular Mycorrhizae in Horticulture

John Larsen
Sabine Ravnskov
Jorn Nygaard Sorensen

Over the past decade, inocula of arbuscular mycorrhizal fungi (AMF) have become commercially available, but we are still far from integrating arbuscular mycorrhiza (AM) into plant production. The role of AMF in plant production has been marginalized in high-input agriculture through the use of pesticides and fertilizers, creating scenarios where the symbiosis may even be causing growth depressions when the cost (carbon drain) of the symbiosis becomes higher than the benefit (phosphorous uptake) (Ryan and Graham, 2002). More recently, the focus has been on areas with little or no AMF inoculum potential, such as plant production in horticulture that is based on soilless growth media. AMF have also been used in the restoration of disturbed landscapes, such as after major construction work (Dodd et al., 2002), in the bioremediation of polluted soils (Joner and Leyval, 2003; Vivas et al., 2003), and in the restoration of desertified ecosystems (Requena et al., 2001).

Since several other reviews on AM and horticulture are available (Linderman, 1986; Gianinazzi, Trouvelot, and Gianinazzi-Pearson, 1990; Azcón-Aguilar and Barea, 1997; Vosatká et al., 1999; Marx, Marrs, and Cordell, 2002), this chapter is not an extensive review on

Mycorrhizae in Crop Production
© 2007 by The Haworth Press, Inc. All rights reserved.
doi:10.1300/5425_04

AMF and horticulture; rather, it is an update presenting the prospects and limitations for integrating AMF into horticultural plant production, with a focus on compatibility between AMF and plant production systems.

ROLE OF AM IN PLANT PRODUCTION

In the early phase of AM research (1960s-1980s), researchers focused on plant nutrition and tried to promote the commercial use of AMF as biofertilizers for plant production. However, this approach did not fit into modern agriculture with its intensive use of fertilizers and pesticides. More recently, other plant benefits, such as increased stress tolerance, have become more important arguments in favor of AM integration. This is especially true in plant-production systems without AMF, either in soil where AMF have been eradicated by soil disinfection or in soilless growth media in greenhouse production, where inoculation with AMF can protect the host plant against both biotic (Rosendahl and Rosendahl, 1990; St-Arnaud et al., 1994; Linderman, 2000; Thygesen, Larsen, and Bødker, 2004; Whipps, 2004) and abiotic stress (Al-Karaki and Hammad, 2001; Cantrell and Linderman, 2001; Davies et al., 2002). In modern horticulture, plants are often grown in soilless growth media with nutrient solutions that are optimal for plant growth. These inert systems are used worldwide in horticulture chiefly because they provide precise plant nutrition and pathogen management and because these media are much lighter than soil and therefore easier to transport and handle in the nursery. Over the past ten years, it has become evident that production systems using inert growth media lack biological buffers against root diseases; pathogens, when they invade these systems, have little competition from other microorganisms. Recent results show that fewer root diseases develop if the inert growth media are reused (Postma et al., 2000), indicating that inert systems can be too clean. Microbial biocontrol agents have therefore become more common in greenhouse systems (Paulitz and Belanger, 2001). A parallel can be found in field-grown systems, where the soil is disinfected to manage pests. Here, biological activation with green manure crops and/or inoculation with AMF have been explored (Haas et al., 1987; Kabir and Koide, 2002).

POTENTIAL AREAS FOR INTEGRATING AMF INTO HORTICULTURE

Most horticultural crops can potentially benefit from AMF, as the vast majority of these crops are known to host AMF. In general, AMF can be integrated into plant production in two different ways, (1) by managing indigenous populations and (2) by inoculating the plant growth substrate with known AMF. The applicability of these methods depends on the individual production system. Indigenous populations of AMF can be managed using various methods for the production of plants in the field, low input of pesticides and P fertilizers, reduced tillage, crop rotations avoiding non-mycorrhizal crops, cover crops, and intercropping. In systems where the indigenous AMF populations have been reduced by soil disinfection, it is possible to reestablish AMF by growing a mycorrhizal cover crop prior to the main crop and/or by adding an AMF inoculum to the soil. However, to reduce the amount of AMF inoculum needed, it may be more feasible to preinoculate transplants with AMF. Preinoculation of transplants may also be useful in non-disinfected soil as a way to "kick start" transplant growth and/or improve plant establishment, especially for high-value crops. In some cases, transplants are produced through micropropagation; in those growth systems, AMF inoculation can be performed aseptically or simply by mixing an inoculum into the growth media during acclimatization (Lovato et al., 1996). AMF can also be applied to soilless growth media in the greenhouse either by preinoculating transplants in mycorrhiza-conducive media or simply by mixing an AMF inoculum into the growth media.

USE OF AMF IN FIELD-GROWN HORTICULTURAL CROPS

Management of Indigenous AMF Populations

Interactions with Other Soil Biota

It is well-known that the soil inoculum potential of AMF is affected by common plant production practices (Hayman, 1982; Hamel, 1996), but our understanding of the impact of other soil biota on the inoculum potential of AMF and how this affects soil/fertility is still limited despite enormous interest in this area (Fitter and Garbaye, 1994; Jeffries et al., 2003; Johansson, Paul, and Finlay, 2004; Leake et al., 2004). This may explain why it has been difficult to validate P transport by AMF and other functions of AMF in the field. AMF spores and mycelium can be subject to parasitism by hyperparasitic fungi (Rousseau et al., 1996) and predation by mycophagous soil animals such as collembolans (Larsen and Jakobsen, 1996). Studies on the impact of macroinvertebrates such as earthworms on AMF are limited, but new results show that earthworms have no effect on root colonization (Wurst et al., 2004) or P transport by an AM fungus (Tuffen, Eason, and Scullion, 2002); in fact, increases in external mycelium growth in root-free compartments as a result of the presence of earthworms have even been reported (Gormsen, Olsson, and Hedlund, 2004). Most studies on the interaction between AMF and soil biota show limited effects on AM fungal growth and P transport (Larsen and Jakobsen, 1996; Green et al., 1999; Ravnskov et al., 1999, 2003). However, most of these studies have been performed under controlled conditions in pots with only a few isolates of AMF and soil biota, demonstrating the need for more field-oriented studies (Schweiger, Spliid, and Jakobsen, 2001) to demonstrate the extent to which interactions with other soil biota affect AM functioning in the field.

Crop Rotation

Some plant species, mainly those belonging to the *Cruciferae* and *Chenopodiaceae* families, do not form symbioses with AMF. The P uptake of an AMF-dependent crop can be reduced in particular during early growth following those plant species or after a fallow period (Thompson, 1994; Gavito and Miller, 1998a). AMF inoculum potential in soils that were previously used to grow a non-mycorrhizal crop can be increased by growing a mycorrhizal green manure cover crop (Boswell et al., 1998). The growth of mycorrhiza-dependent crops such as leeks, onions, lettuce, and carrots can respond to the previous crop (Sorensen, Larsen, and Jakobsen, 2003; Sorensen, Larsen, and Jakobsen, 2005). However, these mycorrhiza-dependent crops often have long production times, so in many situations colonization is simply delayed and impact on yield is not always apparent. In contrast, Bødker and Kristensen (1999) showed that the use of non-mycorrhizal green manure crops *(Raphanus sativus* and *Brassica campestris)* as pre-crops prior to a main crop of peas resulted in an unexpected increase in AMF colonization. Suppression of other root-inhabiting fungi or soil biota, which is antagonistic to AMF during organic matter decomposition, may be responsible for these results; this highlights the importance of ensuring that interactions with other soil biota are not ignored when the effect of plant production methods on AMF inoculum potential is examined. Intercropping can be used as a way to link the roots of a transplanted crop to the AM mycelial network that is already established. This possibility was studied in some detail after it became evident that plants can share the same mycorrhizal network (Graves et al., 1997). In a field experiment, we tested the role of a black medick cover crop where leek transplants were planted.

Tillage

It is well established that tillage reduces the inoculum potential of AMF owing to disruption of the extraradical mycelium (McGonigle and Miller, 1993). The result may be reduced nutrient uptake, growth, and final yield (McGonigle, Evans, and Miller, 1990; Kabir, O'Halloran, Fyles et al., 1998; Mozafar et al., 2000). McGonigle and Miller (1993) suggested that the roots of newly developing plants become connected to the intact mycelium, which then serves as a nutrient acquisition system, obviating the need to develop a new mycelium from primary infections. Disturbing only the top 5 cm soil layer did not affect AMF colonization; this finding might be used to develop a tillage strategy that is compatible with AM (Kabir, O' Halloran, Widdens et al., 1998). Gavito and Miller (1998b) showed that, compared with conventional tillage, no-tillage increases shoot P concentration and AMF colonization, but the yield in no-tillage treatments is lower than in the conventional tillage treatment.

Pest Management Methods

Pests such as weeds, herbivores, and diseases can potentially cause major losses in horticultural plant production. Consequently, growers try to manage these pests using pesticides, soil disinfection, crop rotation, and resistant varieties. These pest management methods are not always compatible with the integration of AMF into plant production. The impact of pesticides on AMF has been studied intensively, and it is clear that AMF are sensitive to a broad range of fungicides (Larsen et al., 1996; Kjoller and Rosendahl, 2000). Since most of the studies have been carried out as pot experiments, under conditions in which it can be difficult to convert the recommended field application dosage, there is a need for more field-oriented work in this area. Schweiger, Spliid, and Jakobsen (2001) examined the effects of benomyl on hyphal P transport by indigenous AMF populations in a pea field. In this study, the recommended dosage of benomyl actually increased AM hyphal P transport, most likely as a result of interactions with saprotrophic fungi, which may have been suppressed more by benomyl than the AMF were. AMF should be more compatible with foliar pesticides than soil pesticides, since foliar pesticides are applied with little runoff to the soil and are not transported from the shoot to the root. However, foliar applications in extremely sandy soil may have a stronger impact on AMF. In addition, persistent pesticides applied to foliage may still be active when crop residues are incorporated into the soil. Few pesticides are used in soil to manage root diseases in field-grown crops, and in most cases these pesticides are used only in limited amounts as a seedcoat. It is well-known that soil fumigation to control weeds and soilborne diseases reduces AMF root colonization; the effects depend on the dosage and depth of penetration of the fumigant (Menge, 1982). Other soil disinfection methods, such as steaming (An, Guo, and Hendrix, 1998) and solarization (BendavidVal et al., 1997), also adversely affect AMF. Control of weed seeds can be achieved by disinfecting the upper 5 cm of the soil, which may not have a strong impact on AMF. Recently, it was found that AMF colonization of *Alnus incana* roots in a forest nursery was not affected by disinfection of the upper 5 cm of the soil surface using dazomet fumigation, steaming, and biofumigation with a *Brassica* green manure (Welc, Ravnskov, and Larsen, unpublished). This finding suggests that this method of managing weed seeds may be compatible with AMF.

Fertilizers

High nutrient input in plant production seems to have adverse effects on AMF. A negative correlation between soil P concentration and AM formation has been clearly demonstrated (Hayman, 1982; Thomson, Robson, and Abbott, 1992; Olsson, Bååth, and Jakobsen, 1997). It has been known for several decades that AM colonization is low in agricultural systems in which there is high availability of P in the soil (Ryan and Graham, 2002). However, the mechanism underlying this reduced colonization in soil with high P is still not fully understood. Little is also known about the correlation between colonization and functioning of AM in terms of P transport and stress alleviation. In addition, the impact of soil P on AM depends on the mycotrophic nature of the host and the P sensitivity of the AM fungus (Sylvia and Schenck, 1983). Some AM symbioses are therefore more strongly affected by high P levels than others. In contrast, organic nutrients seem to affect AMF differently depending on the source of the organic matter in question. The response of AMF to organic matter seems to depend on the carbon/nitrogen ratio of these organic substrates. Ravnskov et al. (1999) examined the impact of various organic substrates with and without nitrogen on the growth of the external mycelium of *Glomus intraradices* and found that compounds containing N, such as bovine serum albumin and bakers' dry yeast, increased mycelium growth in *G. intraradices,* whereas compounds without N, such as cellulose, decreased the growth of the external mycelium. The suppression of mycelial growth in AMF by cellulose seems to be related to increased saprotrophic fungi growth (Ravnskov et al., 1999). Similarly, Gryndler et al. (2002) found that cellulose decreased AM formation in the early phase of plant growth, but increased AM formation after a longer period of growth. AM seem to accelerate the decomposition of organic matter (Hodge, Campbell, and Fitter, 2001), most likely through their modulating effects on soil microbial communities. It is well established that AMF are obligate symbionts and therefore need their host plants to complete their life cycles. However, Hildebrandt, Janetta, and Bothe (2002) suggested that AMF might be able to complete their life cycles without their plant partners by establishing consortia with other soil microbes instead. Mosse (1988) also suggested the possibility of such "saprotrophic" growth of AMF, but such growth of AMF alone or in consortia with other microorganisms remains to be demonstrated.

Preinoculation of Transplants

Many plants, especially high-value crops, are produced using a transplant stage; in such cases, preinoculation with AMF has been shown to increase plant growth and vigor. Post–in vitro inoculation of micropropagated plants with AMF has resulted in better growth and acclimatization as compared to non-inoculated controls (Estaun et al., 1999; Jaizme-Vega et al., 2003). Several studies have shown that pre-inoculation of vegetable transplants with AMF can increase yield not only in low P soil but also in soils with moderate P levels (Regvar, Vogel-Mikus, and Severkar, 2003; Sorensen, Larsen, and Jakobsen, 2003; Ortas, 2003). However a response to pre-inoculation may not be obtained in field soils containing healthy AMF populations. The authors explained the difference as being due to a more abundant and effective indigenous AMF population in the organically cultivated soil. In contrast, Douds and Reider (2003) found no difference in growth response to AMF in preinoculated mycorrhizal green pepper plants grown in high P soil that was fertilized with dairy cow manure and conventional chemical fertilizer, respectively. They tested two kinds of AMF inocula, a single-fungus inoculum consisting of *Glomus intraradices,* and a mixed inoculum combining *Glomus mosseae, Glomus etunicatum,* and *Gigaspora rosea*. The experiment was repeated over three years. In the first year, the yield of marketable fruit was higher from plants inoculated with the mixture of AMF, whereas the yield was lower with single-species inoculation using *G. intraradices* as compared to non-inoculated control plants. In the third year, both mycorrhizal treatments resulted in an increased yield of marketable fruit as compared to non-mycorrhizal plants.

Root Trimming

The roots of transplants are often trimmed to facilitate planting, and in theory this practice seems to be incompatible with AM formation. The use of plug plants instead of bare-root plants would increase the benefit of preinoculation with AMF, because the external mycelium that develops during the transplant production period of 10 weeks or more can start working right after transplanting. However, the impact of trimming transplant roots on AM functioning remains to be examined.

Soil Disinfection

Outplanting preinoculated seedlings seems to be highly compatible with production systems using soil disinfection, in which the indigenous AMF have been eradicated. Linderman and Davis (2001) tested the effect of three AMF on seven grapevine rootstock cultivars grown in fumigated low-P soil. In general, inoculation with AMF resulted in increased shoot growth in the plants, although the effect of AMF on plant growth depended on the combination of plant cultivar and AMF. Similarly, higher yield has been found in several other crops after inoculation with AMF in fumigated soil (Haas et al., 1987; Kapulnik et al., 1994; Koch et al., 1997).

Undesired Effects of Introducing Commercial AMF to Soil Ecosystems

Most commercial AMF inocula contain a mix of only a few AMF, and concern has been raised about whether the introduction of AMF inocula to agroecosystems could have undesired effects on indigenous AMF populations, reducing diversity not only in the agroecosystems but also in neighboring natural soil ecosystems. Very little work has been done in this area to date, but new molecular techniques will make it possible to examine such potential undesired effects of inoculation. One way to avoid the potential effects of introducing AM inocula to the field could be to use indigenous AMF populations as sources of inoculum (Gaur and Adholeya, 2005).

Plant Breeding

In most types of plant production, breeders are constantly providing growers with new varieties that have desired features such as disease resistance, taste, and color. Breeding and/or genetic manipulation of this sort will continue, most likely without consideration being given to whether the new varieties are compatible with AMF or other plant-beneficial microbes. It therefore seems logical to use a mixed inoculum of AMF, as is the case with most commercial AMF inocula, since AMF inoculum producers are unlikely to produce inocula that have the optimal isolates of AMF for every new plant variety. Indeed, one of the main challenges in plant production may be to convince breeders to consider AM compatibility when developing new varieties (Hetrick, Wilson, and Cox, 1993; Zhu et al., 2001).

Economics

The parasitic-mutualistic continuum of root-inhabiting fungi is complex, and even common pathogens have been shown to be mutualistic in terms of plant growth (Johnson, Graham, and Smith, 1997; Redman, Dunigan, and Rodriguez, 2001). It is therefore difficult to perform exact cost-benefit analyses that focus solely on the host plant, making the cost-benefit analyses of integrating AMF into plant production even more difficult. Miller, McGonigle, and Addy (1994) performed an economic analysis that considered the potential environmental benefits of making better use of AMF in field-grown crops. The analysis suggested that the benefit of no-tillage and lower P fertilization practices that give mycorrhizal fungi a more significant role in plant production would be a cleaner environment. This would save the cost of cleaning polluted lakes and rivers after eutrophication caused by fertilizer runoff has occurred.

INTEGRATING AM INTO GREENHOUSE PLANT PRODUCTION

Plant-Beneficial Features of AMF

Greenhouse production of ornamentals and vegetables is to a great extent based on soilless growth media, and it is necessary to inoculate these production systems in order to benefit from AMF. The main benefit of AMF in these systems seems to be increased plant tolerance to stress, since the nutritional issue is not important. This is because plant nutrients can be given at an optimum dosage with minimal environmental problems if the nutrient solution is recirculated. Besides increased stress tolerance, the benefits of AMF in these systems are early flowering, better rooting of plants propagated from cuttings, and biological activation of inert growth media.

Stress Alleviation

AM can alleviate both abiotic and biotic stress. It is well-known that mycorrhizal plants in general are more resistant to and develop fewer diseases caused by root pathogens (reviewed e.g., by Linderman, 1994; Azcón-Aguilar and Barea, 1996; Graham, 2001; Whipps, 2004). The role of AM in alleviating biotic stress caused by plant pathogens is reviewed by St-Arnaud and Vujanovic in Chapter 3 of this book. AM have also been shown to increase plant resistance and tolerance to high levels of salinity (e.g., Cantrell and Linderman, 2001) and drought. The role of AM in drought stress has recently been reviewed by Ruiz-Lozano (2003). Abiotic stress is most likely not as common during plant production in greenhouse systems, since water and nutrient solutions are optimized. However, these features of AM may be valuable in increasing the longevity of pot plants when they are stored before sale in plant nurseries or supermarkets, and in their new homes, where they may lack water.

Early and More Abundant Flowering

AMF have been shown to have an impact on plant flowering. Flowering is important not only because it increases the value of ornamental plants but also because it determines the production period. Early flowering due to AM may shorten the production period. The fact that more abundant flowering is likewise valuable for plant producers means that these benefits of AMF would be highly appreciated. Indeed, AMF have been shown to affect the flowering of both ornamentals and vegetables. *Sparaxis tricolor* that was inoculated with AMF flowered at least one week earlier than non-mycorrhizal controls (Scagel, 2004). Similar results were obtained with *Freesia* x *hybrida* (Scagel, 2003a), *Abutilon theophrasti* (Lu and Koide, 1994), and *Chrysanthemum morifolium* (Sohn et al., 2003). Scagel (2003b) also studied the effect of AMF on the flowering of three cultivars of *Zephyranthes* spp. and found that the effect of AMF varied from delayed to earlier flowering depending on the plant cultivar. Inoculation with AMF has also been shown to prolong flowering (Lu and Koide, 1994) and result in more flowers in *Petunia hybrida, Callistephus chinensis,* and *Impatiens balsamina* (Gaur, Baur, and Adholeya, 2000). In addition, vegetables such as tomatoes *(Lycopersicon esculentum)* have been shown to produce more flowers after inoculation with AMF (Poulton, Koide, and Stephenson, 2001). However, Bryla and Koide (1990) tested the effect of AMF on 10 different tomato cultivars and found that the effect on flowering time, number of flowers, and flowering duration differed from cultivar to cultivar.

Biological Activation of Inert Growth Media

Inert growth systems are deficient in microbial activity, and applying AMF inocula to these systems may result not only in AM formation but also in the introduction of AM-associated bacteria (Mansfeld-Giese, Larsen, and Bødker, 2002), which can promote plant growth and have biocontrol features (Budi et al., 1999). Isolates of *Paenibacillus polymyxa* and *Paenibacillus macerans,* from a cucumber/*G. intraradices* symbiosis, were shown to be antagonistic to the root pathogen *Pythium ultimum* and to promote cucumber growth as well (Larsen, Cornejo, and Barea, unpublished). Similarly, the biological activation of the growth media is most likely also the reason for the growth promotion of *Trifolium subterraneum* in peat after AMF inoculation with no AM formation (Larsen and Ravnskov, unpublished).

Compatibility with Production Systems

Integrating AMF into highly industrialized plant production systems may be difficult because of various AMF-adverse conditions and practices such as the type of growth media and the use of fertilizers and pesticides.

Growth Media

Most greenhouse plant production systems are based on soilless growth media, except in warm areas where field-grown crops are simply protected with plastic covers. Most potted plants are produced in peat-based substrates, and vegetables and cut flowers are produced in inactive growth media such as rockwool, perlite, and LECA. Peat-based substrates have been shown to suppress AM formation (Calvet, Estaun, and Camprudi, 1993; Linderman and Davies, 2003), which may be a result of the presence of an antagonistic microflora or toxic humic substances. However, Corkidi et al. (2004) examined the infectivity of commercial inocula in different growth media and, interestingly, found that three out of six commercial AMF inocula produced higher colonization in maize grown in peat than in maize grown in soil/sand, indicating that some AMF are actually well suited for peat-based substrates. Such differences in AMF performance in different growth media must be taken into consideration by commercial producers of AMF inoculum when screening for isolates with the ability to thrive in the environment in which they are to be applied (Feldmann and Grotkass, 2002).

Fertilizers

Plant nutrition in greenhouse production is managed for optimal plant growth, and there are high P levels in the nutrient solutions, which are incompatible with AMF (e.g., Olsson, Bååth, and Jakobsen, 1997). New results with AM in pot roses and cucumbers from our AM research program suggest that it is possible to reduce the P level in the nutrient solution by 50 percent (from 0.50 mM to 0.25 mM of P), which is compatible with AM development without affecting plant growth (Larsen and Ravnskov, unpublished). Another possibility is to manage P availability by using slow-release nutrients (Graham and Timmer, 1983, 1985) or P buffers, which bind P and release it according to the plant's requirements (Hansen and Petersen, 2004). However, more work needs to be done to determine the compatibility between P buffers/slow-release P products and AM development. It is much easier to manage the P levels in biologically inactive growth media such as rockwool, which has also been one of the driving forces behind the development of these growth media. These systems seem to be compatible with AMF, as P levels can be managed. This is because P is mobile in these inert media, unlike in soil systems. Also, potential antagonism from other microorganisms seems to be limited because of the inert nature of these substrates. It is therefore surprising that so little work has been done to integrate AMF into these production systems.

Pest Management

Fungicides can be used to manage root diseases in greenhouse systems either by mixing the fungicides with the growth media or by applying them as a drench. In general, AMF are less sensitive to fungicides against Peronosporomycetes *(Pythium* and *Phytophthora)*, such as Previcur, Alliette, and metalaxyl (Seymour, Thompson, and Fiske, 1994; Fontanet et al., 1998; Sramek, Dubsky, and Vosatká, 2000), but are highly sensitive to other fungicides against true fungi (e.g., *Fusarium* and *Rhizoctonia*), such as carbendazim (Larsen et al., 1996; Kjoller and Rosendahl, 2000). Seedcoating is also a normal practice in horticulture, but since this is performed only to protect the seed against pre- and postemergence damping-off, it most likely does not affect AMF (Spokes, Hayman, and Kandasamy, 1989). In systems in which the nutrient solution is recirculated, it is common to use disinfectants such as soaps, dihydrogen peroxide, or the release of Cu ions. Little is known about the sensitivity of AMF to these disease-management methods, except that Cu ions have been shown to reduce AM development in maples (Kosuta et al., 2002). Overall, more research is needed in order to develop integrated pest management systems based on the reduced use of pesticides and the increased use of biological control.

Hormones

Potted-plant production often involves hormonal treatments applied to the foliage to produce more compact, bushy plants. To our knowledge, no research has been conducted to study the effect of such hormonal treatments on AM development. Alternative methods aimed at retarding plant elongation, especially low P methods using P buffers and/or slow-release P, look very promising (Petersen and Hansen, 2003) and seem to be compatible with AM development. However, more research is needed to clarify whether AM is compatible with P buffers/slow-release P substrates. Hormones are also used in the production of ornamental plants to promote the rooting of cuttings. Scagel, Reddy, and Armstrong (2003) studied the effect of AMF on the rooting of Hick's yew stem cuttings that were treated with a commercial hormone product. They found that, in general, adding AMF to the growth substrate promoted root initiation in the cuttings, but that the AM inoculum density had a significant impact on the effect. In addition, Scagel (2001) found that the effect of AMF on root initiation in miniature roses depended on the rose cultivar.

OTHER OBSTACLES TO INTEGRATING AM INTO HORTICULTURE

Potential Undesired Effects of Inoculation with AMF

In order to gain a balanced view of the prospects for integrating AMF into horticulture, it is important not only to focus on the benefits but also to consider possible undesired features of AMF.

Increase in Foliar Pests

Most of the work on interactions between AMF and plant pathogens has focused on root pathogens; here, AMF generally suppress disease development and/or induce tolerance (Whipps, 2004). The impact of AM on foliar diseases has received less attention. Nevertheless, the research that is available shows that AM generally increase foliar diseases caused by fungi, bacteria, and viruses (Whipps, 2004), which may be related to increased P levels in AM plants as compared to non-AM plants (West, 1995). Although mycorrhizal plants seem to compensate by increasing disease tolerance (Dugassa et al., 1996; Gernns, von Alten, and Poehling, 2001), this may be a problem in ornamentals with no tolerance to foliar diseases. Mycorrhizal plants have also been reported to be better hosts for aphids (Gange, Bower, and Brown, 1999) than non-AM plants but may experience decreased herbivory by chewing insects (Wamberg, Christensen, and Jakobsen, 2003). Interestingly, AM plants have also been reported to be more attractive to predators than non-AM plants (Lingua et al., 2002). In greenhouse systems producing vegetables in Denmark, pests such as aphids and spider mites are managed entirely with biological control using predators (Eilenberg et al., 2000); this success is mainly due to intensive studies of population biology, which may need to be redefined with the introduction of AM plants. This underscores the need to address pest management development in a more holistic fashion.

Growth Depressions

Are growers willing to make an investment in plant biomass in order to achieve plant protection against diseases? This is a crucial question for some crops, such as cucumbers, in which AM growth depressions are common (e.g., Larsen and Yohalem, 2004). The answer is simple, growers will not accept a parasitic relationship with AM, even if there is an opportunity to protect the plants against diseases. Mycorrhiza-induced growth depressions are more pronounced in growth systems with high P (Ryan and Graham, 2002) and, in most cases, it is possible to reduce the P levels in the nutrient solutions to counteract a potential parasitic AM relation.

Inoculum

The beneficial use of AMF in production systems requires an effective inoculum. Inoculum producers have to perform testing to ensure that the inoculum is functionally effective and free of pathogens, and they also have to inform users about how to store and use the inoculum properly (von Alten et al., 2002; Gianinazzi and Vosatká, 2004).

Compatibility with Other Microbial Biocontrol Agents

Closed systems such as greenhouse areas are ideal for the introduction of microbial biocontrol agents. In some crops, insect management is achieved entirely through biological control; this could also be a future scenario for disease management, as more and more pesticides are being phased out. More than 80 commercial microbial biocontrol agents have been marketed (Paulitz and Belanger, 2001), mostly as "plant strengtheners" and plant-growth promoters, which facilitates product registration. However, these microbial products are used mainly to manage diseases caused by fungi. Products to manage root diseases are mainly based on the fungi *Trichoderma* and *Gliocladium* and the bacteria *Pseudomonas* and *Bacillus*. Developing a strategy that combines biotrophs (AM) and saprotrophs to manage root diseases would be ideal if the combinations are compatible (Nemec, Datnoff, and Strandberg, 1996). Several studies have examined the combination of AMF and *Trichoderma* species and have shown more efficient biocontrol of root pathogens when the fungi were combined (Calvet, Pera, and Barea, 1993; Dubsky, Sramek, and Vosatká, 2002). However, studies on interactions between *G. intraradices* and *Trichoderma harzianum* have shown that the two fungi are mutually inhibitory and most likely compete for inorganic nutrients (Green et al., 1999). In addition, the in vitro parasitism of *G. intra-*

radices spores and mycelium by *T. harzianum* has been demonstrated (Rousseau et al., 1996). Mycorrhiza-associated bacteria from the genus *Paenibacillus* have also been shown to increase disease suppression (Budi et al., 1999). In contrast, several combinations of AMF and other microbial biocontrol agents have shown no additive or even adverse effects on disease control (Vestberg et al., 2004). This is therefore another area in which more research is needed in order to develop biological disease management strategies. Results from our biocontrol research program suggest that combinations of AM and foliar biocontrol agents may increase pathogen control (Larsen, unpublished); there is thus a need for further research on the impact of AM on microbial communities in the phyllosphere.

Breeding Programs

New varieties of greenhouse crops are constantly being developed. A good illustration of the pace at which new varieties are introduced is our project on AMF and cucumber functional compatibility, which began in 2001 and used the five most commonly grown cucumber varieties at the time. Two years later, when we started another project on integrating AM into the production of greenhouse-grown cucumbers and tomatoes, none of these varieties were being grown commercially. We consequently had to develop new screening methods for functional compatibility between commercial AMF and the new varieties of cucumbers in terms of plant growth and colonization. Such a rapid change in cucumber varieties was mainly due to problems with disease resistance to mildew. In ornamentals, however, breeders are looking for new colors and shapes, since potted plants follow current fashion trends. Is it realistic for inoculum producers to keep pace with plant breeders? This does not seem possible; with respect to AMF in field-grown crops, perhaps it makes more sense to focus on AMF mixtures that are compatible with the general practices used in plant production systems.

CONCLUSIONS AND PERSPECTIVES

There is no doubt that AMF affect plant growth and health, but we are still far from fully understanding how to take advantage of these beneficial features in plant production. This is chiefly because we are also far from understanding the ecology of AMF. The main reason for this lack of knowledge is the biotrophic nature of these fungi. We still cannot rule out the possibility that AMF can complete their lives with alternative partners, such as their associated bacteria (Hildebrandt, Janetta, and Bothe, 2002). The role that inoculum-associated bacteria potentially play in the plant benefits that are normally attributed to AMF, such as plant growth promotion and biocontrol of root pathogens (Budi et al., 1999), are also worthy of study. Most often, traditional AM pot experiments are based on disinfected soil and involve introducing a known isolate of an AMF as a crude soil inoculum and giving the non-mycorrhizal control a filtrate of the crude soil inoculum to create a similar microbial background in the treatments with and without AMF. It is important to develop protocols with better control of the background microbial communities, which also play an important role in plant growth and health. Most of all, however, more fieldwork is needed. Jakobsen (1994) provided methods for studying AM hyphal P uptake in the field and also used these methods to study the impact of fungicides on hyphal P transport (Schweiger, Spliid, and Jakobsen, 2001).

In order for AMF to be implemented in horticulture, it is very important that the strengths and limitations of AMF in the plant production systems in question be identified. Grower-based experiments are important for validating the beneficial features of AMF. The main limitation for AMF in horticulture is the conventional use of agrochemicals and peat-based substrates, but slow-release fertilizers and/or P buffers and alternative growth media seem to be compatible with AMF. Although it is not expected that AMF will be the only players in plant protection in horticulture, they should be integrated as an important component for root health that is compatible with biological and chemical disease management methods. It is also important to take the potential undesired features of AMF into consideration, which requires more holistic examinations of AM in different plant production systems. Finally, it is crucial to consider the economics not only in terms of plant productivity but also on a larger environmental scale.

REFERENCES

Al-Karaki, G.N. and R. Hammad (2001). Mycorrhizal influence on fruit yield and mineral content of tomato grown under salt stress. *Journal of Plant Nutrition* 24:1311-1323.

An, Z.Q., B.Z Guo, and J.W. Hendrix (1998). Viability of soilborne spores of glomalean mycorrhizal fungi. *Soil Biology and Biochemistry* 30:1133-1136.

Azcón-Aguilar, C. and J.M. Barea (1996). Arbuscular mycorrhizas and biological control of soil-borne pathogens, an overview of the mechanisms involved. *Mycorrhiza* 6:457-464.

Azcón-Aguilar, C. and J.M. Barea (1997). Applying mycorrhiza biotechnology to horticulture, significance and potentials. *Scientia Horticulturae* 68:1-24.

BendavidVal, R., H.D. Rabinowitch, J. Katan, and Y. Kapulnik (1997). Viability of VA-mycorrhizal fungi following soil solarization and fumigation. *Plant and Soil* 195:185-193.

Bødker, L. and K.T. Kristensen (1999). Effect of green manure crops on root rot and arbuscular mycorrhizal fungi in pea roots. In *Proceedings from an International Workshop, Designing and Testing Crop Rotations for Organic Farming,* J. Olesen, R. Altun, M.J. Gooing, E.S. Jensen, and U. Köpke (eds.). Danish Research Centre for Organic Farming DARCOF Report No 1:337-344.

Boswell, E.P., R.T. Koide, D.L. Shumway, and H.D. Addy (1998). Winter wheat cover cropping. VA mycorrhizal fungi and maize growth and yield. *Agriculture Ecosystems and Environment* 67:55-65.

Bryla, D.R. and R.T. Koide (1990). Regulation of reproduction in wild and cultivated *Lycopersion esculentum* Mill by vesicular-arbuscular mycorrhizal infection. *Oecologia* 84:74-81.

Budi, S.W., D. van Tuinen, G. Martinotti, and S. Gianinazzi (1999). Isolation from *Sorghum bicolor* mycorrhizosphere of a bacterium compatible with arbuscular mycorrhiza development and antagonistic towards soilborne fungal pathogens. *Applied and Environmental Microbiology* 65:5148-5150.

Calvet, C., V. Estaun, and A. Camprubi (1993). Germination, early growth and infectivity of a vesicular-arbuscular mycorrhizal fungus in organic substrates. *Symbiosis* 14:405-411.

Calvet, C., J. Pera, and J.M. Barea (1993). Growth response of Marigold (*Tagetes erecta* L) with *Glomus mosseae, Thrichoderma aureoviride* and *Pythium ultimum* in a peat perlite mixture. *Plant and Soil* 148:1-6.

Cantrell, I.C. and R.G. Linderman (2001). Preinoculation of lettuce and onion with VA mycorrhizal fungi reduces deleterious effects of soil salinity. *Plant and Soil* 233:269-281.

Corkidi, L., E.B. Allen, D. Merhaut, M.F. Allen, J. Downer, J. Bohn, and M. Evans (2004). Assessing the infectivity of commercial inoculants in plant nursery conditions. *Journal of Environmental Horticulture* 22:149-154.

Davies Jr, F.T., V. Olalde-Portugal, L. Aguilera-Gomez, M.J. Alvarado, R.C. Ferraro-Cerrato, and T.W. Boutton (2002). Alleviation of drought stress of chile ancho pepper (*Capsicum annuum* L cv. San Luis) with arbuscular mycorrhiza indigenous to Mexico. *Scientia Horticulturae* 92:347-359.

Dodd, J.C., T.A. Dougall, J.P. Clapp, and P. Jeffries (2002). The role of arbuscular mycorrhizal fungi in plant community establishment at Samphire Hoe, Kent, UK—the reclamation platform created during the building of the Channel tunnel between France and UK. *Biodiversity and Conservation* 11:39-58.

Douds, D.D. and C. Reider (2003). Inoculation with mycorrhizal fungi increases the yield of green peppers in a high P soil. *Biological Agriculture and Horticulture* 21:91-102.

Dubsky, M., F. Sramek, and M. Vosatká (2002). Inoculation of cyclamen *(Cyclamen persicum)* and poinsettia *(Euphorbia pulcherrima)* with arbuscular mycorrhizal fungi and *Trichoderma harzianum. Rostlinna Vyroba* 48:63-68.

Dugassa, G.D., H. von Alten, and F. Schonbeck (1996). Effects of arbuscular mycorrhiza (AM) on health of *Linum usitatissimum* L. infected by fungal pathogens. *Plant and Soil* 185:173-182.

Eilenberg, J., A. Enkegaard, S. Vestergaard, and B. Jensen (2000). Biocontrol of pests on plant crops in Denmark, present status and future potential. *Biocontrol Science and Technology* 10:703-716.

Estaun, V., C. Calvet, A. Camprubi, and J. Pinochet (1999). Long-term effects of nursery starter substrate and AM inoculation of micropropagated peach x almond hybrid rootstock GF677. *Agronomie* 19:483-489.

Feldmann, F. and C. Grotkass (2002). Directed inoculum production—shall we be able to design AMF populations to achieve predictable symbiotic effectiveness? In *Mycorrhiza Technology in Agriculture, from Genes to Bioproducts,* S. Gianinazzi, H. Schuepp, J.M. Barea, and K. Haselwandter (eds.). Switzerland: Birkhaüser, pp. 261-279.

Fitter, A.H. and J. Garbaye (1994). Interactions between mycorrhizal fungi and other soil organisms. *Plant and Soil* 159:123-132.

Fontanet, X., V. Estaun, A. Camprubi, and C. Calvet (1998). Fungicides added to potting substrate affect mycorrhizal symbiosis between a peach-almond rootstock and *Glomus* sp. *Hortscience* 33:1217-1219.

Gange, A., E. Bower, and V. Brown (1999). Positive effects of an arbuscular mycorrhizal fungus on aphid life history traits. *Ecologia* 120:123-131.

Gaur, A. and A. Adholeya (2005). Diverse response of five ornamental plant species to mixed indigenous and single isolate arbuscular-mycorrhizal inocula in marginal soil amended with organic matter. *Journal of Plant Nutrition* 28:707-723.

Gaur, A., A. Baur, and A. Adholeya (2000). Growth and flowering in *Petunia hybrida, Callistephus chinensis and Impatiens balsamia* inoculated with mixed AM inocula or chemical fertilizers in a soil of low P fertility. *Scientia Horticulturae* 84:151-162.

Gavito, M.E. and M.H. Miller (1998a). Early phosphorus nutrition, mycorrhizae development, dry matter partitioning and yield of maize. *Plant and Soil* 199: 177-186.

Gavito, M.E. and M.H. Miller (1998b). Changes in mycorrhiza development in maize induced by crop management practices. *Plant and Soil* 198:185-192.

Gernns, H., H. von Alten, and H.M. Poehling (2001). Arbuscular mycorrhiza increased the activity of a biotrophic leaf pathogen—is a compensation possible? *Mycorrhiza* 11:237-243.

Gianinazzi, S., A. Trouvelot, and V. Gianinazzi-Pearson (1990). Role and use of mycorrhizas in horticultural crop production. *Advances in Horticutural Science* 4:25-30.
Gianinazzi, S. and M. Vosatká (2004). Inoculum of arbuscular mycorrhizal fungi for production systems, science meets business. *Canadian Journal of Botany* 82:1264-1271.
Gormsen, D., P.A. Olsson, and K. Hedlund (2004). The influence of collembolans and earthworms on AM fungal mycelium. *Applied Soil Ecology* 27:211-220.
Graham, J.H. (2001). What do pathogens see in mycorrhizas? *New Phytologist* 149:357-359.
Graham, J.H. and L.W. Timmer (1983). Vesicular-arbuscular mycorrhizal development and growth response of rough lemon in soil and soilless media: Effect of phosphorous source. *Hortscience* 18:581.
Graham, J.H. and L.W. Timmer (1985). Rock phosphate as a source of phosphorous for vesicular-arbsucular mycorrhizal development and growth of citrus in a soilless medium. *Journal of the American Society of Horticultural Science* 110: 489-492.
Graves, J.D., N.K. Watkins, A.H. Fitter, D. Robinson, and C. Scrimgeour (1997). Intraspecific transfer between plants linked by a common mycorrhizal network. *Plant and Soil* 192:153-159.
Green, H., J. Larsen, P.A. Olsson, D.F. Jensen, and I. Jakobsen (1999). Suppression of the biocontrol agent *Trichoderma harzianum* by the external mycelium of the arbuscular mycorrhizal fungus *Glomus intraradices. Applied and Environmental Microbiology* 65:1428-1434.
Gryndler, M., M. Vosatká, H. Hrselova, I. Chvatalova, and J. Jansa (2002). Interaction between arbuscular mycorrhizal fungi and cellulose in growth substrate. *Applied Soil Ecology* 19:279-288.
Haas, J.H., B. Bar-Yosef, J. Krikun, R. Barak, T. Markovitz, and S. Kramer (1987). Vesicular-arbuscular mycorrhizal fungus infestation and phophorus fertigation to overcome pepper stunting after methyl bromide fumigation. *Agronomie Journal* 79:905-910.
Hamel, C. (1996). Prospects and problems pertaining to the management of arbuscular mycorrhizae in agriculture. *Agriculture Ecosystems and Environment* 60: 197-210.
Hansen, C.W. and K.K. Petersen (2004). Reduced nutrient and water availability to Hibiscus rosa-sinensis 'Cairo Red' as a method to regulate growth and improve post-production quality. *European Journal of Horticultural Science* 69:159-166.
Hayman, D.S. (1982). Influence of soils and fertility on activity and survival of vesicular-arbuscular mycorrhizal fungi. *Phytopathology* 72:1119-1125.
Hetrick, B.A.D., G.W.T. Wilson, and T.C. Cox (1993). Mycorrhizal dependence of modern wheat cultivars and ancestors—a synthesis. *Canadian Journal of Botany* 71:512-518.
Hildebrandt, U., K. Janetta, and H. Bothe (2002). Towards growth of arbuscular mycorrhizal fungi independent of a plant host. *Applied and Environmental Mircrobiology* 68:1919-1924.

Hodge, A., C.D. Campbell, and A.H. Fitter (2001). An arbuscular mycorrhizal fungus accelerates decomposition and acquires nitrogen directly from organic material. *Nature* 413:297-299.

Jaizme-Vega, M.C., A.S. Rodriguez-Romero, C.M. Hermoso, and S. Declerck (2003). Growth of micropropagated bananas colonized by root-organ culture produced arbuscular mycorrhizal fungi entrapped in Ca-alginate beads. *Plant and Soil* 254:329-335.

Jakobsen, I. (1994). Research approaches to study the functioning of vesicular-arbuscular mycorrhizas in the field. *Plant and Soil* 159:141-147.

Jeffries, P., S. Gianinazzi, S. Perotto, K. Turnau, and J.M. Barea (2003). The contribution of arbuscular mycorrhizal fungi in sustainable maintenance of plant health and soil fertility. *Biology and Fertility of Soils* 37:1-16.

Johansson, J.F., L.R. Paul, and R.D. Finlay (2004). Microbial interactions in the mycorrhizosphere and their significance for sustainable agriculture. *FEMS Microbiology Ecology* 48:1-13.

Johnson, N.C., J.H. Graham, and F.A. Smith (1997). Functioning of mycorrhizal associations along the mutualism-parasitism continuum. *New Phytologist* 135: 575-586.

Joner, E.J. and C. Leyval (2003). Rhizosphere gradients of polycyclic aromatic hydrocarbon (PAH) dissipation in two industrial soils and the impact of arbuscular mycorrhiza. *Environmental Science and Technology* 37:2371-2375.

Kabir, Z. and R.T. Koide (2002). Effect of autumn and winter mycorrhizal cover crops on soil properties, nutrient uptake and yield of sweet corn in Pennsylvania, USA. *Plant and Soil* 238:205-215.

Kabir, Z., I.P. O'Halloran, J.W. Fyles, and C. Hamel (1998). Dynamics of the mycorrhizal symbiosis of corn (*Zea mays* L.), Effects of host physiology, tillage practice and fertilization on spatial distribution of extra-radical mycorrhizal hyphae in the field. *Agriculture Ecosystems and Environment* 68:151-163.

Kabir, Z., I.P. O'Halloran, P. Widden, and C. Hamel (1998). Vertical distribution of arbuscular mycorrhizal fungi under corn (*Zea mays* L.) in no-till and conventional tillage systems. *Mycorrhiza* 8:53-55.

Kahiluoto, H. and M. Vestberg (1998). The effect of arbuscular mycorrhiza on biomass production and phosphorus uptake from sparingly soluble sources by leek (*Allium porrum* L.) in Finnish field soils. *Biological Agriculture and Horticulture* 16:65-85.

Kapulnik, Y., B. Heuer, N.A. Patterson, D. Sadan, Z. Bar, G. Nir, and B. Kishinevsky (1994). Stunting syndrome in peanuts and agronomic approaches for its release. *Symbiosis* 16:267-278.

Kjoller, R. and S. Rosendahl (2000). Effects of fungicides on arbuscular mycorrhizal fungi, differential responses in alkaline phosphatase activity of external and internal hyphae. *Biology and Fertility of Soils* 31:361-365.

Koch, M., Z. Tanami, H. Bodani, S. Wininger, and Y. Kapulnik (1997). Field application of vesicular-arbuscular mycorrhizal fungi improved garlic yield in disinfected soil. *Mycorrhiza* 7:47-50.

Kosuta, S., C. Hamel, Y. Dalpé, and M. St-Arnaud (2002). Copper release from chemical root-control baskets in hardwood tree production. *Journal of Environmental Quality* 31:910-916.
Larsen, J. and I. Jakobsen (1996). Effects of a mycophagous collembola on the symbiosis between *Trifolium subterraneum* L. and three arbuscular mycorrhizal fungi. *New Phytologist* 133:295-302.
Larsen, J., I. Thingstrup, I. Jakobsen, and S. Rosendahl (1996). Benomyl inhibits phosphorus transport but not fungal alkaline phosphatase activity in a *Glomus*-cucumber symbiosis. *New Phytologist* 132:127-133.
Larsen, J. and D.S. Yohalem (2004). Interactions between mycorrhiza and powdery mildew of cucumber. *Mycological Progress* 3:123-128.
Leake, J.R., D. Johnson, D.P. Donnelly, G.E. Muckle, L. Boddy, and D.J. Read (2004). Networks of power and influence, the role of mycorrhizal mycelium in controlling plant communities and agroecosystem functioning. *Canadian Journal of Botany* 82:1016-1045.
Linderman, R.G. (1986). Managing rhizosphere microorganisms in the production of horticultural crops. *HortScience* 21:1299-1302.
Linderman, R.G. (1994). Role of VAM fungi in biocontrol. In *Mycorrhizae and Plant Health,* F.L. Pfelger and R.G. Linderman (eds.). St. Paul, MN: APS Press, pp. 1-25.
Linderman, R.G. (2000). Effects of mycorrhizas on plant tolerance to diseases, mycorrhiza-disease interactions. In *Arbuscular Mycorrhizas, Physiology and Function,* Y. Kapulnik and D.D. Douds, Jr. (eds.). Dordrecht: Kluwer Academic Publishers, pp. 345-365.
Linderman, R.G. and E.A. Davis (2001). Comparative response of selected grapevine rootstocks and cultivars to inoculation with different mycorrhizal fungi. *American Journal of Enology and Viticulture* 52:8-11.
Linderman, R.G. and E.A. Davis (2003). Soil amendment with different peatmosses affects mycorrhizae of onion. *HortTechnology* 13:285-289.
Lingua, G., G. Berta, M.C. Digilio, N. Massa, and E. Guerriri (2002). Mycorrhizal colonisation enhances plant defenses against aphids in tomato. *Proceedings of the Fourth International Conference on Mycorrhizae,* August 10-15, 2003, Montreal, abstract 499.
Lovato, P.E., V. Gianinazzi-Pearson, A. Trouvelot, and S. Gianinazzi (1996). The state of art of mycorrhizas and micropropagation. *Advances in Horticutural Science* 10:46-52.
Lu, X. and R.T. Koide (1994). The effect of mycorrhizal infection on components of plant growth and reproduction. *New Phytologist* 128:211-218.
Marx, D.H., L.F. Marrs, and C.E. Cordell (2002). Practical use of the mycorrhizal fungal technology in forestry, reclamation, arboriculture, agriculture, and horticulture. *Dendrobiology* 47:27-40.
Mansfeld-Giese, K., J. Larsen, and L. Bødker (2002). Bacterial populations associated with mycelium of the arbuscular mycorhrhizal fungus *Glomus intraradices. FEMS Microbiology Ecology* 41:133-140.

McGonigle, T.P., D.G. Evans, and M.H. Miller (1990). Effect of degree of soil disturbance on mycorrhizal colonisation and phosphorus absorption by maize in growth chamber and field experiments. *New Phytologist* 116:629-636.

McGonigle, T.P. and M.H. Miller (1993). Responses of mycorrhizae and shoot phosphorus of maize to the frequency and timing of soil disturbance. *Mycorrhiza* 4:63-68.

Menge, J.A. (1982). Effect of soil fumigants and fungicides on vesicular-arbuscular mycorrhizal fungi. *Phytopathology* 72:1125-1132.

Miller, M., T. McGonigle, and H. Addy (1994). An economic approach to evaluate the role of mycorrhizas in managed ecosystems. *Plant and Soil* 159:27-33.

Mosse, B. (1988). Some studies relating to independent growth of vesicular-arbuscular endophytes. *Canadian Journal of Botany* 66:2533-2540.

Mozafar, A., T. Anken, R. Ruh, and E. Frossard (2000). Tillage intensity, mycorrhizal and nonmycorrhizal fungi, and nutrient concentrations in maize, wheat, and canola. *Agronomy Journal* 92:1117-1124.

Nemec, S., L.E. Datnoff, and J. Strandberg (1996). Efficacy of biocontrol agents in planting mixes to colonize plant roots and control root diseases of vegetables and citrus. *Crop Protection* 15:735-742.

Olsson, P.A., E. Baath, and I. Jakobsen (1997). Phosphorus effects on the mycelium and storage structures of an arbuscular mycorrhizal fungus as studied in the soil and roots by analysis of fatty acid signatures. *Applied and Environmental Microbiology* 63:3531-3538.

Ortas, I. (2003). Effect of selected mycorrhizal inoculation on phosphorus sustainability in sterile and non-sterile soils in the Harran Plain in South Anatolia. *Journal of Plant Nutrition* 26:1-17.

Paulitz, T.C. and R.R. Belanger (2001). Biological control in greenhouse systems. *Annual Review of Phytopathology* 39:103-133.

Petersen, K.K. and C.W. Hansen (2003). Compact *Campanula carpatica* uniform without chemical growth regulators. *European Journal of Horticultural Science* 68:266-271.

Postma, J., M.J.E.I.M. Willemsen-de Klein, and J.D. van Elsas (2000). Effect of the indigenous microflora on the development of root and crown rot caused by *Pythium aphanidermatum* in cucumber grown on rockwool. *Phytopathology* 90:125-133

Poulton, J.L., R.T. Koide, and A.G. Stephenson (2001). Effects of mycorrhizal infection, soil phosphorus availability and fruit production on the male function in two cultivars of *Lycopersion esculentum. Plant, Cell and Environment* 24: 841-849.

Ravnskov, S., J. Larsen, and I. Jakobsen (2003). No impact of the biocontrol bacterium *Burkholderia cepacia* on hyphal P transport of the arbsucular mycorrhizal fungus *Glomus intraradices. Soil Biology and Biochemistry* 34:1875-1881.

Ravnskov, S., J. Larsen, P.A. Olsson, and I. Jakobsen (1999). Effects of different organic substrates on growth and P uptake of an arbuscular mycorrhizal fungus. *New Phytologist* 141:517-524.

Redman R.S., D.D. Dunigan, and R.J. Rodriguez (2001). Fungal symbiosis from mutualism to parasitism, who controls the outcome, host or invader? *New Phytologist* 151:705-716.
Regvar, M., K. Vogel-Mikus, and T. Severkar (2003). Effect of AMF inoculum from field isolates on the yield of green pepper, parsley, carrot, and tomato. *Folia Geoboatnica* 38:223-234.
Requena, N., E. Perez-Solis, C. Azcon-Aguilar, P. Jeffries, and J.M. Barea (2001). Management of indigenous plant-microbe symbioses aids restoration of desertified ecosystems. *Applied and Environmental Microbiology* 67:495-498.
Rosendahl, C.N. and S. Rosendahl (1990). The role of vesicular-arbuscular mycorrhiza in controlling damping-off and growth reduction in cucumber caused by *Pythium ultimum. Symbiosis* 9:363-366.
Rosseau, A., N. Benhamou, I. Chet, and Y. Piché (1996). Mycoparasitism of the extrametrical phase of *Glomus intraradices* by *Trichoderma harzianum. Phytopathology* 86:434-443.
Ruiz-Lozano, J.M. (2003). Arbuscular mycorrhizal symbiosis and alleviation of osmotic stress. New perspectives for molecular studies. *Mycorrhiza* 13:309-317.
Ryan, M.H. and J.H. Graham (2002). Is there a role for arbuscular mycorrhizal fungi in production agriculture? *Plant and Soil* 244:263-271.
Scagel, C.F. (2001). Cultivar specific effects of mycorrhizal fungi on the rooting of miniature rose cuttings. *Journal of Environmental Horticulture* 19:15-20.
Scagel, C.F. (2003a). Inoculation with arbuscular mycorrhizal fungi alters nutrient allocation and flowering of Freesia x hybrida. *Journal of Environmental Horticulture* 21:196-205.
Scagel, C.F. (2003b). Soil pasteurization and inoculation with *Glomus intraradices* alters flower production and bulb composition of *Zephyranthes* spp. *Journal of Horticultural Science and Biotechnology* 78:798-812.
Scagel, C.F. (2004). Inoculation with vesicular-arbuscular mycorrhizal fungi and rhizobacteria alters nutrient allocation and flowering of harlequin flower. *HortTechnology* 14:39-48.
Scagel, C.F., K. Reddy, and J.M. Armstrong (2003). Mycorrhizal fungi in rooting substrate influences the quantity and the quality of roots on stem cuttings of Hick's yew. *HortTechnology* 13:62-66.
Schweiger, P.F., N.H. Spliid, and I. Jakobsen (2001). Fungicide application and phosphorous uptake by hyphae of arbuscular mycorrhizal fungi into field-grown peas. *Soil Biology and Biochemstry* 33:1231-1237.
Seymour, N.P., J.P. Thomson, and M.L. Fiske (1994). Phytotoxicity of fosetyl-Al and phosphonic acid to maize during production of vesicular-arbuscular mycorrhizal inoculum. *Plant Disease* 78:441-446.
Sohn, K.B., K.Y. Kim, S.J. Chung, W.S. Kim, S.M. Park, J.G. Kang, Y.S Rim, J.S. Cho, T.H. Kim, and J.H. Lee (2003). Effect of the different timing of AMF inoculation on plant growth and flower quality of chrysanthemum. *Scientia Horticulturae* 98:173-183.
Sorensen, J.N., J. Larsen, and I. Jakobsen (2003). Management strategies for capturing the benefits of mycorrhiza in the production of field-grown vegetables. *Acta Horticulturae* 627:65-71.

Sorensen, J.N., J. Larsen, and I. Jakobsen (2005). Mycorrhiza formation and nutrient concentration in leeks *(Allium porrum)* in relation to previous crop and cover crop management on high P soils. *Plant and Soil* 273:101-114.

Spokes, J.R., D.S. Hayman, and D. Kandasamy (1989). The effects of fungicide-coated seeds on the establishment of VA mycorrhizal infection. *Annals of Applied Biology* 115:237-241.

Sramek, F., M. Dubsky, and M. Vosatká (2000). Effect of arbuscular mycorrhizal fungi and *Trichoderma harzianum* on three species of balcony plants. *Rostlinna Vyroba* 46:127-131.

St-Arnaud, M., C. Hamel, M. Caron, and J.A. Fortin (1994). Inhibition of *Pythium ultimum* in roots and growth substrate of mycorrhizal *Tagetes patula* colonized with *Glomus interadices. Canadian Journal of Plant pathology* 16:187-194.

Sylvia, D.M. and N.C. Schenck (1983). Application of super-phosphate to mycorrhizal plants stimulates sporulation of phosphorus-tolerant vesicular arbuscular mycorrhizal fungi. *New Phytologist* 95:655-661.

Thompson, J.P. (1994). Inoculation with vesicular-arbuscular mycorrhizal fungi from cropped soil overcomes long-fallow disorder of linseed (*Linum usitatissimum* L.) by improving P and Zn uptake. *Soil Biology and Biochemcistry* 26:1133-1146.

Thomson, B.D., A.D. Robson, and L.K. Abbott (1992). The effect of long-term application of phosphorus fertilizer on populations of vesicular-arbuscular mycorrhizal fungi in pastures. *Australian Journal of Agricultural Research* 43: 1131-1142.

Thygesen, K., J. Larsen, and L. Bødker (2004). Arbuscular mycorrhizal fungi reduce development of pea root-rot caused by *Aphanomyces euteiches* using oospores as pathogen inoculum. *European Journal of Plant Pathology* 110:411-419.

Tuffen, F., W.R. Eason, and J. Scullion (2002). The effect of earthworms and arbuscular mycorrhizal fungi on growth of and P-32 transfer between *Allium porrum* plants. *Soil Biology and Biochemistry* 34:1027-1036.

Vestberg, M., S. Kukkonen, K. Saari, P. Parikka, J. Huttunen, L. Tainio, D. Devos, F. Weekers, C. Kevers, P. Thonart, M.C. Lemoine, C. Cordier, C. Alabouvette, and S. Gianinazzi (2004). Microbial inoculation for improving the growth and health of micropropagated strawberry. *Applied Soil Ecology* 27:243-258.

Vivas, A., A. Voros, B. Biro, J.M. Barea, J.M. Ruiz-Lozano, and R. Azcon (2003). Beneficial effects of indigenous Cd-tolerant and Cd-sensitive *Glomus mosseae* associated with a Cd-adapted strain of *Brevibacillus* sp. in improving plant tolerance to Cd contamination. *Applied Soil Ecology* 24:177-186.

Von Alten, H., B. Blal, J.C. Dodd, F. Feldmann, and M. Vosatká (2002). Quality Control of arbuscular mycorrhizal fungi inoculum in Europe. In *Mycorrhiza Technology in Agriculture, from Genes to Bioproducts,* S. Gianinazzi, H. Schuepp, J.M. Barea, and K. Haselwandter (eds.). Switzerland: Birkhäuser, pp. 281-296.

Vosátka, M., J. Jansa, M. Regvar, F. Šramek, and R. Malcová (1999). Inoculation with mycorrhizal fungi—a feasible biotechnology for horticulture. *Phyton Annales Rei Botanicae* 39:219-224.

Wamberg, C., S. Christensen, and I. Jakobsen (2003). Interaction between foliar-feeding insects, mycorrhizal fungi, and rhizosphere protozoa on pea plants. *Pedobiologia* 47:281-287.
West, H.M. (1995). Interactions between arbuscular mycorhizal fungi and foliar pathogens, consequences for host and pathogen, multitrophic interactions in terrestrial systems. In *The 36th Symposium of the British Ecological Society, Royal Holloway,* V.K. Brown (ed.), pp. 79-86.
Whipps, J. (2004). Prospects and limitations of mycorrhizas in biocontrol of root pathogens. *Canadian Journal of Botany* 82:1198-1227.
Wurst, S., D. Dugassa-Gobena, R. Langel, M. Bonkowski, and S. Scheu (2004). Combined effects of earthworms and vesicular-arbuscular mycorrhizas on plant and aphid performance. *New Phytologist* 163:169-176.
Zhu, Y.G., S.E. Smith, A.R. Barritt, and F.A. Smith (2001). Phosphorus (P) efficiencies and mycorrhizal responsiveness of old and modern wheat cultivars. *Plant and Soil* 237:249-255.

Chapter 5

Advances in the Management of Effective Arbuscular Mycorrhizal Symbiosis in Tropical Ecosystems

Ramón Rivera
Félix Fernández
Kalyanne Fernández
Luis Ruiz
Ciro Sánchez
Manuel Riera

Numerous scientific publications on the benefits of arbuscular mycorrhizal (AM) symbioses to plant growth and productivity were published during the past few decades (Harley and Smith, 1983; Howeler, 1985; Siqueira and Franco, 1988; Gianinazzi-Pearson and Gianinazzi, 1989; Marschner and Dell, 1994; George, 2000). The vast majority of the experiments reported were conducted under controlled or greenhouse conditions and perhaps without a system perspective. These experiments failed to clarify the mechanisms regulating the AM symbiosis, which could hardly be used to increase agricultural plant productivity.

The importance of mycorrhizal symbiosis in the plant kingdom is obvious; approximately 80 percent of land plants are colonized by some 200 AM fungal species belonging to the Glomeromycota

Mycorrhizae in Crop Production
© 2007 by The Haworth Press, Inc. All rights reserved.
doi:10.1300/5425_05

(Brundrett, 2002). Some authors pointed out an aspect of AM symbioses that is of great significance: the level of specificity existing between AM strains and crop plants is low (Harley and Smith, 1983; Siqueira and Franco, 1988). These authors felt that different species of AM fungi were unspecific to host plants when conditions were conducive to mycorrhizae formation. In the review by Siqueira and Franco (1988) it is shown that this concept of low-specificity was based mainly on the fact that very few AM species (142 reported at the time) could associate with a very large number of plant species (300,000), in contrast to the highly specific ectomycorrhizal symbioses that involves 5,000 fungal species and 2,000 plant species in only a few families of the Gymnosperm and fewer families of the Angiosperm.

This concept of low-specificity is largely correct despite the occurrence of some preference between associating strains and plant species. There has been a lack of study on the preference between strains and crop plants. Systematic trials conducted under tropical conditions with a range of crops could clarify the question of symbiotic preference of arbuscular mycorrhizae and distinguish the factors driving the effectiveness of the associations. The most complete interdisciplinary program made on the management of AM associations was probably that which was conducted in Colombia in the 1980s and published by one of its principal investigators, Sieverding (1991). Important knowledge on the main factors influencing the AM symbiosis was gained from this work.

One important aspect of this work was the management of native species. It is clear that the AM fungal population of a given location is specific and subject to the influence of agronomic practices, hence it is possible to increase native AM fungal populations and change the species composition of these populations. The effective management and evaluation of native populations appears to be very complicated. It would require rapid and precise methods to assess the mycorrhizal potential of soils, to identify the AM fungal species present, and to understand the influence of cropping practices on these species (Sieverding, 1991). The management of the symbiosis in crop production through inoculation of efficacious strains seems simpler than the manipulation of native populations.

One plant species, *Manihot esculentum* Crantz, was used for most of the experiments of the above-mentioned program, and new findings on the effectiveness of AM species or strains and on their adaptability to soil fertility conditions were revealed. It was found that plant inoculation with efficacious AM fungal species could lead to the reduction of fertilizer application, although the need for some fertilizer input was required to ensure adequate yields. The variation in AM fungal strains' effectiveness was revealed in this study, but the influence of soil type on strains' effectiveness remained unknown.

In parallel, research on AM fungi was initiated in the 1980s in Cuba. This work not only provided information on the importance of AM fungi in plants (Herrera et al., 1984a,b), but established the basis for further work on the taxonomy and functionality of AM fungi, and produced a document on the functioning of tropical forests (Herrera et al., 1995).

SPECIFICITY BETWEEN SOIL AND AM FUNGI AND STRAINS SELECTION

In 1990, a research team with the goal of developing cropping practices for the management of the AM symbiosis in crop plants and developing new mycorrhizal products was formed in Cuba. The work of this team was based on three main areas: (1) crop inoculation with efficacious AM fungi, (2) the evaluation of the importance of the soil environment in the selection of efficacious AM strains, and (3) the influence of nutrients on symbiotic effectiveness. The team first sought to understand the relationship between symbiotic activity and soil type and the importance of nutrient availability on the symbiosis. Causal relationships between soil resources and AM fungi functionality were expected based on the fact that AM fungi are soil organisms and that one of the main functions of the AM symbiosis is to enhance plant root system nutrient absorption capacity. A large body of knowledge eventually developed from numerous experiments conducted in Cuba and from validation campaigns carried out in different Latin American countries. The results of selected research works, which developed knowledge in the three target areas mentioned above, are presented in this chapter. The FAO-ISRIC and ISSS

(1998) international soil classification system is used throughout the chapter to refer to the soils used in the different experiments reported.

Coffee Transplants

Some researchers concentrated their effort on the development of AM inoculation technologies for the production of coffee *(Coffea arabica)* transplants from 1990 to 2000. They studied the effects of AM fungal strains in twelve different soil types under production conditions (Table 5.1). The soils used covered a large range of fertility levels spanning from very low, as indicated by very low cation exchange capacity (CEC) like that seen in Haplic Acrisols, to rich Eutric Cambisols with CEC of 45 $cmol_c\ kg^{-1}$.

TABLE 5.1. Main chemical characteristics of soils used in the different experiments.

			P_2O_5	K_2O	Ca	Mg	Al	
			$mg\ 100\ g^{-1}$		$cmol_c\ kg^{-1}$			
Haplic Acrisol[a]	4.9	1.35	2.80	6.8	1.5	1.30	2.3	Very low
Distric Nitisol[b]	4.7	2.48	4.00	5.8	2.7	1.30	–	Low
Chromic Luvisol[b]	5.7	3.22	10.10	12.2	7.3	1.60	–	Fair
Chromic Cambisol[a]	5.8	3.00	13.00	17.7	5.2	1.50	–	Fair
Humic Chromic Cambisol[a]	6.9	3.60	15.20	25.8	8.1	1.30	–	Fair
Gleyic Cambisol[b]	6.0	3.45	17.60	19.7	8.1	2.30	–	Fair
Eutric Ferralsol[c]	6.0	3.00	4.30	5.0	12.0	1.10	–	Fair
Eutric Nitisol[d]	6.2	1.25	12.10	16.9	18.4	1.85		Fair
Eutric Cambisol[a]	6.3	3.25	25.00	24.0	14.8	2.00	–	High
Humic Eutric Cambisol[a]	6.8	4.10	38.90	37.4	23.5	1.30	–	Very high
Calcaric Cambisol[d]	7.8	2.00	3.51	25.7	46.5	4.25		Very high
Eutric Cambisol[a]	7.3	3.80	57.00	52.8	32.0	7.90	–	Very high

[a]Soils studied in the coffee transplants program from 1990 to 1994.

[b]Soils studied in the coffee transplants program from 1994 to 1998.

[c]Soils studied in the coffee transplants program from 1997 to 2000.

[d]Soils studied in the program of root and tuber crops.

A positive response to inoculation was obtained under all soil conditions. This response, however, varied depending on three factors (1) the AM fungal strain, (2) the soil, and (3) the fertilization level, which determined the extent of symbiotic development. The consideration of these three intricately related factors determining inoculation effectiveness lead to the initiation of further work to define the influence of the soil condition on AM strains' effectiveness. In these experiments, the soil condition was determined by both the soil type and the vermicompost used as fertilizer.

All the AM strains tested triggered different growth responses under different soil conditions, indicating that AM strains' effectiveness was soil specific. Soil type appeared to be the fundamental criteria to determine which strains or species would be effective. A very thorough research program was conducted on this topic using Distric Nitisols, Chromic Luvisols, and Gleyic Cambisols. The performance of 15 AM fungal strains was evaluated in each soil over three years with vermicompost used in a 5:1 (soil:compost) proportion for fertilization. This proportion was found to be most adequate for AM development in these soils (Sánchez, 2001).

Principal component analysis revealed a different effect of crop inoculation with different AM fungal strains, and this, in all soil types used. Results from the Distric Nitisols are shown in Figure 5.1. Strains are clustered in four groups. The first group included very efficacious strains, which enhanced growth indicators (leaf area, nutrient uptake, and leaf tissue concentrations, height, leaf number, and AM root colonization level) as compared to non-inoculated control plants (T14), which were given larger amounts of vermicompost (3:1 ratio) and consequently, more nutrients. This result indicates the more effective nutrient uptake ability of coffee plants colonized by Group I strains. Group II included strains that lead to the production of plants larger or similar to control plants T14. The remaining two groups of AM fungal strains triggered little growth enhancement, at times even producing traits similar to the non-mycorrhizal and non-fertilized control plants T15.

In addition to the occurrence of specificity between soil type and AM strain, the level of effectiveness achieved by the symbioses after inoculation with the best strains for each soil was dependent on the fertility of that soil. The best effect, expressed as an index of effec-

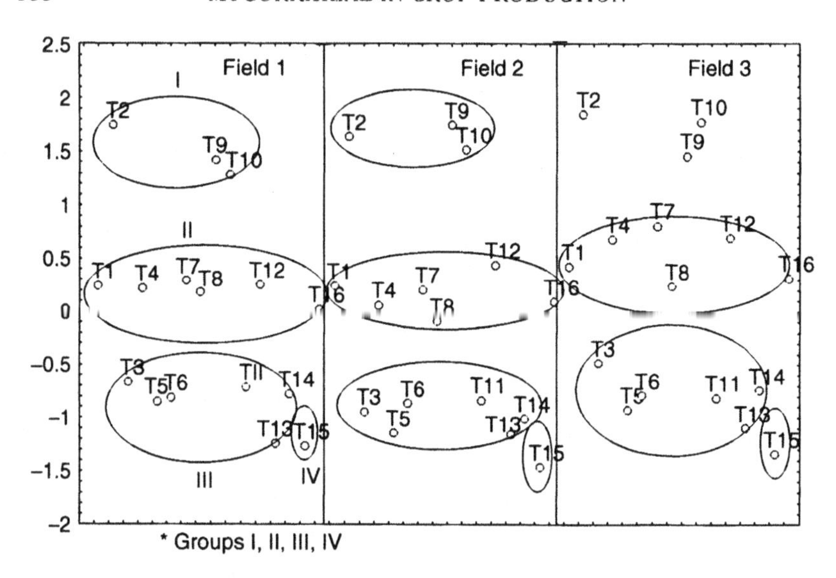

FIGURE 5.1. Effect of inoculation with one of 15 strains of AM fungi on the growth of coffee transplants grown in a Distric Nitisol. Strains higher up have a stronger positive effect. The experiment was repeated three times (Fields 1 to 3).

tiveness (percent of increase in inoculated plants as compared to the non-inoculated control), was reached in the Distric Nitisols (72 percent, 68 percent, and 67 percent), which were among the least fertile in this study. *Glomus clarum, G. intraradices,* and *Acaulospora scrobiculata* were the most effective strains in these soils. In Chromic Luvisols, the most effective AM fungi were strains of *G. fasciculatum* and two ecotypes of *G. mosseae,* which reached about the same index of effectiveness, 65 percent and 67 percent. *G. intraradices* was most effective in Gleyic Cambisols, along with one *G. mosseae* ecotype and *G. fasciculatum,* which had effectiveness indices between 5 percent and 55 percent. The analysis of these results (Sánchez et al., 2000) clearly indicated the positive effect of inoculation with AM strains and the dependence of this effect on soil type (Figure 5.1). These effects were repeatedly obtained, indicating the high reproducibility of positive AM inoculation effects in coffee production.

In the initial work conducted from 1990 to 1994 (Fernández, 1999) in Haplic Acrisols, Chromic Cambisols, and Eutric Cambisols, *Glomus* species were found to perform well. Tables 5.2 and 5.3 illustrate

TABLE 5.2. Effect of inoculation with effective AM fungal strains and of the proportion of soil: worm compost on the production of leaf surface area and Index of Effectiveness (I.E.) of coffee transplants, using Haplic Acrisols.

Experiment No.	AM fungi	Proportion of soil:worm compost	Leaf surface area (cm^2)	Index of effectiveness[a] (%)
1	*Glomus clarum*	3:1	138.8a*	140
		5:1	33.6d	−42
	Acaulospora scrobiculata	3:1	119.6ab	107
		5:1	71.7c	24
	Control	3:1	57.7B	
	Standard error		6.32***	
2	*Glomus clarum*	3:1	210.3a	143
	Glomus sp.1	3:1	213.0a	147
	Control	3:1	86.4A	
	Standard error		8.30***	

Source: Data selected from Fernández (1999).

[a]Index of effectiveness = (value with AM fungi − value without fungi)/value without fungi × 100.

*Means with the same higher case or lower case letter are not significantly different according to Duncan Multiple Range test $p < 0.001$.

***Significant at $P < 0.001$.

this fact with the most efficacious strains under each condition studied. The species of the genus *Glomus,* according to Barros (1987) and Sieverding (1991), possess a good functional stability in soils with high or fair fertility where they are competitive and effective. Results obtained on coffee indicate that their effectiveness extends to conditions of low to very low fertility. The performance of *G. clarum* was particularly good in these low fertility soils under the various conditions studied. Its application not only produced positive effects on leaf area under low soil fertility, but also under fair fertility levels (Tables 5.2 and 5.3), with indices of effectiveness between 80 percent and 257 percent. Inoculation with *G. clarum* was always effective except in highly fertile Humic Eutric Cambisols with percentage of base saturation of 25 $cmol_c\ kg^{-1}$ to 30 $cmol_c\ kg^{-1}$.

Another AM fungal species that triggered very good responses was *G. fasciculatum.* This AM fungus was studied only in Humic Eutric Cambisols and in Eutric Cambisols in these initial trials (Fer-

TABLE 5.3. Effect of inoculation with effective AM fungal strains and of the proportion of soil: worm compost on the production of leaf surface area and Index of Effectiveness (I.E.) of coffee transplants, using Humic Eutric Cambisols (experiments 4 and 5) and Eutric Cambisols (experiments 6, 7, and 8).

Experiment No.	AM fungi	Proportion of soil:worm compost	Leaf surface area (cm²)	Index of effectiveness[a] (%)
3	*Glomus clarum*			
	Glomus fasciculatum			
	Control	3:1	166.20d	
	Standard error		5.67*	
4	*Glomus fasciculatum*			
	Native strains			
	Control	3:1	420.1c	
	Standard error		2.10*	
5	*Glomus fasciculatum*			
	Glomus sp. 2			
	Control	3:1	297.0c	
	Standard error		12.08*	
6	*Glomus fasciculatum*			
	Glomus sp. 2			
	Control	3:1	295.5c	
	Standard error		11.90*	
7	*Glomus fasciculatum*			
	Glomus mosseae			
	Control	3:1	341.5bc	
	Standard error		8.10*	

Source: Selected from Fernández (1999).

Note: Means with the same letter are not significantly different according to Duncan Multiple Range test $p < 0.001$.

[a]Index of effectiveness = (value with AM fungi − value without fungi)/value without fungi × 100.

*Significant at $P < 0.001$.

nández, 1999). *G. fasciculatum* always provided a positive and stable response, with increases in leaf surface area ranging from 15 to 86 percent (Table 5.3) as compared to the non-inoculated control. Joao (2002) would later show the higher effectiveness of *G. fasciculatum* as compared to *G. clarum* in Eutric Ferrasols with CEC of about 20 $cmol_c\ kg^{-1}$.

Isolated native species *Glomus* sp. 1 and *Glomus* sp. 2 produced positive results when inoculated under edaphic conditions similar to those of the soil where they originate, but their effect was not as good as those of AM ecotypes from collections. Inoculation with *A. scrobiculata* (Table 5.2) was also assessed. The best response obtained with this species was observed in Haplic Acrisols with high aluminum levels (2.3 cmol Al kg^{-1} to 2.5 cmol Al kg^{-1} of soil). This concurs with the results of Barros (1987) who recovered this species from soils rich in aluminum.

Table 5.4 contains a summary of the results obtained from these experiments. The results show the high level of specificity existing between soil type and strain effectiveness in the AM symbioses with coffee plants. Soil type appears as the basic criterion for the selection of efficacious strains in a given agroecosystem. A regular transition of efficacious strains can be observed when soils are ranked according to their fertility level or fertility indicator such as the sum of exchangeable Ca^{2+} and Mg^{2+}. The fact that strains were effective in a large range of soils also needs to be pointed out. It appears that with only two or three efficacious species, very good results can be obtained with AM inoculation in coffee transplant production using practically all soils growing this crop in Cuba.

TABLE 5.4. Recommendation of effective AM fungal strains by soil types, for the production of coffee transplants.

Soil type	Species and strains of AM fungi recommended	Ca + Mg $cmol_c\ kg^{-1}$
Haplic Acrisols	*G. clarum, Glomus* sp. 1 and *Acaulospora scrobiculata*	2.8
Distric Nitisols	*G. clarum, G. intraradices* and *Acaulospora scrobiculata*	4.0
Humic Chromic Cambisols and Chromic Cambisols	*G. clarum y Glomus* sp. 2	6.7-9.4
Chromic Luvisols	*G. fasciculatum, G. mosseae (5)*[a] and *G. mosseae (8)*[a]	8.7
Gleyic Cambisols	*G. intraradices, G. mosseae (5)*[a] and *G. fasciculatum*	10.4
Eutric Ferralsols	*G. fasciculatum*	12-15
Eutric Cambisols and Humic Eutric Cambisols	*G. fasciculatum*	16.8-39.9

Source: Rivera and Fernández (2003).

[a]Ecotypes of *G. mosseae* from the collection of the Instituto de Ecología y Sistemática, Cuba.

According to our experience, AM inoculation should be done using "simple" inoculants, which contain only one strain. This eliminates the possibility of competition between strains. Because these strains are not effective under all edaphic conditions, their successful colonization of roots under improper soil conditions would reduce the establishment of the most effective strains, and consequently, AM effectiveness in crops.

The success of AM inoculation is not only related to the infectivity and effectiveness of the strains applied, but also to the number and type of native propagules in a soil (Dodd and Thompson, 1994). Most reports on the abundance of native propagules in soil dedicated to coffee nursery have been realized in Brazil and have indicated low concentrations of AM fungi propagules (Lópes, Díaz, and Costa, 1986; Siqueira and Colozzi-Fhilo, 1986; Siqueira et al., 1987). These reports were corroborated in our conditions by Sánchez (2001) who reported approximately 50 spores per 100 g^{-1} of soil.

Roots, Tubers, and Vegetable Crops

A series of experiments (Ruiz, 2001) with objectives similar to those for coffee transplant production were conducted from 1993 to 1999 with two soils, Eutric Ferrasols and Calcaric Cambisols (Table 5.1). They involved a large variety of crops: cassava *(M. esculentum),* potato (*Solanum tuberosum* L.), sweet potato (*Ipomoea batatas* [L.] Lam.), yam (*Discorea* sp.), and two types of malanga (*Xanthosoma* sp. and *Colocasia* sp.). Tomato (*Lycopersicon esculentum* L.), cucumber (*Cucumis sativus* L.), and micropropagated plantain *(Musa* × *paradisiaca)* were later included in the study, but only grown in Calcaric Cambisols. The main difference between these trials and those reported earlier is that the source of nutrients applied was mineral fertilizers supplying N, P, and K.

Figures 5.2 and 5.3 show the results of a principal component analysis conducted to evaluate the effectiveness of AM fungal strains in the root and tuber crops grown in the two soils. In both cases, only one component was necessary to explain 76.55 and 85.85 percent of the variability that existed in the trials with Calcaric Cambisols and Eutric Nitisols, respectively. Among the most important results obtained was the good performance of the strains in both soils. All

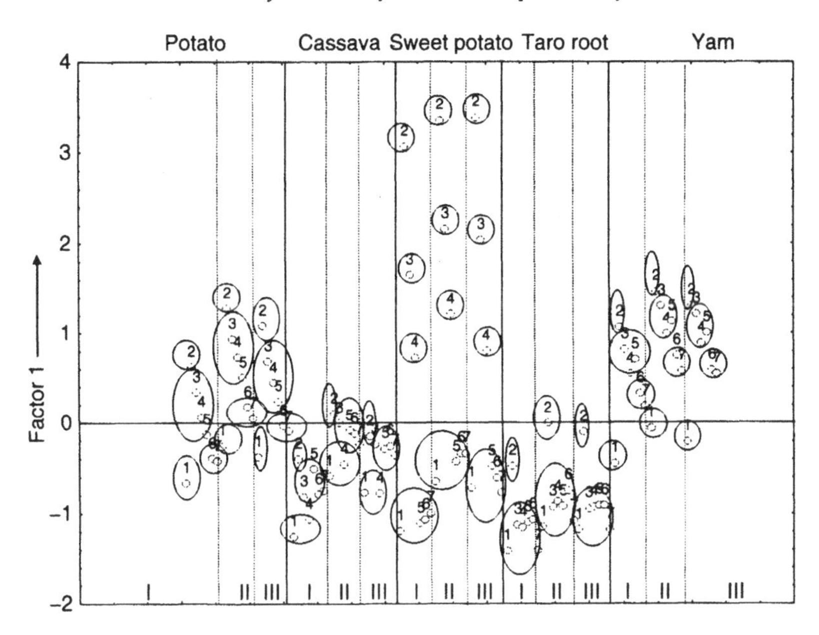

FIGURE 5.2. Graphic representation of the effect of AM strains inoculated on different root and tuber crops growing in Calcaric Cambisols, on the principal component C1 (Ruiz, 2001). Strains higher up have a stronger positive effect. I, II, and III = first, second, and third crops. 1, control; 2, *G. intraradices;* 3, *G. fasciculatum;* 4, *G. mosseae;* 5, *G. manihotis;* 6, *G. occultum;* 7, *A. scrobiculata.*

strains enhanced plant development as compared to the controls. But the most interesting result was that independently of the crop, there was one superior strain with which the best growth response was obtained. Thus, two effects were seen: large soil-strain specificity and low crop plant-strain specificity. These results were repeatedly obtained in the different years of the experiments in both soils despite the fact that crop species varied, as did their nutritional requirements and the availability of soil P and K, which also changed from year to year (Ruíz, 2001). This low-specificity between AM strains and crop species has important consequences and was very promising for the management of AM symbioses in these crops. Low-specificity facilitates the selection of efficacious strains and reduces selection criteria to one condition, the soil type, which appears to be the fundamental factor determining a strain's effectiveness.

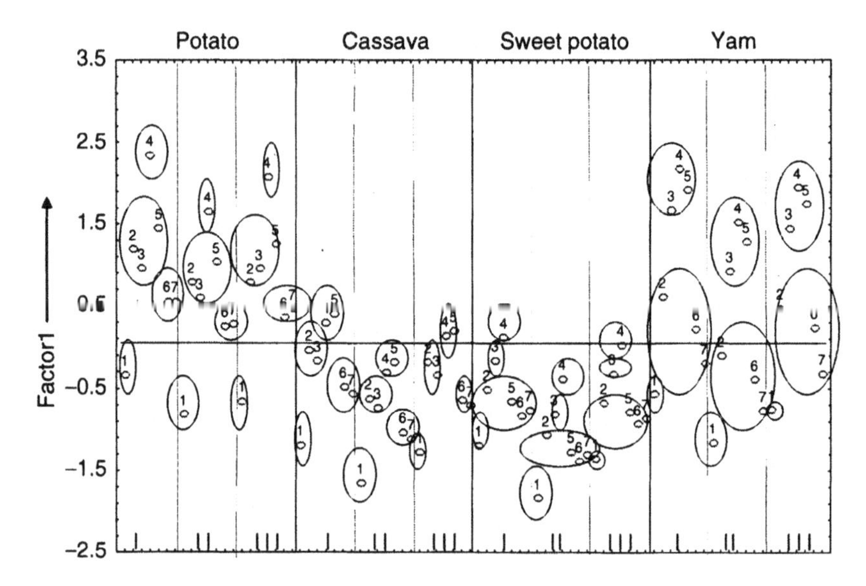

FIGURE 5.3. Graphic representation of the effect of AM strains inoculated on different root and tuber crops growing in Eutric Nitisols, on the principal component C1 (Ruiz, 2001). Strains higher up have a stronger positive effect. I, II, and III = first, second, and third crops. 1, control; 2, *G. intraradices;* 3, *G. fasciculatum;* 4, *G. mosseae;* 5, *G. manihotis;* 6, *G. occultum;* 7, *A. scrobiculata.*

The most effective strain for all crops grown in Calcaric Cambisols was *G. intraradices*. It is worth noting, however, that inoculation with *G. fasciculatum* was also very effective on tuber and root crops in Calcaric Cambisols, albeit less so than *G. intraradices*. Similar results were found in Eutric Nitisols, but in this case, *G. mosseae* was the most effective, followed closely by *G. clarum*.

SPECIFICITY BETWEEN EFFICACIOUS STRAINS AND CROP SPECIES

Results obtained from root and tuber crops support those obtained from coffee. AM fungal strains have a functional specificity for soil types. Furthermore, this last series of experiments indicated that a strain selected for its performance in a soil effectively enhances growth in all crop species that can form an AM symbiosis.

There were differences between the responses in different crops. In Calcaric Cambisols, for example, several strains were effective, with *G. intraradices, G. fasciculatum,* and *G. mosseae* triggering large growth increments in sweet potato, while in taro root, only one AM fungal strain, *G. intraradices,* increased growth as compared to the non-inoculated control. In cassava, on the other hand, *G. clarum* enhanced growth similarly in both soils and was as effective as the strains that were most effective in each of the two soils. In other words, the level of effectiveness of strains effective in a given soil type, was influenced by the crop species. From these results, which were highly reproducible across years and soil conditions, it could be concluded that root and tuber crops have a similar response to inoculation, which is dominated by the AM fungal strain specificity for soil type, supporting the general concept of low AM fungi/host plant specificity proposed by Siqueira and Franco (1988).

Experiments involving other crops were then undertaken in Calcaric Cambisols. These experiments had a similar design and tested almost the same AM fungal species. *G. spurcum,* which seemed to possess desirable attributes for inoculation purposes, was also studied.

The results obtained form the univariate statistical analysis of yield and biomass data, transformed as relative effectiveness indices, are presented in Table 5.5. The similarity of results obtained for root and tuber, and vegetable crops despite differences in nutritional requirements between these two crop groups, strongly supports the fact that while crop-strain specificity has relatively little importance, selecting strains for their effectiveness in given soil types is crucial.

The inclusion of *G. spurcum* in the trial revealed the good function of this AM species in Calcaric Cambisols. The superior effectiveness of *G. intraradices* was observed once more and the performance of *G. fasciculatum* was again very good. Tomato studies showed how the specificity between crop and strain is expressed. *G. intraradices* was the most effective strain for tomato as well as for all other crops grown in this soil, but *G. mosseae* also performed very well with this crop. *G. mosseae* was found particularly effective in Eutric Ferrasols (Medina, 1994; Hernández, 2000).

The results of both research programs indicate that some AM fungal strains are adequately effective on a large group of crops in given

TABLE 5.5. Index of effectiveness of AM fungal species inoculated on different crops in Calcaric Cambisols.

	Index of effectiveness (%)[a]								
AM fungal strains	**Potato**	**Cassava**	**Sweet potato**	**Malanga**	**Yam**	**Plantain**	**Tomato**	**Pepper**	**Cucumber**
G. intraradices	43.9a	48.8a	397.6a	110.0a	47.8a	68.0a	148.5a	77.7a	74.2a
G. fasciculatum	31.2ab	27.4bc	319.5b	6.6bc	39.8b	56.3a	28.3c	38.2b	44.5b
G. mosseae	24.7bc	1.1d	186.5c	20.0b	29.5c	10.5cd	92.1b	37.6b	9.4c
G. clarum	18.0bc	38.0a	7.3d	3.3bc	35.4bc	29.2bc	23.2c	26.1c	18.7c
G. occultum	5.4c	29.8bc	3.6d	18.3b	22.5d	17.9cd	–	–	–
A. scrobiculata	1.8d	20.2c	0.0d	10.0c	17.7d	45.2ab	–	–	–
G. aggregatum	–	–	–	–	–	–	89.8b	32.8bc	43.0b
G. spurcum	–	–	–	–	–	–	130.0a	73.1a	71.2a
cv %	12.8	7.1	6.9	8.6	3.5	12.6	6.35	4.95	11.31

Source: Ruiz and Rivera (2001).

Note: Means with the same letter are not significantly different according to Duncan Multiple Range test $P < 0.001$.

[a]Index of effectiveness = (value with AM fungi − value without fungi)/value without fungi × 100.

soil conditions. The expression "adequately effective" is a more general criteria than "effective," which was used to refer to the most effective strains in a given soil type. Table 5.4 shows how the transition between a strain's effectiveness as a function of soil type is a gradual phenomenon in which strains are effective or adequate in a range of more or less similar soil types, depending on the genotype of the strain. For example, *G. fasciculatum* was effective in soils of average to high fertility, *G. clarum* in soils of low to average fertility, and *A.scrobiculata* only in soils of low fertility. *G. intraradices* is effective in a wide range of soils, but no one strain was effective in all soil types.

Subsequently, we found that *G. fasciculatum* functions adequately on soils with high base saturation and high salinity (1,300 μS cm^{-1}-2,400 μS cm^{-1}) associated with vegetable crops and pastures. These results concur with general reports on the beneficial effect of the AM symbiosis in saline environments (Azcón-Aguilar and Barea, 1997; Subba Rao and Dommergues, 1998) and indicate that efficacious AM fungal strains can establish effective symbioses in such environments even if they were not isolated from a saline soil, as reported by Tian et al. (2004).

Sieverding (1991) classified strains on a scale ranging from high to low effectiveness, notwithstanding how soil type influence strain effectiveness. The results presented here provide a better criterion for a more precise selection of strains for crop inoculation purposes. One strain will be effective in a certain range of soils, which is larger or narrower depending on the strain.

More knowledge on the functional specificity of AM strains would be useful to optimize taro root production or that of other crop species with strong preference for AM fungal strains. But the general scheme of "effective" and "adequately effective" strains in a soil type will remain valid and sufficient for the effective inoculation of most crops.

INFLUENCE OF NUTRIENT AVAILABILITY ON MYCORRHIZAL EFFECTIVENESS

An important factor for the management of effective AM associations in crop production is the availability of nutrients. These nutrients may be derived from the intrinsic fertility of soils or from the application of organic or mineral fertilizers that are required to completely fulfill plant growth requirements.

Influence of Organic Fertilizers on Mycorrhizal Effectiveness in Coffee

The nutrient source studied in the coffee transplant production program was vermicompost. The influence that the amount of compost has on the effectiveness of AM fungal strains was evaluated. Figure 5.4 and Tables 5.2 and 5.3 present selected results obtained by soil type and AM strain. In order to correctly interpret these results, it is important to know that the amount of vermicompost recommended for coffee plant production is three parts of soil for one part of compost (Rodríguez, 1992). Figure 5.4 shows that lower effectiveness of inoculation with different efficacious strains was obtained with this proportion in all soil types (Distric Nitisol, Chromic Luvisol, and Gleyic Cambisol). Better AM effectiveness was obtained with a proportion of 5:1. Lower additions of nutrients, like in a soil: vermicompost proportion 7:1, were also inadequate in these soils. A similar effect was observed by Fernández (1999) with Chromic Cambisols

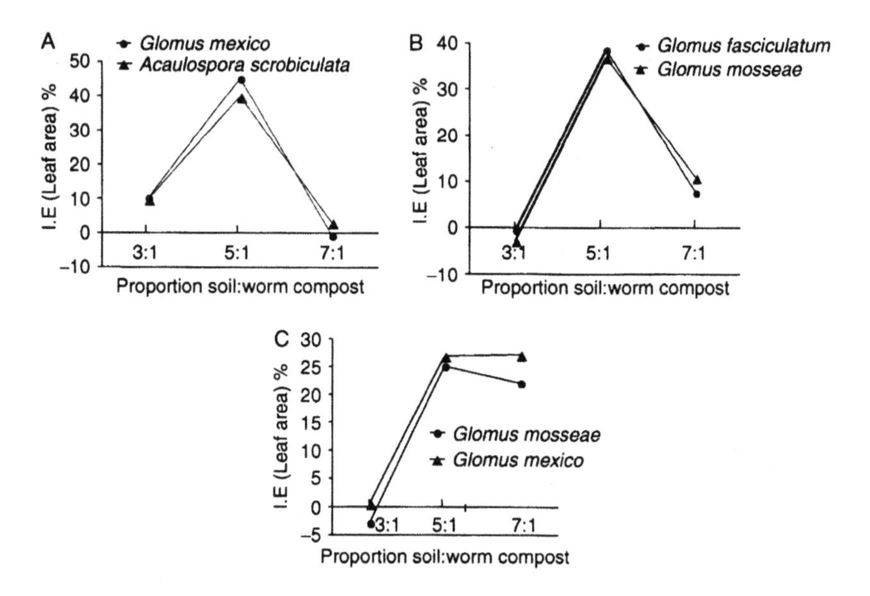

FIGURE 5.4. Influence of the proportion of soil and worm compost on the effectiveness of symbioses formed in coffee transplants inoculated with effective AM fungal strains, in soils of different types. Coffee plants were grown in 1,000 cm^3 to 1,200 cm^3 bags. *A*, Distric Nitisol; *B*, Chromic Luvisol; *C*, Gleyic Cambisols (Adapted from Sánchez, 2001).

and Humic–Chromic Cambisols where growth enhancement was optimal with a proportion of 5:1 for inoculated coffee plants, but decreased with a proportion of 3:1. In Eutric Cambisols with high fertility (CEC > 25 $cmol_c$ kg^{-1}), a proportion of 5:1 is not conducive to the establishment of effective AM symbioses and a proportion of 7:1 must be used.

The results indicate that for an AM symbiosis to be effective, nutrient availability level must be lower than that required for optimal production of non-mycorrhizal coffee plants. This corresponds to general information on the relationship between AM effectiveness and nutrient availability (Pacovsky et al., 1986; Siqueira and Franco, 1988). The only soils where a proportion of three parts of soil to one part of compost was appropriate for the creation of effective AM symbioses were Haplic Acrisols with low fertility. The small growth response obtained in these soils suggests that the proportion 3:1 was still too low. Obviously, the production of transplants with optimal vigor produced using less organic fertilizer is attributable to the increased nutrient absorption effectiveness in AM inoculated plants resulting in a better nutrient use efficiency.

High nutrient availability reduces the development of mycorrhizal structures in roots (Fernández, 1999; Sánchez, 2001), expressed in terms of endophytic mass or in percentage of root colonization (Phillips and Hayman, 1970), which indicates that the reduction in AM effectiveness is a consequence of a malfunction or inhibition of symbiosis development. Again, it is important to stress that the amount of fertilizer material for optimal AM effectiveness with inoculated crops depends on the soil's inherent fertility.

Criteria to interpret measurements of endophytic mass—an indicator to evaluate the intensity of the symbiosis—were established. Values indicative of effective AM symbiosis were relatively similar and ranged from 18 mg g^{-1} to 22 mg g^{-1} in all soils with average and high fertility levels, increased to 30 mg g^{-1} to 32 mg g^{-1} in Distric Nitisols of low fertility, and increased even more in infertile Acrisols (37 mg g^{-1}-39 mg g^{-1}). This suggests that in soils with low fertility, or when some nutritional factors are limiting, a greater abundance of fungal structures is required to ensure the proper function of the symbiosis.

Table 5.6 presents the results obtained from these fertilization trials. It appears that mycorrhizal plants require the application of an optimal level of nutrients—defined here by the proportion of soil and compost in the substrate mix in relation to the inherent fertility of the soil used—in order to develop effective mycorrhizae, favor the good functioning of the AM symbiosis, and optimize coffee transplants growth. Suboptimal compost application rates resulted in slow growing transplants bearing fewer AM fungal structures and consequently, with reduced AM function.

It appears clear from these results that AM effectiveness does not only depend on the selection of efficacious AM fungal strains, but also on the level of fertilizer applied and the soil type used. Here again, the soil type appears as the determinant factor in effective management of AM symbioses through inoculation. The soil type determined both the strain's effectiveness and the amount of organic fertilization (soil:vermi compost proportion) required for the development of effective symbioses.

TABLE 5.6. Proportion of soil:worm compost permitting optimal AM effectiveness of inoculated coffee transplants grown in 1,000 cm^{-3} to 1,200 cm^{-3} bags, and some chemical properties of these soils.

Soil type	Ca + Mg $cmol_c\ kg^{-1}$	Al $cmol_c\ kg^{-1}$	Optimal proportion soil: worm compost	Optimal endophytic mass $mg\ g^{-1}$
Acrisols	2.5-2.8	>2.3	3:1	37-39
Distric Nitisols	4.0-6.5	0.5-1.0	5:1	29-31
Chromic Luvisols	8.0-10.0	–	5:1	20-22
Chromic Cambisols	7.0-9.5	–	5:1	21-22
Gleyic Cambisols	10.5-12.0	–	5:1	21-22
Eutric Ferralsols	13. 0-15.0	–	5:1	22
Eutric Cambisols	16.0-17.0	–	5:1-7:1	18-20
Humic Eutric Cambisols	25.0	–	7:1	18-20
Humic Eutric Cambisols	40.0	–	7:1	19

Source: Rivera and Fernández (2003).

We recently found that the size of the growth bag and the rate of nutrient uptake by the crop plant are other factors to be considered in the establishment of the optimal proportion of soil and compost (Rivera, Ruíz, and Calderón, 2006). In 1000 cm^3 bags, micropropagated banana requires a soil:compost proportion of 3:1 for optimal AM development, but this proportion increases to 1:3, in plants placed in 180 cm^3 containers at the adaptation stage. Larger proportions of compost are required in smaller containers because nutrient absorption by plants and AM effectiveness respond more to the amount of available nutrients than to the concentration of available nutrients.

Influence of Mineral Fertilizers on Mycorrhizal Effectiveness in Root, Tuber, and Vegetable Crops

The experiments selected to illustrate this theme were conducted in Calcaric Cambisols. In these experiments, the influence of mineral N, P, and K fertilization on the effectiveness of *G. intraradices* was studied. *G. intraradices,* as seen earlier, is the most effective in Calcaric Cambisols. In each experiment, the optimal fertilization rates, which were derived from the "Research Program on the Nutrition and Mineral Fertilization of Root and Tuber Crop" (Portieles et al., 1982; Portieles, Ruíz, and Sánchez, 1983; Ruíz and Portielles, 1985; Ruíz, Milián, and Portieles, 1990), were used as positive controls. Here, these treatments are referred to as the "100% NPK" treatments.

The different crops studied responded in a similar way. The responses of cassava and sweet potato are presented in Figures 5.5 and 5.6, where five important observations can be made: (1) A positive response to inoculation with the efficacious strain was expressed by an increase in AM root colonization levels and in yields as compared to non-inoculated controls; (2) Lower fertilizer rates improved symbiotic effectiveness, as expressed by increased AM colonization and yield in all crops. The optimal fertilization rates, that is, those producing the best yields in plants inoculated with efficacious strains, were lower than the rates required to produce similar yields in non-inoculated plants; (3) The application of fertilizer rates higher than optimal for inoculated plants reduced mycorrhizal colonization, and consequently, symbiotic development, up to nearly complete inhibition with the 100 percent NPK rate. Yields did not decline, however, indicating that plants were fulfilling their nutritional requirements al-

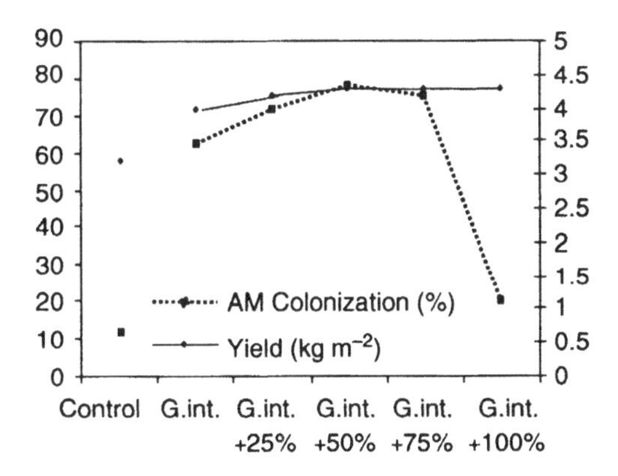

FIGURE 5.5. Influence of mineral fertilization with N, P, and K on the AM symbioses of cassava inoculated with *G. intraradices* and grown in Calcaric Cambisols (Adapted from Ruiz, 2001).

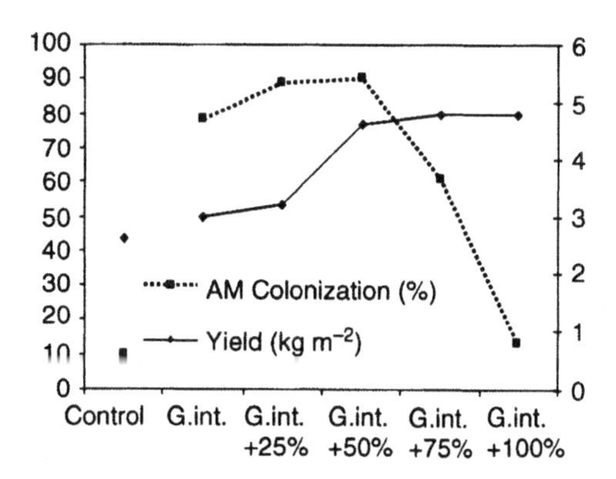

FIGURE 5.6. Influence of mineral fertilization with N, P, and K on the AM symbioses of sweet potato inoculated with *G. intraradices* and grown in Calcaric Cambisols (Adapted from Ruiz, 2001).

though less efficiently than with better mycorrhizal development; (4) the optimal fertilization rates for inoculated plants depended on the crop species; and (5) optimal nutrient availability allows an efficacious strain to function effectively.

The AM symbiosis appears as a mechanism allowing plants to fulfill their nutritional requirements and to reach their yield potential, which is determined by their genotype, the climate, and cultural practices. Depending on the resources of the soil or growth substrate, larger or lesser amounts of organic or mineral fertilizer are required.

The influence of mineral fertilization (N, P, and K) on the AM effectiveness of 10 crops is presented in Tables 5.7 and 5.8. These tables show the yields of the non-inoculated controls, the 100 percent NPK controls, and the inoculated treatments receiving the N, P, and K level that produced the best AM effectiveness and yields, which were similar to those obtained with 100 percent NPK. The good response to inoculation of crops with the efficacious strain selected for this soil, *G. intraradices,* and the high reproducibility of the effects through time are worth pointing out. In all crops, fertilization was necessary for the expression of optimal AM effectiveness, although the necessary amounts varied with crop species. Only 25 percent of the recommended N, P, and K rates were required by inoculated cassava, whereas sweet potato, malanga (*Colocasia* and *Xanthosoma* spp.), yam, and cucumber required about 50 percent of the recom-

TABLE 5.7. Optimal rates of mineral fertilizer (N, P, K) for different inoculated field-grown crops, in Calcaric Cambisols.

	First season		Second season	
	t ha^{-1}	IE %[a]	t ha^{-1}	IE %
Control	35.5c[b]	–	37.06c	–
AMF + 25% NPK	48.8a	37.80	52. 5a	41.7
100% NPK	50.3a	41.60	52.0a	40.3
cv %	9.9	–	4.4	–
Control	18.8c	–	27. 2c	–
AMF + 50% NPK	36.1a	92.60	40.1a	47.4
100% NPK	38.3a	99.36	41.4a	47.6
cv %	4.4	–	7.0	–
Control	25.5c	–	29.0	–
AMF + 50% NPK	36.7a	43.90	40.9a	41.0
100% NPK	38.3a	50.10	41.4a	42.7
cv %	5.9	–	4.7	–
Control	17.9c	–	23.2d	–
AMF + 50% NPK	40.1a	124.20	42.7a	84.1
100% NPK	40.3a	125.10	43.2a	86.2
cv %	4.3	–	5.7	–
Control	17.8c	–	25.1d	–
AMF + 50% NPK	26.9a	51.00	33.0a	31.5
100 % NPK	27.9a	56.40	32.8a	30.7
cv %	6.8	–	4.3	–
Control	5.9c	–	7.0c	–
AMF + 50% NPK	13.1a	122	15.43a	120.4
100% NPK	13.35a	126.70	15.56a	122.3
cv %	3.88	–	2.52	–

Source: Ruiz (2001).

[a]Index of effectiveness = (value with AM fungi – value without fungi)/value without fungi × 100.

[b]Means with the same letter are not significantly different according to Duncan Multiple Range test $p < 0.001$.

mended rates for non-inoculated plants. The rate of 75 percent N, P, and K was required by other crops (potato, tomato, pepper [Capsicum annuum L.], and plantain).

Thus, while improving plant nutrient uptake, the use of AM inoculation could lead to reduction in fertilizer use and to the reduction of nutrient seepage into the environment, an important potentially nega-

TABLE 5.8. Optimal rates of mineral fertilizer (N, P, K) for different inoculated field-grown crops, in Calcaric Cambisols.

	First season		Second season	
	t ha^{-1}	IE %[a]	t ha^{-1}	IE %
Control	18.0d[b]	–	20.5d	–
AMF + 75% NPK	29.2a	61.7	36.8a	79.6
100% NPK	30.1a	66.9	37.2a	81.2
cv %	7.5		8.6	–
Control	18.6e	–	20.1e	–
AMF + 75% NPK	40.9a	119.6	41.9a	108.9
100% NPK	41.1a	120.9	42.0a	109.0
cv %	4.8	–	4.4	–
Control	15.9d	–	17.2d	–
AMF + 75% NPK	37.3a	134.0	41.6a	141.8
100% NPK	37.4a	134.0	41.9a	143.4
cv %	3.6	–	–	–
Control	15.6d	–	–	–
AMF + 75% NPK	29.6a	89.7	–	–
100% NPK	29.7a	89.7	–	–
cv %	6.9	–	–	–

Source: Ruiz (2001).

[a]Index of effectiveness = (value with AM fungi – value without fungi)/value without fungi × 100.

[b]Means with the same letter are not significantly different according to Duncan Multiple Range test $p < 0.001$.

tive impact of agriculture. The maintenance of high yield despite reduction of fertilizer application is achieved through improvement of crop nutrient uptake capacity, or in other words, increased nutrient use efficiency.

The influence of fertilization on the AM effectiveness of a range of crops inoculated with an efficacious AM fungal strain were similar to those obtained with coffee transplants, although in the latter, an organic fertilizer was used. These results underline the compatibility of AM inoculation with mineral fertilization. Plant response depends on the type of soil, on nutrient availability, and on the yield potential of crops. These are key factors to the effective management of AM symbioses in productive agroecosystems.

AM FUNGI-RHIZOBACTERIA COINOCULATION

In Cuba, research on the use of microbial inoculant containing rhizobacteria has been very successful. This work evaluated the influence of nitrogen-fixing rhizobacteria such as *Rhizobium* and *Bradyrhizobium* species for legumes, and plant-growth promoting rhizobacteria (PGPR) including *Azotobacter chroococcum* for vegetable crops, species of *Azospirillum* for rice (*Oryza sativa* L.), maize (*Zea mays* L.), sorghum (*Sorghum bicolor* [L.] Moench.), *Burkholderia cepacia* for maize, and *Pseudomonas fluorescens* for various crops. Commercial forms of these inoculants exist under the names Biofert®, Azofert®, Dimargon®, Fosforina®, and others.

PGPRs have different modes of action. *B. cepacia* controls plant pathogens, *P. fluorescens* solubilizes phosphorus, and *A. chroococcum* and *Azospirillum* species fix atmospheric nitrogen. These rhizobacteria also stimulate plant growth.

Research work to evaluate the effects of coinoculation of different rhizobacteria and AM fungi was initiated in the early 1990s, in Cuba. This work was based on the observations of these microorganisms in the rhizosphere of AM plants and on reports of their mutualistic relationships (Fitter and Garbaye, 1994; Höflich, Wiehe, and Kuhn 1994; Gryndler, 2000). Furthermore, PGPRs have been found in spores of *Gigaspora* (Bianciotto et al., 2000; Minerdi et al., 2001) and most recently in *G. clarum* (Mirabal, Ortega, and Rodes, 2002). In the early program of coffee transplant production, we observed that the sterilization of the substrate before inoculation trials reduced the effectiveness of inoculation (Rivera et al., 1997). This lower effectiveness in sterilized soil could be due to the elimination of rhizospheric organisms associated with coffee, which may have contributed somehow to AM effectiveness.

The first results from this program of coinoculation of PGPRs and AM fungi were presented by Rivera et al. (1997) and Fernández (1999). Coinoculation of *A. chroococcum* with efficacious AM fungal strains was tested in Eutric Cambisols. The rhizobacterium significantly enhanced the effect of inoculation with one AM strain without increasing the development of fungal structures in plants, suggesting that the two organisms have complementary modes of action. In Haplic Acrisols of low fertility, the impact of coinoculation was always negative, with worse effects than inoculation with any of the microorganisms alone. These results were explained on the basis of the low nutrient availability in this soil, which could limit the development of AM fungal structures in plants. Based on these results coinoculation of *A. chroococcum* was recommended only in soils of fair to high fertility and with one part of vermicompost for five parts of soil.

The best and more consistent results were probably obtained in the tripartite symbiosis AM fungi-Rhizobium-legume. Symbiotic relationships seemingly provide better exchange between partners resulting in larger plant growth promotion than relationships based on nonsymbiotic associations. The plant symbiosis with both Rhizobium and AM fungi mobilizes N_2 through biological fixation on the one hand and enhancement of P uptake, an element very important for effective nitrogen fixation and plant growth on the other hand. The international literature (Pacovsky, Bethlenfalvay, and Paul, 1986) and the local literature alike (Corbera and Hernández, 1997; Corbera, 1998) contain numerous reports on the benefits of such coinoculations that are still being demonstrated in large production areas in Bolivia (INCA, 1999).

Based on the previous results and those related to AM fungi coinoculation with *Azotobacter* spp. in maize, sunflower (*Helianthus annuus* L.), tomato, and bean (*Phaseolus vulgaris* L.) (Terry et al., 2002), with *A. brasilense* in maize and sorghum (Medina et al., 1999), with *A. lipoferum* in tomato and sorghum (Terry, Pino, and Medina, 1998), and with *B. cepacia* in potato (Hernández-Zardón, 2001) in which coinoculation was better than single inocula, it was appropriate to recommend the practice of coinoculation. Coinoculation is considered as a practice for the management of AM associations in crop production. Coinoculation practices always take into account the specificity that exists between the different crop species and the rhizobacteria, and a bacterium is added with AM fungal inoculum only for the inoculation of receptive crops.

AM INOCULATION AND CROP ROTATION

Although experiment results on the persistence of inoculation effects in crop sequences or crop rotation either in research plots or in the field are rare, the persistence of inoculation effects is a very important subject. It is also a very promising subject considering the low level of specificity existing between efficacious AM strains and crop species.

In the extensive review by Sieverding (1991), there are only two reports of research work conducted to assess the possible residual effects of AM inoculation. No residual effect of inoculation was reported although inoculation had initially triggered positive growth response in inoculated crops. Different results were obtained in the research program on root and tuber production conducted by Ruíz (2001) in Calcaric Cambisols. The residual effect of *G. intraradices*, a highly effective AM strain in this soil type, was studied in a crop rotation composed of potato-sweet potato-cassava-sweet potato grown in microplots, in the field. Two cycles of the rotation were completed. Table 5.9 shows the yields and AM root colonization percentages averaged over the two cycles of the trial. Inoculation always significantly enhanced both variables as compared to the non-inoculated

TABLE 5.9. Persistence of the effect of *Glomus intraradices*, an effective strain inoculated in the yield and percentage of AM root colonization of the crops of a rotation established on Calcaric Cambisols.

	Potato		Sweet potato		Cassava		Sweet potato	
Treatment	Yield ($kg\ m^{-2}$)	AM roots (%)	Yield ($kg\ m^{-2}$)	AM roots (%)	Yield ($kg\ m^{-2}$)	AM roots (%)	Yield ($kg\ m^{-2}$)	AM roots (%)
Control	2.0b[a]	3c	2.1b	5c	2.1c	7c	1.8b	4c
All crops inoculated	2.6a	76ab	2.8a	75a	3.1a	81a	2.5a	79a
Inoculation of every 2 crops	2.6a	77ab	2.5a	70b	3.1a	79a	2.2a	71a
Inoculation of every 3 crops	2.6a	75b	2.5a	71b	2.7ab	69b	2.3a	79a
Inoculation of every 4 crops	2.6a	79a	2.5a	70b	2.6b	67b	2.0b	34b
cv %	18.1	2.9	18.3	4.1	16.6	5.0	17.7	3.2

Source: Ruiz (2001).

[a]Means with the same letter are not significantly different according to Duncan Multiple Range test $p < 0.001$.

control, confirming the positive effect of inoculation with *G. intraradices* in all crops growing in this soil.

At the end of the experiment there was no significant difference in yield or AM root colonization percentages whether inoculation was done in all rotation phases, every other phase, or every third phase of the rotation. When inoculation was only performed every fourth phase (i.e., once per rotation), the yield (2 kg m^{-2}) and percentage of colonization (34 percent) obtained in the last phase were significantly less than those obtained with more frequent inoculation, but was still better than the values obtained in the non-inoculated control. These results reveal the contribution of an inoculum to two or three subsequent crops as a consequence of the low level of specificity between efficacious AM strains and crop species. Thus, when an efficacious strain is introduced into a soil, it is able to colonize different crops, reproduce and build a population of propagules large enough to colonize subsequent crops.

The persistence of AM inoculation effects is certainly linked to a range of soil management and cropping practice-related factors. Fertilization and pesticide application, crop species used, and their mycorrhizal dependency, as well as planting pattern are all factors that may influence the production of propagules of the efficacious strain introduced, and thus, the capacity of the latter to colonize a subsequent crop.

The "self-repair" capacity of the extraradical mycelium, with which areas of the mycelium can be reconnected to the main hyphal network through a phenomenon of hyphal fusion known as anastomosis (see Chapter 2 of this book), is most developed in *Glomus* species (Providencia et al., 2005). This mechanism may also contribute to the good functionality of the symbiosis and may also be a factor involved in the persistence of inoculation effects in crop sequences.

Riera (2003) conducted some rotation experiments under field conditions, in Eutric Ferrasols. The effectiveness of inoculation with *G. clarum,* expressed by yield enhancement, was evaluated over two rotation cycles (Table 5.10). Results generally concurred with the previous ones except that in this case, the effects of inoculation only persisted up to the second rotation phase. The density of mycorrhizal structures observed in roots (Trouvelot, Kough, and Gianinazzi-Pearson, 1986), an indicator of the extent of AM functionality, was

TABLE 5.10. Effect of field inoculation of different rotation crops with *Glomus clarum* on the yield and percentage of density of AM structures (DV%) in the roots of these crops grown in rotations, on Eutric Ferrasols.

	Soybean		Maize		Sweet potato	
	Yield ($t\ ha^{-1}$)	DV (%)	Yield ($t\ ha^{-1}$)	DV (%)	Yield ($t\ ha^{-1}$)	DV (%)
Control	1.83b[a]	1.4b	5.15b	2.6b	20.3b	1.9c
1st crop inoculated	2.00ab	3.7a	5.37ab	2.8b	20.9b	2.5bc
1st and 2nd crops inoculated	2.02ab	4.4a	5.62a	4.8a	23.4a	3.4b
All crops inoculated	2.12a	4.4a	5.67a	5.1a	24.95a	5.7a
Standard error	0.06	0.2	0.11	0.35	0.59	0.39
	Soybean		Sunflower		Sorghum	
Control	1.91b	1.6b	1.71b	2.8c	2.64c	0.6c
1st crop inoculated	2.08ab	4.9a	1.77ab	5.7b	2.72b	1.5bc
1st and 2nd crops inoculated	2.10a	4.9a	1.91a	7.9a	2.73b	2.2ab
All crops inoculated	2.16a	4.6a	1.94a	8.4a	2.82a	2.9a
Standard error	0.05	0.25	0.05	0.30	0.02	0.30
	Rice		Dry bean		Sweet potato	
Control	1.09b	0.9c	1.59b	0.5b	22.6c	1.4c
1st crop inoculated	1.13ab	1.5bc	1.60b	0.7b	23.3c	2.3bc
1st and 2nd crops inoculated	1.18a	2.1ab	1.77a	2.1a	25.5b	4.2ab
All crops inoculated	1.19a	2.3a	1.78a	2.1a	27.3a	5.7a
Standard error	0.02	0.19	0.03	0.21	0.5	0.57

Source: Riera (2003).

[a]Values are the means of two years. Means with the same or lower case letter are not significantly different according to Duncan Multiple Range test $p < 0.001$.

always reduced in absence of inoculation, although densities observed in the subsequent crop were close to those measured in the inoculated crop.

The information available on the persistence of inoculation effects through crop sequence is limited. The literature that does exist refers to the influence of cropping practices on the conservation of native AM populations (Primavesi, 1990; Sieverding, 1991; Chu and Diekmann, 1994; Sivila and Hervé, 1994). The results presented here are the first to report on the persistence of AM inoculation effects through the phases of crop rotations.

The work of Riera (2003) also evaluated the dynamics of the AM spore population, as influenced by crop species and treatments, and showed that inoculation significantly increased spore number in all rotation phases. The extent of this increase depended on the frequency of inoculation, the genotype of the crop species inoculated, and the genotype of the subsequent crop. Spore density could reach 800 to 900 spores g^{-1} of soil under inoculated sweet potato and sunflower, while initial population densities were approximately 50 spores g^{-1} of soil.

INFLUENCE OF EFFICACIOUS AM ON SOIL AGGREGATION

In Cuba, pioneering work on the influence of AM inoculation on soil aggregation was conducted by Riera (2003) as a part of his research on crop rotations in Eutric Ferrasols. He found that the treatments favoring AM effectiveness also increased the proportion of larger class size soil aggregates and their stability. These treatments reduced the soil coefficient of dispersion, which prevented the progressive decline in soil physical quality associated with continuous cropping. Increases in the proportion of soil bound into aggregates was positively correlated to the indicators of AM functionality such as AM root colonization, suggesting the involvement of enhanced AM development in soil structural quality. These results concur with current understanding of the role of AM fungi in the formation of soil aggregates (Miller and Jastrow, 1992; Rillig and Wright, 2002) in relation to their production of glomalin, a glycoprotein excreted by AM fungi that is involved in the formation and stabilization of aggregates (Wright et al., 2001).

A study was recently conducted in Cuba (Morell, 2005) to characterize the degradation of Eutric Nitisols. This study revealed better soil aggregation, larger amounts of glomalin, and an AM fungal activity more intense in soil profiles showing less degradation, giving greater support to the fact that AM fungi have a beneficial influence on soil physical properties.

GREEN MANURE USE AND MYCORRHIZAL EFFECTIVENESS

A series of experiments was conducted from 1995 to 1998 (Sánchez, 2001) to define the use of green manure crops as a component of coffee transplant production systems. These crops are grown to be incorporated into the soil as a source of nutrients and organic matter. Green manure production could be used to reduce or replace the use and transport of traditional sources of organic fertilizers (vermicompost and manure) and to reduce farmers' dependence on off-farm inputs.

The following work was conducted in Gleyic Cambisols, Chromic Luvisols, and Distric Nitisols with characteristics similar to those presented in Table 5.1. Each species of green manure had different characteristics (Table 5.11). Sorghum and crotalaria (*Crotalaria juncea* L.) presented interesting characteristics both exhibiting high dry matter and nutrient yields. Crotalaria is richer in N and sorghum contains more P and K. Canavalia (*Canavalia ensiformis* [L.] DC.) had lower nutrient and dry matter yields, which were nevertheless better than dolicho (*Lablab purpureus* L.). The soil type studied did not have a different influence on the productivity of these crops

Table 5.12 presents a summary of the results obtained from three crops of coffee transplants grown in Gleyic Cambisols. Results, which were reproducible in time, revealed marked differences be-

TABLE 5.11. Characteristics of the green manure crops under study.

Green manure crop	Fresh biomass (t ha^{-1})	Dry biomass (t ha^{-1})	N (kg ha^{-1})	P (kg ha^{-1})	K (kg ha^{-1})
Dolicho[a]	14.69d[b]	2.60d	85.93c	9.65d	31.12d
Canavalia	28.6-33.4	4.2-4.64	143.7-160.4	13.0-15.2	49.8-56.8
Crotalaria	37.8-47.1	6.46-7.38	165.0-201.9	17.17-18.7	98.9-110.0
Sorghum	60.1-70.5	8.95-9.53	145.6-164.8	24.2-30.5	151.6-167.3

Source: Sánchez (2001).

[a]Only used in Gleyic Cambisols.

[b]Numbers are means of three growth cycles and three soil types.

TABLE 5.12. Effect of green manure crops and inoculation with *Glomus mosseae* ecotype 1 (AM) on the vigor, percentage of AM root colonization, and AM endophytic biomass in roots of coffee transplants produced in Gleyic Cambisols.

	1995-1996	1996-1997			1997-1998		
Treatments	**Leaf surface area (cm^2)**	**Leaf surface area (cm^2)**	**AM col. (%)**	**AM endophyte ($mg\ g^{-1}$)**	**Leaf surface area (cm^2)**	**AM col. (%)**	**AM endophyte ($mg\ g^{-1}$)**
Soil	[illegible][a]	[illegible]	[illegible]	7.0d	[illegible]	17.7f	[illegible]
Soil + AM	240c	255d	28.0cd	8.1d	276de	24.0e	9.1c
Dolicho	250c	270c	24.05d	12.0c	286d	18.0f	13.6b
Dolicho + AM	307b	346b	31.05c	15.2b	346c	38.3c	16.0b
Canavalia	285bc	281c	32.35c	13.0c	295d	30.0d	15.3b
Canavalia + AM	372a	424a	50.7a	19.6a	396ab	48.0b	19.0a
Crotalaria	312b	341b	33.0c	15.5b	338c	37.3c	15.9b
Crotalaria + AM	381a	443a	58.3a	20.0a	417a	58.3a	19.2a
Sorghum	322b	430a	42.0b	18.4a	380b	44.7bc	19.4a
Sorgo + AM	379a	433a	52.3a	19.5a	413a	57.0a	20.2a
3:1 (soil:compost)[b]	385a	451a	24.3d	12.0c	415a	20.0f	13.4b
3:1 (soil: compost) + AM	389a	448a	25.3d	12.2c	404a	25.0e	14.1b
cv %	9.54	10.01	8.15	6.24	5.66	5.66	9.66
Standard error	12.34**	11.28**	1.15**	0.49*	11.95**	0.87**	0.88**

Source: Sánchez (2001).

[a]Means with the same or lower case letter are not significantly different according to Duncan Multiple Range test $p < 0.001$.

[b]3:1 (soil:compost), recommended fertilization in conventional production of coffee transplants.

*Treatment effect is significant according to ANOVA $p < 0.01$ level.

**Treatment effect is significant according to ANOVA $p < 0.001$ level.

tween treatments. When dolicho or canavalia green manure crops were used, a slight improvement in coffee growth was seen, but coffee development was still much less than when grown with the conventional production method (the use of a 3:1 soil:vermi compost mix [Rodríguez, 1992]). Crotalaria showed a better performance, although it was still inferior to that produced by the conventional method. Green manure treatments combined with soil inoculation with an efficacious AM fungal strain markedly enhanced coffee transplant growth, and this was true with crotalaria or canavalia. The performance of the green manures + AM inoculation treatment did not differ significantly from that of the conventional production control treatment, indicating that the green manure crop could fulfill coffee transplants' nutritional requirement. Incorporation of sorghum produced better results than those found with the other green manures. The transplants obtained with sorghum green manure had as much as 84 percent to 95 percent of the leaf surface area as those with conventional production control treatment and AM inoculation with sorghum green manure treatment produced plants similar to conventional production control treatment.

The difference between the contrasting effects of inoculation with green manure crops on coffee growth was explained by two indicators of AM functionality: the concentration of AM endophyte in roots and the percentage of coffee transplant root colonization. Surprisingly, the production and incorporation of green manure in the absence of inoculation enhanced mycorrhizal development, as measured by the amount of AM endophyte and percentage of AM colonization of coffee plant roots. These effects were directly related to green manure crop growth and dry matter yield, indicating that larger green manure yields were followed by more extensive AM development in non-inoculated coffee transplants, in other words, the greater effectiveness of native AM endophytes. Sorghum green manure increased the AM endophytic mass of coffee plants to as much as 18 mg g^{-1} to 20 mg g^{-1} of roots, levels that were associated with high AM symbiotic effectiveness. Inoculation with sorghum green manure did not increase further AM development and symbiotic effectiveness.

Other important conclusions from these experiments were that conventional organic fertilizers could be replaced by green manure crops used alone or in combination with AM inoculation, which resulted in optimal coffee transplant development. Green manure crops could entirely fulfill the nutritional requirements of the mycorrhizal coffee transplants and support optimal AM functionality. This was the case for canavalia, crotalaria, and sorghum, but not for dolicho.

Sorghum green manure increased the AM endophytic mass from less than 10 mg to almost 20 mg g^{-1} of roots, a level associated with high AM symbiotic effectiveness, and inoculation did not further increase AM development. This indicates that although there were differences in the amounts of nutrients brought by the different green manure crops, the main reason for the difference seen in the interaction between green manure crops and AM inoculation was the effect of the green manure crops on native AM fungi. Green manure crops may influence the production of propagules by native AM fungi differently. For example, sorghum is one of the host plants most used for the propagation of AM fungal inoculum (Sieverding, 1991) and the data suggests that sorghum, and to a lesser extent canavalia and crotalaria may have raised the level of native inoculum in soil. The concentration of AM spores increased from 60 to 90 spores 100 g^{-1} of soil up to 450 spores 100 g^{-1} of soil with sorghum or crotalaria (Sánchez, 2001). Although in two out of the three years of the study coffee growth did not respond to inoculation with sorghum green manure, it is safe to recommend inoculation of coffee transplants with efficacious AM strains in production systems using green manure, in Gleyic Cambisols.

The effect of green manure in coffee transplant production in the lower fertility Chromic Luvisols and in Distric Nitisols (Table 5.13) was similar to that observed in Gleyic Cambisols, with regard to their positive effects on coffee growth and stimulation of native AM fungal populations. However, effects were smaller and probably related to the lower fertility of these soils. Sorghum and crotalaria, when used alone, produced the largest AM endophytic biomass in coffee roots, but this biomass was lower than what is considered indicative of an effective AM symbiosis, that is, 20 mg g^{-1} to 22 mg g^{-1} in Chromic Luvisols and 30 mg g^{-1} to 32 mg g^{-1} in Distric Nitisols (Table 5.6). This indicates that the production and incorporation of green manure does not ensure the effectiveness of native AM fungi on coffee transplants. Similarly, coffee transplants showed reduced growth in AM inoculated soil amended with green manure as compared to control plants under conventional production management, indicating that green manure crops could not completely fulfill the needs of inoculated coffee transplants.

The results obtained from other experiments where green manure treatments were complemented with different amounts of organic fertilizers revealed the reason for different treatment effects in differ-

TABLE 5.13. Effect of green manure crops (GM) and inoculation with *Glomus intraradices* (AM) on the vigor, percentage of AM root colonization, and AM endophytic biomass in roots of coffee transplants produced in Distric Nitisols.

	1996-1997			1997-1998		
Treatments	**Leaf surface area (cm^2)**	**AM col. (%)**	**AM endophyte ($mg\ g^{-1}$)**	**Leaf surface area (cm^2)**	**AM col. (%)**	**AM endophyte ($mg\ g^{-1}$)**
Soil	141e[a]	23f	6.1h	153f	24g	5.8f
Soil + AM	144e	25f	6.1h	154f	31e	7.9e
Soil + GM Canavalia	223d	27de	7.2f	215e	30ef	9.8d
Soil + GM Canavalia + AM	251d	40c	12.1e	261de	47b	14.1c
Soil + GM Crotalaria	243d	33d	13.1d	257de	36d	12.3cd
Soil + GM Crotalaria + AM	290c	56a	17.0b	297bc	51a	17.0b
Soil + GM Sorghum	285c	46b	13.0d	266cd	40c	16.8b
Soil + GM Sorghum + AM	321b	55a	21.2a	330b	49ab	19.2a
3:1 (soil:compost)[b]	410a	27de	16.9b	416a	28f	16.0b
3:1 (soil:compost) + AM	432a	28de	19.4ab	440a	30ef	18.8a
cv %	6.37	7.72	4.06	9.08	5.85	6.93
Standard error	10.08**	1.19**	0.279**	14.64**	0.905*	0.519**

Source: (Sánchez, 2001).

[a]Means with the same or lower case letter are not significantly different according to Duncan Multiple Range test $P < 0.001$.

[b]3:1 (soil:compost), recommended fertilization in conventional production of coffee transplants.

*Treatment effect is significant according to ANOVA $p < 0.01$ level.

ent soils. In Distric Nitisols (Table 5.14) the comparison of the best inoculated treatment (7:1, soil + green manure:vermi compost + AM inoculation) and AM inoculation with green manure only indicated that, effectively, in these soils a green manure crop alone could not fulfill mycorrhizal coffee transplants' nutritional requirements. Fertilization with vermicompost in addition to green manure and AM inoculation produced a much better growth response in coffee and a more

TABLE 5.14. Effect of a green manure crop of sorghum (GM), worm compost addition (C), and inoculation with *Glomus intraradices* (AM) on the vigor, percentage of AM root colonization, and AM endophytic biomass in roots of coffee transplants produced in Distric Nitisols.

	1996-1997	1997-1998		
Treatments	**Leaf surface area (cm^2)**	**Leaf surface area (cm^2)**	**AM col. (%)**	**AM endophyte ($mg\ g^{-1}$)**
Soil	141d[a]	153g	24e	5.8g
Soil + AM	144d	157g	31d	7.9f
Soil + GM	285c	286f	40c	16.8d
Soil + GM + AM	321c	330de	49b	19.2c
(Soil + GM): $C_{9:1}$[b]	313c	306ef	36c	14.3e
(Soil + GM): $C_{9:1}$ + AM	360b	357cd	50ab	23.3b
(Soil + GM): $C_{7:1}$	374b	381bc	39c	13.9e
(Soil + GM): $C_{7:1}$ + AM	421a	446a	54a	30.4a
(Soil + GM): $C_{5:1}$	430a	441a	38c	16.7d
(Soil + GM): $C_{5:1}$ + AM	434a	445a	46b	17.4d
Conventional Soil-$C_{3:1}$	410a	417ab	28de	16.0d
Soil-$C_{3:1}$ + AM	432a	440a	30d	23.4b
c.v %	6.68	6.63	7.20	5.20
Standard error	13.07*	13.27*	1.21*	0.48*

Source: Sánchez (2001).

[a]Means with the same or lower case letter are not significantly different according to Duncan Multiple Range test $p < 0.001$.

[b]3:1 proportion of soil to worm compost; 5:1, 7:1, and 9:1, proportion of (soil + GM) to worm compost.

*Treatment effect is significant according to ANOVA $p < 0.01$ level.

**Treatment effect is significant according to ANOVA $p < 0.001$ level.

functional AM symbiosis. It was necessary to complement this soil with organic fertilizer in a 7:1 (soil + green manure:compost) proportion to obtain optimal coffee transplant growth.

The amount of fertilizer required for mycorrhizal coffee transplant production was reduced from 5:1 to 7:1 through the use of green manure crops. This production system allows for less off-farm inputs as compared with the 3:1 soil:compost mix traditionally recommended for coffee transplant production with these soils. Furthermore, the high response of AM inoculation to the presence of green manure and vermicompost (7:1 soil:compost) suggests either that green manure crops did not effectively stimulate the production of native AM propagules or that the native AM fungi were ineffective. Nevertheless, it was possible in these soils to reduce organic fertilizer application by about half using AM inoculation with efficacious strains and still obtain optimal growth. This last experiment also showed that the use of more than one part of vermicompost with seven parts of soil concurrently with green manure did not further increase leaf area, but reduced AM endophytic occupation of roots. This negative effect of excessive nutrient availability on mycorrhizal effectiveness was reported in earlier experiments.

In Chromic Luvisols, which have a fertility level between that of Gleyic Cambisols and Distric Cambisols, experimental results showed that the use of one part of vermicompost in nine parts of soil with green manure crops was appropriate to optimize the effect of inoculated AM strains and insure vigorous plant growth. Thus, approximately 60 percent of the amount of organic fertilizer previously recommended could be replaced by green manure and AM inoculation in coffee transplant production. Also, in these soils, higher amounts of compost reduced the functionality of the AM fungi introduced through inoculation.

The growth and incorporation of sorghum and crotalaria green manure crops prior to coffee production offers large economic benefits and is a practice that can be perfectly integrated in the production of AM coffee transplants. The use of green manure should not be limited to the inoculation of coffee with efficacious AM strains at seeding. It may be better to inoculate the green manure crop and to obtain in this way additional benefits in the multiplication of efficacious AM strains propagules, thus reducing the cost of AM transplant production.

USE OF THE AM SYMBIOSIS ON A PRODUCTION SCALE

This section presents some results derived from the effective management of the AM symbiosis in agriculture using the inoculant EcoMic®, a product formulated for the application of 1 kg ha^{-1} to 10 kg ha^{-1} of the product as a seedcoating (INCA, 1999). This product can be used in high-input agricultural systems as well as in low-input traditional family farming systems. In all cases presented, *G. fasciculatum* was used based on previous results showing the effectiveness of this strain in soils of fair to high fertility.

High-Input Agriculture

Examples from high-input systems were taken from results obtained from the Bolivian province of Santa Cruz de la Sierra in Eutric Regosols (INCA, 1999). In high-input systems, the biofertilizer is mechanically applied on seeds and this seedcoating is compatible with mechanical seeding.

All crops responded to inoculation with increased productivity in the range of 16 percent to 78 percent and averaging 43.3 percent (Table 5.15). The seedcoated inoculant was applied at rates from 6 to 10 percent of the seed weight. It is important to note the range of crops used in the test, which included cotton, wheat, soybean, sunflower, bean, maize, sorghum, hot pepper, and lentil, verifying once more the low level of specificity between efficacious AM strains and crops species.

Crops were produced in high-input systems on land areas as large as 1,000 ha. This demonstrates the compatibility of the AM inoculation practice with large-scale production systems. Furthermore, coinoculation of the AM strain and Rhizobia was successfully done in

TABLE 5.15. Results of experiments testing the inoculant EcoMic® *(Glomus fasciculatum)* applied as seedcoating at a rate of 6 percent to 10 percent of seed weight in different crops and soil types under high-input production systems.

Crop-Country	Soil type	Validation Area ha	AM $t\ ha^{-1}$	Control $t\ ha^{-1}$	Yield increase (%)
Rice-Colombia	Eutric Fluvisol	16	4.80	2.70	77.7
Cotton-Bolivia	Eutric Regosol	94	0.94	0.69	38.0
Maize-Cuba	Eutric Ferralsol	16	2.84	2.34	21.3
Maize-Cuba	Eutric Ferralsol	16	5.41	3.04	77.9
Maize-Bolivia	Eutric Regosol	150	2.92	2.16	35.1
Maize-Bolivia	Eutric Regosol	150	3.12	2.51	24.3
Wheat-Bolivia	Eutric Regosol	50	3.19	2.75	16.0
Wheat-Bolivia	Eutric Regosol	50	3.12	1.82	71.4
Soybean-Bolivia	Eutric Regosol	150	2.73	1.94	40.7
Soybean-Bolivia	Eutric Regosol	150	2.20	1.78	23.5
Soybean-Bolivia	Eutric Regosol	1,000	2.93	2.32	26.3
Bean-Bolivia	Eutric Regosol	11	1.71	0.96	78.1
Sunflower-Bolivia	Eutric Regosol	40	1.23	0.86	43.1

Source: Rivera and Fernández (2003).

soybean and bean. This shows that efficacious AM inoculation can be advantageously integrated with established high-input production systems.

Low-Input Agriculture

Examples of the successful use of EcoMic® on low-input family farms come from small farms on Eutric Fluvisols in Casanares, Colombia (Sosa, 1999), on Eutric Regosols in Santa Cruz de la Sierra, Bolivia (Hernández-Zardón, 2005) and from farms on Cuban Eutric Ferrasols. Soils were highly fertile and grew various crops including dry bean, cassava, maize, and rice. In all cases, yield increases between 12 and 78 percent and averaging 46.9 percent were attributable to AM inoculation with EcoMic® seedcoating (Table 5.16). These results are similar to those obtained under high impact production conditions. Coinoculation of soybean and bean with the AM strain and *Rhizobium* or *Bradyrhizobium* species was also successful. Under low-input agriculture AM inoculation was highly ef-

TABLE 5.16. Results of experiments testing the inoculant EcoMic® *(Glomus fasciculatum)* applied as seedcoating at a rate of 10 percent of seed weight in different crops and soil types under low-input production systems.

Crop-Country	Soil type	Area ha	AM $t\ ha^{-1}$	Control $t\ ha^{-1}$	Yield increase (%)
Rice-Cuba	Petroferric Gleysol	1.0	6.80	4.60	47.8
Rice-Colombia	Eutric Fluvisol	2.0	2.15	1.30	65.3
Rice-Colombia	Eutric Fluvisol	2.0	2.40	1.40	71.4
Cotton-Colombia	Molic Gleysol	1.5	2.60	2.20	18.2
Cotton-Colombia	Molic Gleysol	1.5	2.50	1.90	31.5
Maize-Colombia	Eutric Fluvisol	1.0	2.96	1.64	80.5
Bean-Colombia	Eutric Fluvisol	1.0	0.50	0.29	72.4
Bean-Cuba	Eutric Ferralsol	1.0	1.00	0.70	42.8
Potato-Bolivia[a]	Eutric Cambisol	0.5	80.00	50.00	60.0
Soybean-Cuba	Eutric Ferralsol	0.5	2.63	1.50	75.3
Maize-Cuba	Eutric Ferralsol	3.0	2.83	2.44	12.4
Peanut-Cuba	Eutric Ferralsol	1.2	1.12	1.00	12.0
Tomato-Cuba	Eutric Ferralsol	2.3	27.30	21.80	20.0

Source: Rivera and Fernández (2003); Hernandez-Zardón (2005).

[a]Seed potatos were inoculated with 1 kg EcoMic® 50 kg^{-1}.

fective producing higher yield through enhanced nutrient uptake, but also probably through improved tolerance to drought stress, which is commonly encountered in these systems.

In low-input systems, seedcoating and seeding are done manually. Seedcoating is done in a container or on a sheet of polyethylene, depending on the amounts of seeds that need to be treated. It is recommended to immerse vegetative propagules like those of sweet potato or cassava in a thin paste made with EcoMic® and water.

The positive results obtained in the testing of EcoMic® in both high- and low-input systems were based on the fundamental principles driving AM effectiveness: the use of AM strains adapted to soil types, the low level of specificity existing between AM strains and crops, the maintenance of optimal AM effectiveness through adequate fertilization, the coinoculation of AM strains and rhizobium inoculants, as well as the effective and practical seedcoating formulation of EcoMic®.

CROPPING SYSTEMS WITH EFFICACIOUS AM STRAINS

The importance of the AM symbiosis in plant growth was demonstrated throughout this chapter. The AM symbiosis increases plant nutrient and water uptake capacity and allows them to better tolerate nutritional and water stresses, but this does not mean that this symbiosis can be advantageously used only under stressful conditions. It is clear that the AM symbiosis is as useful to plant growth as rhizobia are for legumes, and enough information is available for the design and application of cropping systems based on the model of effective AM symbiosis. We have seen that such systems can be adapted to low-input as well as to highly mechanized, large-scale production. In both cases soil biological quality is enhanced, and the negative impacts of drought on crop plants are reduced, as are the risks of environmental impacts linked to crop fertilization. Results presented in this chapter are a guide for the successful management of the AM symbiosis in crop production, which is based on the selection of efficacious strains according to the type of soils, their inoculation, and the adoption of cropping practices that optimize AM effectiveness.

The observation of growth response upon inoculation with efficacious strains of AM fungi in a large range of soils indicates the prevalence of low numbers of effective native AM propagules in the cropping systems studied; this demonstrates the importance of the AM symbiosis beyond conditions of low soil fertility and degraded or eroded soils. Inoculation with efficacious strains is necessary to ensure the sufficient abundance of efficacious AM propagules.

The benefits obtained under experimental and production conditions using recommended AM strains, which were mostly extraneous, not only provide a strong basis for the elaboration of sound research objectives, but also provide basic principles for the effective management of AM symbioses.

The low level of specificity existing between AM strains and crop species simplifies management and allows efficient use of AM inoculation in crop rotations. The effect of inoculation on one crop can persist into a subsequent crop. Inoculation of green manure can be employed to enhance its productivity and thus its input as a nutrient and organic matter source, but these crops may also be used as a means to propagate efficacious AM strains for the benefit of a subsequent crop.

A new concept of efficient soil fertility management for mycorrhizal plants was established. Efficient fertilization of AM crops leads to the production of high yields using mineral or organic fertilizers at lower rates than those required for the production of similar yields in non-inoculated plants. The rates recommended for non-inoculated plants limit AM development and functionality. This stems from two important considerations: (1) the compatibility between the AM symbiosis and the proper management of soil fertility, and (2) the need to redefine the fertilizer requirements of efficiently mycorrhized crops and the critical levels of soil nutrients.

The concept of rhizosphere expands to a concept of hyphosphere and mycorrhizosphere establishing mutually beneficial relationships with different microbial groups (Fitter and Garbaye, 1994; Gryndler, 2000). This is the basis on which the tripartite associations established through coinoculation with AM fungi and *Rhizobium* in legumes; *Azospirillum* spp. in rice, tomato, and maize; *B. cepacia* in potato; and *A. chroococcum* in coffee, tomato, maize, and vegetable crops in general, provided good results under experimental and large scale production conditions (Rivera, 1993; Medina, 1994; Corbera and Hernández, 1997; Rivera et al., 1997; INCA, 1999; Sanchez et al., 1999; Hernández, 2000; Terry et al., 2002).

From the universality of the AM symbiosis (Harley and Smith, 1983; Brundrett, 2002) and from different results obtained emerges the concept of efficient AM cropping systems (Rivera, 2000), which are constructed around the model of the effective mycorrhizal plant, through inoculation with efficacious strains and management practices optimizing their efficiency. This symbiosis has always been present in the life of plants (Pate, 1994; Fitter and Moyersoen, 1996), even in agricultural soils where it is not efficient and almost always unnoticed. The proposition is to recognize and support this phenomenon effectively as a basis in crop production. The productive basis is effectively mycorrhized plants, and around them, cultural practices developed to support and optimize the system. Finally, it is important not to consider the inoculation of AM fungi as another agricultural input, but as a concept central to plant production. An agroecological and conservationist agriculture that protects the environment, sustains the soil resources, and ensures high yields corresponds to the economical, environmental, and social expectations that the twenty-first century demands.

REFERENCES

Azcón-Aguilar, C. and J.M. Barea (1997). Applying mycorrhiza biotechnology to horticulture: significance and potentials. *Sciencia Horticulturae* 68:1-24.

Barros, A. (1987). *Micorrizas Vesículo Arbusculares em Cafeiros da Regiao Sul do Estado de Minas Gerais.* M.Sc. Thesis. Lavras: Minas Gerais. p. 97.

Bianciotto, V., E. Lumini, I. Lanfranco, D. Minerdi, P. Bonfante, and S. Perotto (2000). Detection and identification of bacterial endosymbionts in arbuscular mycorrhizal fungi belonging to the family Gigasporaceae. *Applied and Environmental Microbiology* 66:4503-4509.

Brundrett, M.C. (2002). Coevolution of roots and mycorrhizas of land plants. *New Phytologist* 154:275-326.

Chu, E.Y. and U. Diekmann (1994). Efeito das actividades agricolas em populacao de fungo endomicorrízico nativo do solo da Amazonia Orental. *Resúmenes V REBRAM* Florianópolis, S.C. Brazil: Univ. Federal Sta. Catarina, p. 11.

Corbera, J. (1998). Coinoculación *Bradyrhizobium japonicum*—Micorriza vesículo arbuscular como fuente alternativa de fertilización para el cultivo de la soya. *Cultivos Tropicales* 19:17-20.

Corbera, J. and A. Hernández (1997). Evaluation of the Rhizobium—MVA association on the growth and development of the cultivation of the soybean (*Glycine max* [L.] Merril). *Cultivos Tropicales* 18:10-12.
Dodd, J.C. and B.D. Thompson (1994). The screening and selection of inoculant arbuscular mycorrhizal and ectomycorrhizal fungi. *Plant and Soil* 159:149-158.
FAO-ISRIC and ISSS (1998). *World Reference Base for Soil Resources. World Soil Resources Report 84*. Vienna: ISSS-ISRIC-FAO.
Fernández, F. (1999). *Manejo de las Asociaciones Micorrízicas Arbusculares sobre la Producción de Posturas de Cafeto (C. arabica L. var. Catuaí) en Algunos Tipos de Suelos*. Ph.D. Thesis, Ciencias Agrícolas. La Habana: INCA, 102 p.
Fitter, A.H. and J. Garbaye (1994). Interactions between micorrhizal fungi and others soil organisms. *Plant and Soil* 159:123-132.
Fitter, A.H. and B. Moyersoen (1996). Evolutionary trends in root-microbe symbioses. *Philosophical Transactions of the Royal Society of London* B351:1367-1375.
George, E. (2000). Nutrient uptake. In *Arbuscular Mycorrhizas: Physiology and Function,* Y. Kapulnik and D.D. Douds, Jr. (eds). Dordrecht: Kluwer Academic Publishers, Dordrecht, pp. 307-344.
Gianinazzi-Pearson, V. and S. Gianinazzi (1989). The physiology of improved phosphorus nutrition in mycorrhizal plants. In *Arbuscular Mycorrhizas: Physiology and Function,* Y. Kapulnik and D.D. Douds, Jr. (eds). Dordrecht: Kluwer Academic Publishers, pp. 101-109.
Gryndler, M. (2000). Interactions of arbuscular mycorrhizal fungi with other soil organisms. In *Arbuscular Mycorrhizas: Physiology and Function,* Y. Kapulnik and D.D. Douds, Jr. (eds). Dordrecht: Kluwer Academic Publishers, pp. 239-262.
Harley, J.L. and S.E. Smith (1983). *Mycorrhizal Symbiosis*. New York: Academic Press. 483 pp.
Hernández, M.I. (2000). *Nutrición de las Plantas y Biofertilizantes. Las Micorrizas Arbusculares y las Bacterias Rizosféricas como Complemento de la Nutrición Mineral del Tomate (Lycopersicum esculentum Mill.)*. M.Sc. Thesis. La Habana: INCA, 65 p.
Hernández-Zardón, A. (2001). Manejo agronómico integral de sustratos, métodos de siembra y biofertilización en la producción sostenible de tubérculos-semilla de papa por semilla sexual. *Cultivos Tropicales* 22:21-27.
Hernández-Zardón, A. (2005). *Resultados de la Aplicación del Biofertilizante Micorrízico EcoMic® en el Cultivo de la Papa en el Departamento de Santa Cruz de la Sierra, Bolivia*. Informe de Investigaciones. La Habana: INCA. 15 p.
Herrera, R.A., R.L. Ferrer, M.O. Orozco, G. Hernández, and V. Vancura (1984a). Fertilización y micorrizas VA. I. Efectos del nitrógeno, el fósforo y el potasio sobre el crecimiento y las micorrizas de la majagua (*Hibiscus elatus* Sw.). *Acta Botánica Cubana* 20:93-110.
Herrera, R.A., R.L. Ferrer, M.O. Orozco, G. Hernández, and V. Vancura (1984b). Fertilización y micorrizas VA. II. Análisis del balance de macroelementos en varios experimentos. *Acta Botánica Cubana* 20:111-142.
Herrera, R.A. et al. (1995). *Estrategia de Funcionamiento de las Micorrizas VA en un Bosque Tropical. Biodiversidad en Iberoamérica: Ecosistemas, Evolución y Procesos sociales.* Programa Iberoamericano de Ciencia y tecnología para

el desarrollo. Subprograma XII, Diversidad Biológica. Mérida: Maximina Monasterio.
Höflich, G., W. Wiehe, and G. Kuhn (1994). Plant growth stimulation by inoculation with symbiotic and associative rhizosphere microorganism. *Experientia* 50: 897-905.
Howeler, R.H. (1985). *Aspectos Prácticos de la Investigación de Micorrizas Vesículo-Arbusculares Demostrados en el Cultivo de la Yuca*. Cali: CIAT, pp. 44-61.
INCA (1999). *Efecto de las Aplicaciones del Biofertilizante Ecomic® (HMA) en Cultivos de Interés Económico, Durante el Periodo 1990-1998*. Informe de Investigaciones. La Habana: INCA, 15 p.
Joao, J.P. (2002). *Nutrición de las Plantas y Biofertilizantes: Efectividad de la Inoculación de Cepas de HMA en la Producción de Posturas de Cafeto sobre Suelos Ferralítico Rojo Compactado y Ferralítico Rojo Lixiviado de Montaña*. M.Sc. Thesis. La Habana: INCA, 85 p.
Lópes, E.S., R. Díaz, and A.M. Costa (1986). Problemas no desenvolvimiento e na colonizacao micorrízica natural de mudas de café em viveiro. In *Reuniao Brasileira Sobre Micorrizas*. Lavras, 156 p.
Marschner, H. and B. Dell (1994). Nutrient uptake in mycorrhizal symbiosis. *Plant and Soil* 159:89-102.
Medina, N. (1994). Evaluación agronómica de diferentes biofertilizantes en la nutrición mineral del Tomate (*Lycopersicum esculentum*, Mill.) In *Resúmenes IX Seminario Científico*. La Habana: INCA.
Miller, R.M. and J.D. Jastrow (1992). The role of mycorrhizal fungi in soil conservation. In *Mycorrhizae in Sustainable Agriculture*. Special Publication 54. Madison: ASA.
Minerdi, D., R. Fani, R. Gallo, A. Boarino, and P. Bonfante (2001). Nitrogen fixation genes in an endosymbiotic *Burkholderia* strain. *Applied and Environmental Microbiology* 67:725-732.
Mirabal L., E. Ortega, and R. Rodes (2002). Influencia de bacterias endospóricas de *Glomus clarum* en la germinación de semillas de arroz (*Oryza sativa* L.). *Revista Chapingo, Serie Ingeniería Agropecuaria* 4:135-146.
Morell, F. (2005). *Degradación de las Propiedades Agrobiológicas de los Suelos Ferralíticos Rojos Lixiviados por el Cambio del Uso de la Tierra y su Respuesta Agroproductiva al Mejoramiento del Componente Suelo del Agroecosistema*. M.Sc. Thesis. La Habana: INCA, 75 p.
Pacovsky, R.S., G.J. Bethlenfalvay, and E.A. Paul (1986). Comparison between P-fertilizer and mycorrhizal plants. *Crops Science* 26:151-156.
Pacovsky, R.S., G. Fuller, A.E. Stafford, and E.A. Paul (1986). Nutrient and growth interactions in soybeans colonized with *Glomus fasciculatum* and *Rhizobium japonicum*. *Plant and Soil* 92:37-45.
Pate, J.S. (1994). The mycorrhizal association: Just one of many nutrient acquiring specializations in natural ecosystems. *Plant and Soil* 159:1-10.
Phillips, J.M. and D.S. Hayman (1970). Improved procedures for clearing roots and staining parasitic and vesicular arbuscular mycorrhizal fungi for rapid assessment of infection. *Transactions of the British Mycological Society* 55:158-161.

Portieles, J.M., L. Ruíz, and E. Sánchez (1983). Estudio del consumo y coeficientes de aprovechamiento de los fertilizantes y el suelo en el cultivo de la yuca *(Manihot esculenta). Ciencias Técnicas en la Agricultura Viandas Tropicales* 6: 85-100.

Portieles, J.M., L. Ruíz, A. de la Nuez, and V. Gutierrez (1982). Estudio del consumo y los coeficientes de aprovechamiento de los fertilizantes y el suelo en el cultivo de la malanga isleña *(Colocasia esculenta). Ciencias Técnicas en la Agricultura Viandas Tropicales. Suplemento.* pp. 33-44.

Primavesi, A. (1990). *Manejo Ecológico do Solo. A. Agricultura em Regioes Tropicais.* Sao Paulo: Livraria Novel S.A., pp. 164-197.

Providencia, I., F.A. de Souza, F. Fernandez, N. Sejalon-Delmas, and S. Declerck (2005). Comparison of anastomosis formation and healing mechanisms in two distinct arbuscular mycorrhizal fungi phylogenic groups. *New Phythologist* 165:261-271.

Riera, M. (2003). *Manejo de la Biofertilización con Hongos Micorrízicos Arbusculares y Rizobacterias en Secuencias de Cultivos Sobre Suelo Ferralítico Rojo.* Ph.D. Thesis, Ciencias Agrícolas. La Habana: INCA, 102 p.

Rillig, M. and Sara Wright (2002). The role of arbuscular mycorrhizal fungi and glomalin in aggregation: Comparing effects of five plant species. *Plant and Soil* 238:325-333.

Rivera, R. (1993). Efecto de la coinoculación *Azospirillum brasilense* y hongos micorrizógenos va en el cultivo del arroz. In *Informe del Trabajo Anual de 1992 sobre Biofertilizantes.* Documento Interno. La Habana: INCA, 15 p.

Rivera, R. (2000). Disponibilidad de nutrientes y fertilización en los sistemas agrícolas micorrizados: resultados en la producción de posturas de cafeto y de raíces y tubérculos. *XII Seminario Científico del INCA. Libro de Resúmenes.* La Habana: INCA.

Rivera, R. and K. Fernández (2003). In *El manejo efectivo de la simbiosis micorrízica, una vía hacia la agricultura sostenible: Estudio de caso El Caribe,* R.Rivera and K. Fernández (eds.). La Habana: INCA.

Rivera, R., L. Ruíz, and A. Calderón (2006). Influencia del tamaño del recipiente y las fuentes de abono orgánico sobre la micorrización de vitroplantas de plátano. *Cultivos Tropicales* 27 (In press).

Rivera, R., F. Fernández, C. Sánchez, C. Bustamante, and M. Ochoa (1997). Efecto de la inoculación con hongos (MVA) y bacterias rizófericas sobre el crecimiento de las posturas de cafetos. *Cultivos Tropícales* 18:15-23.

Rodríguez, I. (1992). *Certificado de Introducción del Resultado Científico Técnico "Utilización del Humus de Lombriz en la Producción de Posturas de Cafeto."* La Habana: Dirección Nacional de Café y Cacao. 1p.

Ruíz, L. (2001). *Efectividad de las Asociaciones Micorrízicas en Especies Vegetales de Raíces y Tubérculos en Suelos Pardos con Carbonatos y Ferralíticos Rojos de la Región Central de Cuba.* Ph.D. Thesis, Ciencias Agrícolas. La Habana: INCA, 117 p.

Ruíz, L. and J.M. Portieles (1985). Estudio de la interacción de NPK sobre los rendimientos del boniato (*Ipomoea batatas* Lam.) en la época de primavera. *Ciencia Técnica en la Agricultura Viandas Tropicales* 8:53-69.

Ruíz, L. and R. Rivera (2001). La importancia del tipo de suelo en la selección de especies eficientes de HMA en la horticultura tropical. *Horticultura Mexicana* vol. 8.

Ruíz, L., J.O. Milián, and J.M. Portieles (1990). Clima, suelo y fertilización en el cultivo del ñame (*Dioscorea* spp.). *Boletín de Reseñas Viandas Tropicales* 6:15-25.

Sánchez, C. (2001). *Uso y Manejo de los Hongos Micorrizógenos y Abonos Verdes en la Producción de Posturas de Cafeto en Algunos Suelos del Macizo Guamuhaya.* Ph.D. Thesis, Ciencias Agrícolas. La Habana: INCA, 105 p.

Sánchez, C., C. González, C. Cupull, and C. Pérez (1999). Efecto de la aplicación de micorrizas y Azotobacter sobre la germinación y el desarrollo de las posturas de cafeto. *Centro Agrícola* 26:41-49.

Sánchez, C., R. Rivera, C. González, R. Cupull, R. Herrera, and M. Varela (2000). Efecto de la inoculación de hongos micorrizógenos (HMA) sobre la producción de posturas de cafetos en tres tipos de suelos del macizo montañoso Guamuhaya. *Cultivos Tropicales* 21:5-13.

Sieverding, E. (1991). *Vesicular Arbuscular Mycorrhiza in Tropical Agrosystem.* Eschborn, Federal Republic of Germany: Deutsche Gesellsschaft fur techniische Zusammenarbeit (GTZ) GMBH, 371 p.

Siqueira, J.O. and A. Colozzi-Fhilo (1986). Micorrizas vesiculo a arbusculares en mudas de cafeeiro. II. Efeito do fosforo no establecimento e funcionamento da simbiosis. R. Bras. Ci. Solo. *Campinas* 10:31-38.

Siqueira, J.O. and A.A. Franco (1988). *Biotecnología do solo. Fundamentos e Perspectiva.* Brasilia, D.F.: MEC-ESAL-FAEPE-ABEAS, 235 p.

Siqueira, J.O., A. Colozzi-Filho, E. de Oliveira, A.B. Fernandes, and M.L. Florence (1987). Micorrizas vesiculo arbusculares en mudas de cafeeiro produzidas no sul do estado de Minas Gerais. *Pesquisas Agropecuárias Brasileras* 22:31-38.

Sivila, R.C. and D. Hervé (1994). Comportamiento microbiológico del suelo en terrenos agrícolas en descanso. *Resúmenes V REBRAM.* Florianópolis, SC, Brazil: Universidad Federal Sta. Catarina, p. 12.

Sosa, J. (1999). *Informe Final de Resultados de Validación del Biofertilizante Biomonte (Ecomic®) en Municipios del Departamento de Córdoba, Colombia.* Fundación San Isidro-Comercial Mercadu SA, 63 p.

Subba Rao, N. and Y.R. Dommergues (2000). Biological amelioration of salt-affected soils. In: *Microbial Interactions in Agriculture and Forestry, vol. 1.* Enfield, NH: Science Publishers, pp. 21-238.

Terry, E., M.A. Pino, and N. Medina (1998). Efectividad agronómica de Azofert y EcoMic® en el cultivo del tomate (*Lycopersicum esculentum*, Mill). *Cultivos Tropicales* 19:33-37.

Terry, E., Z. Terán, R. Martínez-Viera, and M.A. Pino (2002). Biofertilizantes, una alternativa promisoria para la producción hortícola en organopónicos. *Cultivos Tropicales* 23(3):43-46.

Tian, C.Y., G. Feng, X.L. Li, and F.S. Zhang (2004). Different effects of arbuscular mycorrhizal fungal isolates from saline or non-saline soil on salinity tolerance of plants. *Applied Soil Ecology* 26:143-148.

Trouvelot, A., J. Kough, and V. Gianinazzi-Pearson (1986). Mesure du taux de mycorhization VA d'un système radiculaire. Recherche de méthodes d'estimation ayant une signification fonctionnelle. *Proc. 1st Eur. Symp. on Mycorrhizae: Physiological and Genetical Aspects of Mycorrhizae*. Dijon: INRA, pp. 217-221.

Wright, S.E., K.A. Nichols, L. Jawson, L.F. Mckenna, and A. Almendras (2001). Glomalin—a Manageable Soil Glue. Soil Science Society of America Special Publication Book Chapter. <http://www.ars.usda.gov/research/publications/publications.htm?SEQ_NO_115=123063> January 12, 2006.

Chapter 6

Mycorrhizae in Tropical Agriculture

Alejandro Alarcón
Jesús Pérez-Moreno
Ronald Ferrera-Cerrato

The most widely distributed mycorrhizal symbiosis in tropical agroecosystems is the arbuscular mycorrhizal symbiosis. This symbiosis is established between plant root systems and fungi (Figure 6.1) which belong to the phylum Glomeromycota (Schübler, Schwarzott, and Walker, 2001). The name of this symbiosis is based on the presence of specific fungal structures, called arbuscules, which colonize the cortical cells of roots. Based on current root analysis, it is considered that arbuscular mycorrhizal symbiosis is present in more than 80 percent of all known terrestrial plants. The importance of arbuscular mycorrhizal (AM) fungi is related not only to their beneficial effect on plant growth and nutrition but to the fact that these fungi appear to have played a role in plant evolution and adaptation (Malloch, Pyrosinsky, and Raven, 1980; Taylor et al., 1995). According to fossil evi-

The authors sincerely thank Chantal Hamel for the kind invitation to write this contribution, and are indebted to C. Plenchette, Associate Editor, and an anonymous referee for their very valuable comments and corrections of previous manuscripts. They also wish to thank Alicia Franco-Ramírez and Mauricio Ivan Andrade Luna, MSc students at the Colegio de Postgraduados, Mexico, for kindly providing the photographs for Figure 6.1A, C, and D. Financial support from the Mexican government through Project SEMARNAT-CONACyT 2004-01-45 *"Los hongos silvestres comestibles del Parque Nacional Izta-Popo, Zoquiapan y Anexos"* is also acknowledged.

Mycorrhizae in Crop Production
© 2007 by The Haworth Press, Inc. All rights reserved.
doi:10.1300/5425_06

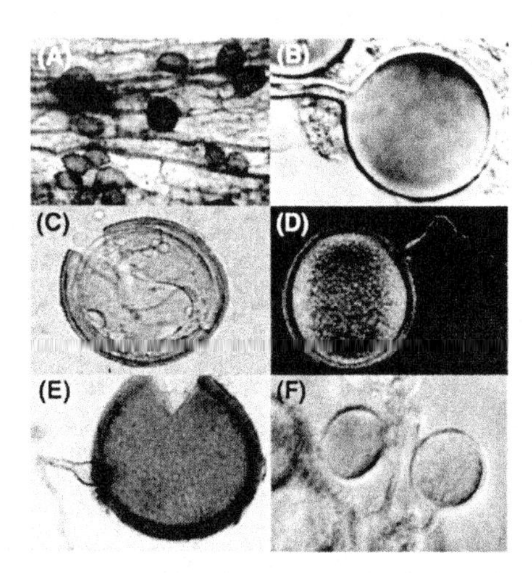

FIGURE 6.1. (A) Abundant vesicles of *G. intraradices* in inoculated roots of the tropical fruit tree *C.a papaya* ("papaya"); (B) *Glomus* sp. spore from a degraded arid tropical zone in Puebla, Mexico; (C, D) Mycorrhizal spores associated with tropical grasses in petroleum-contaminated areas of Mexico; (C) *Acaulospora* aff. *mellea* from Villahermosa; (D) *Gi. margarita* associated with *Panicum* sp. in Minatitlán; (E, F) Mycorrhizal spores associated with the tropical fruit tree Anonna muricata ("soursoap") in Veracruz, Mexico; (E) *Scutelospora* sp.; (F) *Glomus aggregatum.*

dence (Redecker, Kodner, and Graham, 2000) and DNA molecular clock estimates (Berbee and Taylor, 1993), these fungi could be similar in age to the first land plants, that is, around 460 million years old (Brundrett, 2002). AM fungi are an important component of soil biota and could account for 5 to 50 percent of the total microbial biomass in agricultural soils (Olsson et al., 1999). Survival and dispersion of AM fungi in soils are commonly dependent on their establishment in the root system, and these fungi are considered obligate symbionts, requiring carbon compounds synthesized by the host through photosynthesis. Only a few species, such as *Glomus intraradices* (Figure 6.1A), *Glomus versiforme,* and *Gigaspora margarita* (Figure 6.1D), have been successfully propagated using in vitro cultures with host roots (Bago et al., 1998; Bécard and Fortin, 1988; Declerck et al., 1996; Mohammad and Khan, 2002). This physiological limitation on AM fungi poses a challenge to the large-scale production of low-cost inoculum for agricultural purposes.

The importance of tropical ecosystems in providing biodiversity, carbon reservoirs, and regulating weather has been clearly demonstrated at the global scale (Lovelock, 1995a,b). Despite this importance, the study of arbuscular mycorrhizal symbiosis in tropical areas is still in its infancy. Paradoxically, the study of mycorrhizal associations in tropical plants is as old as the word "mycorrhiza" itself. The same year that Frank (1885) coined the word mycorrhiza, Treub (1885) recorded vesicular arbuscular mycorrhizal association in sugar cane in Java (Redhead, 1980). Eleven years later, Janse (1896) carried out an extensive survey of the occurrence of mycorrhizal associations in tropical plants and found that 69 of 75 plant species studied, including all the woody species, were characteristically colonized by AM fungi. More than 30 years later, Pyke (1935) and Laycock (1945) reported the presence of AM fungi in cocoa (*Theobroma cacao* L.) roots. Finally, working in Trinidad in 1949, Johnston found that 80 of 93 tropical plant species, including 13 species of forest trees, also presented AM colonization. Most of the studies conducted in the tropics are more recent than these detailed pioneering surveys and were concentrated in the last four decades of the past century.

In this chapter, a number of basic and applied aspects related to AM fungi in tropical agriculture are discussed. Tropical environments are characterized by the prevalence of poor soil conditions (particularly a critical lack of phosphorus), extremely high biodiversity and lack of dormant biological stages, and the cultural, economic, and social systems in such areas differ from those of other regions. As a result, aspects such as the visible spread of diseases, soil erosion, and land pollution are of critical importance to long-term crop production in the tropics. In this context, the chapter discusses the possible implications of AM fungi for crop growth, phosphorus acquisition, and increased resistance to diseases, and presents an initial analysis of the relevance of AM fungi in sustainable tropical agriculture. In addition, because soil degradation and pollution are factors in long-term crop production in various tropical regions, and because there is a growing recognition that AM fungi could play an important role in "soil health" (Bethlenfalvay and Linderman, 1992; Wright, 2004), the importance of AM fungi in soil conservation and restoration is also discussed.

MYCORRHIZAE AND SUSTAINABLE TROPICAL AGRICULTURE

The dilemma of how to increase crop productivity in tropical areas while preserving the environment has highlighted the importance of sustainable food production. The establishment of agroecosystems geared not only to high productivity but also to sustained production and environmental benefits (e.g., agroecosystems that enhance soil fertility and structure) is an urgent requirement in tropical regions. So-called "low external input agroecosystems" are good examples and have been maintained for centuries in various developing countries. As opposed to conventional agroecosystems, in low external input systems, the levels of disturbance are minimized because the systems are characterized by high genetic and cultural diversity, multiple use of resources, and efficient nutrient and material recycling (Altieri, 1987). The great abundance and importance of AM fungi in these kinds of agroecosystems has been reviewed (Pérez-Moreno and Ferrera-Cerrato, 1997).

Tropical cropping systems are established in areas previously occupied by ecosystems rich in AM fungi species, including mainly: (1) lowland rainforests, (2) savannas, (3) montane cloud forests, and (4) degradation stages of these ecosystems (Janos, 1980; Lovelock et al., 2003; Muthukumar, Udaiyan, and Manian, 1996; Onguene and Kuyper, 2002; Schmidt and Stewart, 2003). As is the case in temperate areas (Helgason et al., 1998), native AM fungal diversity generally appears to decrease when natural ecosystems are converted to agroecosystems (Álvarez-Solis and Anzueto-Martínez, 2004; Sieverding, 1990; Siqueira, Colozzi-Filho, and Oliveira, 1989). This pattern in native AM fungal communities could be related to: (1) reduced diversity in plant communities (Bethlenfalvay and Linderman, 1992; Oehl et al., 2003), (2) influence of cultural practices (Johnson, 1993; Johnson and Pfleger, 1992), (3) duration of fallow or grazing (Duponnois et al., 2001), and (4) modification of soil properties (Brundrett, Jasper, and Ashwath, 1999). However, a number of exceptions to the pattern have been reported. For example, in Nicaragua and Costa Rica, AM fungal diversity did not significantly decline following conversion of lowland evergreen forests to pastures (Picone, 2000), and in Venezuela, diversity of AM fungal communities was reported to be similar to that in natural undisturbed ecosystems (Cuenca and Meneses, 1996).

In a pattern similar to that seen when natural ecosystems are converted to agroecosystems, diversity of AM fungi in tropical areas also appears to decrease as the intensity of agricultural practices increases. In Brazil, Zaire, and Mexico, it has been shown that low external input agroecosystems have greater diversity of AM fungi than comparable agroecosystems with high external inputs (Álvarez-Solis and León-Martínez, 1997; Sieverding, 1990; Trejo-Aguilar, 1997). Because isolates of AM fungi differ enormously in their effects on plants, from mutualistic to parasitic (Johnson, Graham, and Smith, 1997), a highly diverse community of AM fungi may be desirable to increase possible options while fulfilling multifunctional purposes in a context of sustained productivity. In general, it has been shown that AM fungal spore populations (Eason, Scullion, and Scott, 1999; Galvez et al., 2001; Kurle and Pfleger, 1994) and colonization levels (Dann et al., 1996; Douds, Janke, and Peters, 1993; Mäder et al., 2000) are higher in low external input agroecosystems than in those conventionally managed, with high inputs. Interest in the development of less intensively managed systems is also presenting new opportunities for adapting agricultural production systems to enhance the benefits of the networks established by the AM fungal mycelium. These mycorrhizal networks can make an important contribution to sustainability, primarily by increasing nutrient uptake efficiencies, reducing diseases caused by root pathogens, and improving soil-aggregate stability and soil physical properties (Leake et al., 2004). For example, a positive correlation has been demonstrated between hyphal lengths and P-uptake efficiency (Schweiger and Jakobsen, 2000), and in certain cases, even crop yield (Kabir and Koide, 2002). Various practices commonly associated with sustainable agriculture in tropical areas have resulted in increased "soil health" and enhancement of AM functioning in field crops. As reviewed elsewhere (Johnson and Pfleger, 1992), these practices include no-tillage or reduced tillage, crop rotation, cover crops, intercropping, manure addition, and, in some cases, reduced use of fertilizers or pesticides. Although the current state of knowledge of AM fungi in tropical agroecosystems makes it difficult to generalize accurately, information obtained mainly in nontropical areas could be used to make assumptions about the potential benefits for tropical crops of certain agricultural practices affecting AM fungal populations.

For example, it has been demonstrated that soil tillage is generally related to reduced mycorrhizal colonization, and in some cases, is also related to decreased mineral nutrition and lower crop yields in the field (Anderson, Milner, and Kunishi, 1987; Mulligan, Smucker, and Safir, 1985; O'Halloran, Miller, and Arnold, 1986). Mechanical tillage may limit root colonization by reducing the frequency of contact between inoculum sources and new roots, or by disrupting AM fungal networks, thereby reducing nutrient translocation to plants (Evans and Miller, 1988; Vivekanandan and Fixen, 1991). Conversely, no-tillage or reduced tillage increases the length of AM fungal networks associated with field crops (Kabir, O' Halloran, Fyles et al., 1998; Kabir, O'Halloran, Widden et al., 1998).

In addition to the yield benefits and soil fertility enhancement achieved by crop rotation, positive benefits for AM colonization and spore production have been reported in various cropping sequences (Dodd et al., 1990a,b; Harinikumar and Bagyaraj, 1988; Sieverding and Leihner, 1984; Vivekanandan and Fixen, 1991). Selecting crop rotations that are beneficial to AM fungi can improve productivity, for example, in maize (Karasawa, Karahara, and Takebe, 2002). Some cover crops or intercropping combinations can substantially improve soil health and AM fungi functioning in field crops (Guzmán, Ferrera-Cerrato, and Bethlenfalvay, 1992; Kabir and Koide, 2002; Quiroga-Madrigal, 1996). It has even been proposed that some sustainable cropping systems could manage AM fungal inoculum densities and species composition through the use of cover crops and intercropping strategies that are more responsive to AM fungal populations (Dodd et al., 1990a,b; Johnson, Pfleger et al., 1991; Johnson, Zak et al., 1991; Kormanik, Bryan, and Schultz, 1980). Because crop productivity can be differentially affected by AM fungal species, selection of appropriate cover crops or intercropping combinations would be of critical importance to sustainable production in the tropics.

In general, addition of manure significantly increases crop yield, and in certain cases, such increases have been related to AM fungal populations (Matías-Crisóstomo and Ferrera-Cerrato, 1993; Muthukumar and Udaiyan, 2002). It has been shown that addition of manure significantly increases AM colonization; however, when manure is applied with nitrogen (N), phosphorus (P), and potassium (K) fertilizers, mycorrhizal colonization is significantly reduced (Vejsadová, 1992). AM fungal populations seem to be differentially affected by manure amendments. In India, Muthukumar and Udaiyan (2002) found that the addition of sheep manure significantly increased the populations of some AM fungal species, such as *Glomus aggregatum* and *Scutellospora calospora,* but dramatically reduced others, including *Glomus sinuosum* and *Acaulospora scrobiculata.* In Mexico, Matías-Crisóstomo and Ferrera-Cerrato (1993) found that in the reclamation of tepetates ("hardened soil layers") with polycultures, the addition of cow dung significantly increased AM colonization and yields of *Zea mays, Phaseolus vulgaris,* and *Vicia faba*. Maximum colonization rates (up to 65 percent) where seen in *V. faba.* In Rwanda, Heizemann, Sieverding, and Diederichs (1992) found that the combined application of rock phosphate and manure significantly enhanced the proliferation of AM fungal spore populations, mainly those of *G. fasciculatum, G. aggregatum,* and *G. geosporum.* Another practice, used frequently in certain tropical agroecosystems, is terracing. In African tropical highlands, Heizemann, Sieverding, and Diederichs (1992) also found that the formation of terraces differentially affected AM fungal population, enhancing *Glomus callosum* and *G. aggregatum* and suppressing *G. fasciculatum* and *G. occultum.*

EFFECTS OF AM FUNGI ON PLANT GROWTH, PHOSPHORUS UPTAKE, AND INCREASED RESISTANCE TO DISEASE IN TROPICAL AGRICULTURE

Plant Growth

Initial interest in and practical use of mycorrhiza began in the late 1800s and, until the 1970s, was concentrated mainly in forest production (Mikola, 1980). More recently, however, knowledge related to plants of agricultural and horticultural importance, including those in tropical areas, has developed at a rapid pace (Bethlenfalvay and Lindermann, 1992). AM fungi have been experimentally managed in tropical crop production systems by the deliberate addition of fungal propagules to soils. The goal of inoculation has been to enhance crop production by using beneficial strains of AM fungi to improve mycorrhizal colonization. The success of inoculation efforts appears related to initial soil fertility, population densities of indigenous AM fungi, the ability of inoculant fungi to rapidly form extensive mycorrhizal associations, mycorrhizal soil receptiveness (Plenchette, 2000), and the ability of the fungal inoculant to persist in soil (Abbott and Robson, 1981, 1982; Abbott, Robson, and Hall, 1983). It has been demonstrated that, in certain cases, inoculation with AM fungi dramatically improves yields of nursery tropical plants and specialty crops such as citrus (Eissenstat et al., 1993; Kleinschmidt and Gerdemann, 1972), papaya (*Carica papaya* L.) (Sánchez-Espindola et al., 1996) (Figure 6.1A), and guava *(Psidium guajava)* (Chacón and Cuenca, 1998). It has also been reported that vesicular arbuscular mycorrhizal fungi (VAM) inoculation has resulted in up to triple yields of field crops such as maize in Nigeria and Pakistan (Khan, 1972; Mosse and Hayman, 1980), rice in Nigeria (Sanni, 1976), beans (Aguilar et al., 2000; Ferrera-Cerrato, 1995) and papaya in Mexico (Escalona et al., 2000), and cassava in Colombia (Howler, Sieverding, and Saif, 1987). Furthermore, it has been demonstrated that inoculation with AM fungi can potentially increase pasture production (Cuenca, Andrade, and Escalante, 1998). AM fungi produce dramatic responses in the growth of some tropical plants; however, many basic ecological and physiological aspects of mycorrhizal systems need to be better understood before AM fungi can be successfully applied as inoculants on a large-scale.

Although some AM fungi can tolerate certain agronomic practices carried out in intensively managed fields on an exceptional basis (Sieverding, 1991), in general, AM fungal populations in high-input agroecosystems are either suppressed or, if they proliferate, are inferior mutualists (Johnson, 1993; Johnson, Graham, and Smith, 1997). As a result, in these high-input agroecosystems, there may be little scope for AM fungi involvement in agricultural production, since most fungal species are potentially eliminated (Daniell et al., 2001; McGonigle and Miller, 2000; Ryan and Graham, 2002). In heavily fertilized banana and coffee plantations in tropical areas, the AM fungi present could be expected to play only a marginal role in terms of benefiting plants or agroecosystems. For example, in studying coffee plantations in Eastern Mexico, Trejo-Aguilar (1997) and Trejo-Aguilar et al. (2000) found very low diversity in AM fungal populations, and colonization rates ranging from 8 percent to 18 percent in heavily fertilized plantations. Conversely, communities were much more diverse and colonization rates ranged from 50 to 65 percent in comparable low-input coffee plantations grown in the shade of trees (mainly legumes such as *Inga*).

The advantages of mycorrhizal inoculation of plants are often related to increased exploitation of the soil volume beyond the root nutrient–depletion zone, enhanced uptake and assimilation of nutrients with low solubility and mobility, such as phosphorus, calcium, and zinc (Smith and Read, 1997), and improved usage of fertilizer when applied in low doses in plant production systems (Bethlenfalvay and Linderman, 1992). Additional benefits of mycorrhizal fungi have been related to inducing plant tolerance to root diseases, drought and salinity stress, and heavy metals and organic pollutants in the soil (Cabello, 2001; El-Tohamy et al., 1999; Jakobsen, Smith, and Smith, 2002; Meharg, 2001). Because aspects of these benefits are also of enormous agricultural interest, it can now be assumed, with a degree of certainty, that AM fungi are not only important to plants but also to the soil itself, or in other words, to what has been called "the soil-plant system" (Bethlenfalvay, 1992).

Use of Mycorrhizae in Tropical Nurseries and Plantations

In the eighteenth century, truffles were added to the planting holes of oak seedlings in new plantations to enhance natural truffle production. These first attempts to form mycorrhizas on seedlings were made long before the term "mycorrhiza" was coined (Malencon, 1938; Trappe, 1977). More recently, spores of ectomycorrhizal fungi have been used in tropical areas (Brundrett et al., 1996; Mikola, 1970, 1980; Theodorou and Bowen, 1970) in a variety of forms, including ground fresh or dried fruitbodies, to infest soils or inoculate seeds or seedlings in nurseries and containers. Basidiomycetous fungi, including species in the genera *Rhizopogon, Scleroderma,* and *Pisolithus,* have been successfully used because of the massive quantities of spores that can be obtained from their fruitbodies (from 10^6 to 10^9 spores per gram of ground fruitbody) (Martínez-Reyes et al., 2005;

Marx and Bryan, 1975). In addition, pure mycelial inoculum of ectomycorrhizal fungi, including *Suillus, Rhizopogon,* and *Pisolithus,* has been used successfully in tropical plant production (Brundrett et al., 1996; García-Rodríguez et al., 2005; Marx, 1980; Mikola, 1980; Vozzo and Hackslayo, 1971).

The large-scale application of AM fungi under field conditions has been limited, in part, by the current lack of extensive ecological and physiological knowledge about these microorganisms (Bagyaraj, 1992). Under tropical conditions, which are commonly characterized by poor soils, extreme temperatures, high humidity or drought, and the presence of many pathogens, the use of AM fungi can be especially beneficial to perennial crops because: (1) most perennial plants in these areas establish arbuscular mycorrhizal symbiosis (Smith and Read, 1997); (2) most perennial tropical plants are propagated from seeds or cuttings in nurseries before they are planted in the field; and (3) it has been demonstrated that AM fungi not only improve the growth of various tropical crops, but also have the potential to increase the host's resistance to root pathogens and to reduce the severity of foliar disease (Declerck et al., 2002; Elsen, Declerck, and De Weale, 2001; Jaizme-Vega et al., 1997, 1998; Schwob, Ducher, and Coudret, 1999); (4) mycorrhizal plants seem to have reduced sensitivity to transplantation stress, which is of particular importance to nurseries and to micropropagation from tissue cultures (Hooker, Jaizme-Vega, and Atkinson, 1994); and (5) plants can be produced using a small amount of fertilizers (Bethlenfalvay and Linderman, 1992). AM fungi can produce three kinds of propagules: (1) spores (Figure 6.1), (2) hyphae emerging from dead root fragments, and (3) hyphae associated with living plants (Friese and Allen, 1991). AM fungal inoculum in soil containing pieces of roots, spores, and hyphae, maintains its effectiveness for long periods when stored in cold rooms under slightly moist conditions. For example, Howeler, Sieverding, and Saif (1987) reported that AM fungal inoculum maintained under these conditions was found to be effective even after three years of storage. Studies involving tropical plants have shown that AM fungal species vary widely in terms of mycorrhizal colonization and spore production. Native mycorrhizal colonization can range from 20 percent to 80 percent of fine root length (Alexander, 1989; Howeler et al., 1987; Redhead, 1980; St John, 1980), and some AM fungi can produce large numbers of spores (Johnson and Wedin, 1997; Louis and Lim, 1987; Lovelock, Andersen, and Morton, 2003). At low levels of P, mycorrhizal dependence of tropical crops is highly variable, ranging, for example, from 95 percent in cassava *(M. esculenta),* 72 percent in beans *(Phaseolus vulgaris),* and 26 percent in maize *(Z. mays)* (Howeler, Sieverding, and Saif, 1987). In general, it appears that certain plants with thick, fleshy roots and few root hairs, such as cassava (Howeler, Sieverding, and Saif, 1987), citrus (Alarcón and Ferrera-Cerrato, 1999; González-Chávez and Ferrera-Cerrato, 1994), and papaya (Sánchez-Espíndola et al., 1996), frequently have high mycorrhizal dependency. It is important to note, however, that the efficiency of AM fungal species is highly variable and is not always correlated with high colonization rates. In certain tropical plants, inoculation with certain AM fungal species did not produce positive effects in terms of plant growth or P uptake. For example, Howeler et al. (1987) found that *Acaulospora laevis, A. mellea, A. morrowae,* and *Gigaspora heterogama* were not effective in the inoculation of cassava plants. These ecophysiological differences could be important in the selection of strains of AM fungi for large-scale inoculation.

A variety of tropical plants are highly responsive to inoculation with AM fungi in nurseries; these include soursop (Franco-Ramírez et al., 2004) (Figure 6.1E, F), avocado (Reyes, 2000), pineapple (Guillemin, Gianinazzi, and Trouvelot, 1992; Lovato, Guillemin, and Gianinazzi, 1992), papaya (Sánchez-Espíndola et al., 1996), citrus (Alarcón and Ferrera-Cerrato, 1999; González-Chavez and Ferrera-Cerrato, 1994), cacao (Azizah-Chulan and Ragu, 1986; Cuenca, Herrera, and Meneses, 1990; Ezeta and Santos, 1981), and coffee (Fernández et al., 1992; González-Chavez and Ferrera-Cerrato, 1996; Howeler, Sieverding, and Saif, 1987; Lopes et al., 1983). In general, one of the main limitations on the use of AM fungi is the high cost of inoculum production (Bagyaraj, 1992). Despite this limitation, in a number of tropical countries, such as Colombia and Brazil (Feldmann and Idczak, 1992; Sieverding, 1991), AM fungi are already an important factor in certain commercial cultivation systems, diminishing the cost of plant production. Feldemann and Idczak (1992) successfully conducted an interesting inoculum production experiment using expanded clays in the humid tropical region of Amazonas,

Brazil. They found that the use of these expanded clays as inoculum carriers facilitated the following: (1) maintenance of inoculum purity, (2) transportation and distribution of large quantities (because of low weight), (3) long-term storage, and (4) application (because of the nature of the material). The inoculum was successfully applied to the following perennial tropical plants: *H. brasiliensis* (rubber, seringueira), *C. papaya* (papaya, mamao), *M. esculenta* (cassava, maníoca), *Citrus* spp. (orange, laranja), *Theobroma cacao* (cacao), *Theobroma grandiflorum* (cupuacu), and *Guilielma gassipaes* (pupunha).

AM fungal inoculum production systems have now evolved from relatively simple technologies to more sophisticated ones. They range from soil plots in nurseries, containers with various substrates, aeroponic systems, and propagation of in vitro roots (Gianinazzi and Vosátka, 2004). Before application, fungal propagules are placed in a range of carriers, primarily vermiculite, peat, sand, clays, and perlite. Optimum AM fungal inoculation is a function of the crop, its planting system, the growth cycle, and the availability of technological tools (Sieverding, 1985). In short-season, row-planted crops, AM fungal inoculum can be applied in a continuous band under the seed at the time of planting. Fertilizers can be side-banded at the time of planting or after germination. Small seeds can be pelleted with inoculum. In tree crops, the inoculum can be placed under the stake or seedling at time of planting (Howeler, Sieverding, and Saif, 1987). Inoculum can be mixed with sand to increase its spatial distribution below the plants and to reduce the costs of transporting bulky inoculum if sand is available locally.

Another important area of concern is AM fungal inoculation in tropical fruit plants (Pérez-Moreno and Ferrera-Cerrato, 1993) obtained by micropropagation and other techniques (e.g., seeds and cuttings). It has been demonstrated that, in micropropagated plants, inoculation with AM fungi initially increases nutrition and growth, and subsequently enhances survival after transplantation and adaptation to stress (Alarcón et al., 2000; González-Chávez et al., 2000; Lovato et al., 1996). These enhancements have been demonstrated in various micropropagated tropical plants, including banana (Declerck, Plenchette, and Strullu, 1995; Jaizme-Vega et al., 2002; Yano-Melo et al., 1999), pineapple (Guillemin, Gianinazzi, and Trouvelot, 1992; Lovato, Guillemin, and Gianinazzi, 1992), papaya (Alarcón et al.,

2002; Quiñones-Aguilar et al., 1998; Sánchez-Espindola et al., 1996), guava (Chacón and Cuenca, 1998; Estrada-Luna, Davies, and Egilla, 2000), kiwi (Calvet et al., 1989), citrus (Alarcón, González-Chávez, and Ferrera-Cerrato, 2003; Hernández-Meza et al., 1998), avocado (Salazar-García, 2002), and coffee (González-Chávez and Ferrera-Cerrato, 1993).

Phosphorus Uptake

Formation of AM symbiosis is important for P acquisition in most tropical plants (Smith and Read, 1997). However, a variety of factors affect acquisition of this nutrient, including: soil fertility levels, variations within plant species, cultivars and within isolates of AM fungi (Clark and Zeto, 2000). For example, it has been shown that, in soils with high P content, *G. intraradices* is even able to induce growth depression in citrus (Graham and Eissenstat, 1998; Peng, Eissensat, and Graham, 1991), and *Sclerocystis* spp. quickly disappeared in limed and P-fertilized areas converted from natural forests to agroecosystems (Sieverding, 1990). In general, the adverse effect of high soil P levels on AM formation is well documented (Abbott, Robson, and DeBoer, 1984; Bååth and Spokes, 1989; Boddington and Dodd, 1998; Menge et al., 1978; Mosse, 1973). It has been shown that high P levels inhibit AM fungi directly by reducing spore germination and hyphal growth from germinated spores (Miranda and Harris, 1994; Nagahashi, Douds, and Abney, 1996). A high-affinity P transporter is expressed in the extraradical mycelium of the AM fungus *G. versiforme* (Harrison and Van Buuren, 1995). In *G. intraradices,* a similar P transporter is regulated by P availability in the external medium and possibly also by the P status of the host root (Maldonado-Mendoza, Dewbre, and Harrison, 2001). Like high P concentrations, very low concentrations of P also appear to inhibit the benefits of AM fungi for associated plants. With the application of 50 kg of P ha^{-1} to 100 kg of P ha^{-1}, Howeler and Sieverding (1983) reported increased yields of cassava *(M. esculenta)* as a result of AM fungi inoculation under field conditions in Colombia. However, no significant yield increases were observed with applications of 0 kg of P ha^{-1} or 50 kg of P ha^{-1}. AM fungal species and strains have been reported to dramatically differ in their effectiveness in terms of increasing P uptake and plant growth in tropical fruit trees (Chacón and Cuenca, 1998; Declerck, Plenchette, and Strullu, 1995).

Increased Resistance to Disease

A number of studies have shown that, in addition to having a beneficial effect on nutrient acquisition and increasing resistance to such abiotic stresses as aluminum toxicity (Rufyikiri et al., 2000), AM fungi may improve resistance to biotic stresses in tropical plants. Bananas have shown increased resistance, for example, to nematodes (Elsen, Declerck, and De Weale, 2001, 2002; Jaizme-Vega et al., 1997), *Fusarium oxysporum* f.sp. *cubense*, the causal agent of Panama disease (Jaizme-Vega et al., 1998), and the root pathogen *Cylindrocladium spathiphylli* (Declerck et al., 2002). In rubber tree plantations in Brazil, Schwob, Ducher, and Coudret (1999) found an inverse relationship between AM fungi in *Hevea brasiliensis* roots and populations of the root-knot nematode *Meloidogyne exigua*. This inverse relationship was also seen in the surrounding soil. In addition, microscopic observations of roots showed mutual exclusion with histological specificity of the two organisms. Since mycorrhizal rubber tree roots are more lignified than non-mycorrhizal roots, and because it has been demonstrated that lignification can protect roots against other pathogens (Morandi, 1996), the authors proposed that AM fungal populations in roots of rubber trees are able to indirectly affect parasitic nematode populations. They also found that certain cover crops, such as *Brachiaria decumbens* and *Vernonia* sp., commonly intercropped with *H. brasiliensis*, harbor significant AM fungal populations and therefore have an indirect detrimental effect on parasitic nematode populations. Working in Sri Lanka, Waidyanatha (1980) had previously reported that legumes of the genera *Pueraria, Centrosema, Calopogonium, Desmodium,* and *Stylosanthes* grown as ground cover under rubber plantations were heavily colonized by AM fungi. Like Schwob, Ducher, and Coudret (1999), Waceke, Waudo, and Sikora (2001) found significantly lower nematode (*Meloidogyne hapla* Chitwood) gall indices and fewer females, eggs, and juveniles on pyrethrum plants (*Chrysanthemum cinerariefolium* Vis.) in Kenya inoculated with a variety of AM fungi. In general terms, the possible mechanisms involved in increased resistance to disease as a result of AM fungi could range from: (1) increased P and N nutrition (Declerck et al., 2002); (2) activation of the plant defense system and disease resistance (Benhamou et al., 1994; St. Arnaud et al., 1994); (3) direct or indirect competition in the rhizosphere (St-Arnaud et al., 1994) and enhancement of protective rhizosphere microorganisms (Linderman, 1988, 1992, 1993); (4) biochemical changes in the plant, and anatomical changes in the roots (Benhamou et al., 1994; Hooker, Jaizme-Vega, and Atkinson, 1994) (e.g., increased production of phenolic compounds or lignin [Dehne and Schonbeck, 1979; Suresh, Bagyaraj, and Reddy, 1985] or alteration of protein metabolism leading to increases in amino acids with nemastatic properties, such as serine, phenylalanine, and arginine [Ingham, 1988]); (5) competition for host resources in root tissues (Hooker, Jaizme-Vega, and Atkinson, 1994); and (6) altered chemotactic attraction to plant roots brought about by quantitative or qualitative modification of root exudates (Francl, 1993).

AM FUNGI AND REHABILITATION OF DEGRADED SOILS IN TROPICAL AREAS

AM Fungi and Soil Structure

AM fungi develop extensive networks that grow away from the roots into the soil and constitute an important component of the plant-soil system (Leake et al., 2004). These networks enhance soil structure (Miller and Jastrow, 1992a; Tisdall, 1991) and water relations (Augé, 2001), which are important to soil aggregation. Aggregate stability is a prerequisite for healthily managed agroecosystems. This is particularly true in the case of soils prone to erosion (Miller and Jastrow, 1992a), which are especially abundant in tropical regions. Soil aggregation may be attributed to physical binding by roots and AM fungal hyphae, to gluing by microbially produced polysaccharides, or to physical-chemical interactions between clay surfaces and partially decomposed organic matter (Strickland et al., 1988; Tisdall and Oades, 1982). In many situations, loss of soil aggregation and structure reduces the capacity of soil to store nutrients and water (Rose, 1988; Schimel, 1986), thereby increasing its erosion potential

(White, 1985). AM fungal networks vary in their ability to enhance soil stability, and it has been proved that this ability is more important than root length or bacterial populations in stabilizing soil aggregates (Schreiner et al., 1997). It has been demonstrated that when soils are fallowed, or virgin soils are cultivated, the result is a loss of soil carbohydrates (Dalal and Henry, 1988), aggregates (Perry et al., 1989), and AM fungal populations (Thompson and Wildermuth, 1988). Thomas et al. (1986) and Thomas, Franson, and Bethlenfalvay (1993) demonstrated that, in contrast to non-arbuscular mycorrhizal plants, arbuscular mycorrhizal plants can increase the abundance of water-stable macroaggregates. Therefore, by improving soil structure, AM fungi may also affect plant growth and hence crop yield, particularly in tropical regions (Miller and Jastrow, 1992b). Past emphasis has been on plants, but interest is currently also growing in the effects of AM fungi on soil and their influence on so-called soil nutrition (Bethlenfalvay, 1993).

A number of recent studies have shown that AM fungi produce a recalcitrant glycoprotein (or group of closely related compounds) known generically as glomalin, which acts as an insoluble glue to stabilize soil aggregates (Wright, 2000; Wright and Upadhyaya, 1996, 1999; Wright, Upadhyaya, and Buyer, 1998; Wright, Starr, and Paltineaunu, 1999; Wright et al., 1996) and may also represent a recalcitrant pool of C in some soils (Rillig, Wright, and Evanir, 2002; Rillig, Wright, and Torn, 2001). It has been demonstrated that aggregate stability and glomalin are linearly correlated (Wright and Anderson, 2000; Wright and Upadhyaya, 1998). AM fungal species and soil types vary in their production of glomalin, ranging from 2.8 mg g^{-1} to 14.8 mg g^{-1} of soil (Rillig and Allen, 1999; Rillig, Wright, and Tom, 2001; Wright and Upadhyaya, 1998). Although information in tropical areas is very scarce (Lovelock et al., 2003), it can be assumed that agricultural management practices that disturb AM fungal networks, and therefore reduce glomalin production, will negatively affect aggregate stability. Conversely, enhancement of glomalin production by AM fungal hyphae, and hence increased aggregate stability, could be achieved by promoting agricultural practices that influence the development of appropriate AM fungal populations.

Rehabilitation of Disturbed Soils

The combined effect of anthropogenic activities such as intensive agricultural practices, induced forest fire, land management, overgrazing and mining can generate problems related to desertification, salinization, and contamination by heavy metals. These problems are of particular importance in tropical areas where human pressure on natural ecosystems is a growing problem because of high birth rates. Such negative processes bring about drastic changes in plant communities, which lose their fertility and productive capacity. In order to restore and rehabilitate them, the main disturbance factor and the original composition of the natural plant communities must be identified. In general, this information is fundamental to the introduction of specific reforestation programs in which plants that establish arbuscular mycorrhizal symbiosis could be of critical importance (Carpenter et al., 2001; Kyllo, Velez, and Tyree, 2003; Rashid et al., 1997).

The selection of plants to be propagated and utilized is considered a keystone of the soil restoration process. Plants that are likely to form AM symbiosis are the most appropriate, because it is well-known that the establishment of AM fungi in root systems enhances plant adaptation to and survival under adverse soil conditions (Cairney and Meharg, 1999; Van Duin, Griffioen, and Ietswaart, 1991), including the following: (1) compaction (Dew et al., 2003; Nadian et al., 1997), (2) drought (Azcón and Tobar, 1998; Davies et al., 2002; Henderson and Davies, 1990; Sylvia and Williams, 1992), (3) heavy metal toxicity (through avoidance, tolerance [Davies et al., 2001; Griffioen and Ernst, 1989; Meharg and Cairney, 2000; Perotto and Martino, 2001; Yang and Goulart, 2000] or metal binding by glomalin [González-Chávez et al., 2004]), (4) pesticide (Smith, Harnett, and Rice, 2000) and petroleum hydrocarbon toxicities (Hernández-Acosta et al., 2000; Joner and Leyval, 2001, 2003; Leyval and Binet, 1998), and (5) salinity (Cantrelli and Linderman, 2001; Copeman, Martin, and Stutz, 1996; Ruíz-Lozano, Azcón, and Gómez, 1996). Plant inoculation with AM fungi (Figure 6.1B) would ensure the long-term success of soil restoration efforts and could significantly stimulate microbial activity, thereby enhancing the chemical and physical characteristics of the soil.

An evaluation of the intrinsic infectivity potential of native AM fungi is required for any site to be restored. In situ AM fungi characterization can be used to manage various strategies for stimulating the activity of native fungi or for introducing competitive AM fungi under adverse soil conditions. Land recuperation can be enhanced by specific cultural practices, including the following: (1) rational use of fertilizers and pesticides to help preserve some of the physical, and chemical characteristics of the soil (Douds and Miller, 1999; Jeffries et al., 2003), (2) amendments of organic matter, (3) crop coverage, and (4) crop rotations (Johnson and Pfleger, 1992). Legume plants are an excellent choice for inclusion in a soil restoration project because of their ability to make a significant contribution through biological nitrogen fixation and microbiological changes in their rhizosphere. Because legumes tend to form a double symbiosis with nitrogen-fixing bacteria, such as *Rhizobium* and AM fungi, they can play a fundamental role in the assimilation and cycling of both nitrogen and phosphorus in plants and soil (Gardezi and Ferrera-Cerrato, 1989; Gardezi, Ferrera-Cerrato, and Lara, 1988). In general, practices such as no-tillage or reduced tillage, amendments of organic matter from animal manure or plants, and crop rotation and intercropping, not only reduce physical soil losses but also help maintain niches that allow the conservation of AM fungi and many other beneficial microorganisms (Bethlenfalvay and Linderman, 1992; Jeffries et al., 2003; Pérez-Moreno and Ferrera-Cerrato, 1996).

AM fungal inoculation with species in the genera *Acacia* (Martin-Laurent et al., 1999; Michelsen, 1993; Founoune et al., 2002), *Casuarina* (Duponnois et al., 2003), and *Eucalyptus* (Brundrett et al., 1996; Chen, Brundrett, and Dell, 2000) has been attempted with promising results in the rehabilitation of tropical areas of Africa and Australia. In Mexico, 20 tropical plant species (including a number of the genera mentioned above) with potential for use in the rehabilitation of degraded lands were shown to be highly responsive to native strains of AM fungi (Table 6.1). Field inoculation with mycorrhizal propagules from pot cultures was shown to benefit revegetation, with an introduced grass, of degraded tropical lands in Venezuela (Cuenca, Andrade, and Escalante, 1998), and prairie environments in other areas (Smith, Charvat, and Jacobson, 1998). Cuenca, Andrade, and Escalante (1998) found that rehabilitation of degraded lands in La Gran Sabana was not possible with the application of chemical fertilizers alone. It was evident that mycorrhizas were required to achieve

TABLE 6.1. Plant and fungal combinations that are highly responsive to mycorrhizal inoculation in terms of plant growth and nutrient content, and have potential for use in the restoration of disturbed lands in tropical and subtropical areas of Mexico.

Plant species	Inoculated arbuscular mycorrhizal fungi	Experimental condition	References
Legumes			
Erytrhina americana	Consortium of *Glomus* spp. (Zac-19[a])	Addition of organic matter and rock phosphate amendments	(Gardezi et al., 1995, 1996)
Acacia saligna	Consortium of *Glomus* spp. (Zac-8[b])	Three levels of P fertilization and inoculation with double symbiosis (including *Rhizobium* sp.), in acid soils	(Gardezi and Ferrera-Cerrato, 1989)
Acacia cyanophylla	Consortium of *Glomus* spp. (Zac-8[b])	P fertilization and inoculation with double symbiosis (including *Rhizobium* sp.), in acid soils	(Gardezi, Ferrera-Cerrato, and Lara, 1988)
Acacia farnesiana	Two consortia of *Glomus* spp.[b]	Three different soil types	(Gardezi et al., 1990)
Pithecellobium dulce	*Glomus fasciculatum* and several consortia of Glomus spp.[b]	P-fertilization	(Gardezi, Guzmán-Plazola, Ferrera-Cerrato, 1991)
Caesalpinia cacalaco	*Glomus fasciculatum* and several consortia of *Glomus* spp.	Low-fertility soils	(Gardezi and Ferrera-Cerrato, 1992)
Leucaena leucocephala	*Glomus intraradices* and consortia of *Gigaspora* spp. and *Glomus* spp. (including *Glomus Zac-19*)	Reduced defoliation during transplanting, application of rock phosphate and dual inoculation, and higher mycorrhizal dependency in acid soils with fertilizer application	(Guzmán-Plazola et al.,1984, 1987, 1988; Lara-Fernández and Ferrera-Cerrato, 1986)
Eysenhardtia polystacya, Mimosa biuncifera, and *Dodonaea viscosa*	Consortium of *Glomus* spp. (Zac-3[b])	Increased survival when established in Tepetate (hardened volcanic ash-derived soils)	(Rey et al., 1992)

TABLE 6.1 *(continued)*

Plant species	Inoculated arbuscular mycorrhizal fungi	Experimental condition	References
Acacia schaffneri	*Glomus aggregatum* and several consortia of *Glomus* spp.[b]	Tepetate soil and organic matter application	(Ríos-Garay, 1994)
Non-legumes			
Casuarina equisetifolia	*Glomus aggregatum, Glomus intraradices* and consortium *Glomus* spp. (Zac-2[b])	Forest and agricultural soil	(Alarcón and Ferrera-Cerrato, 1996)
Cupressus lindleyi	*Glomus fasciculatum* and consortium of *Glomus* spp. (Zac-19[a])	Fertilization with 15 kg N and 10 kg P ha^{-1} per year	(Ferrera-Cerrato and Jaen, 1989)
Stemmadenia denell-smithii, Heliocarpus appendiculatus, Poulsenia armata, and *Nectandra ambigens*	Native fungi from mexican tropical regions in Veracruz[b]	Differential responses between pioneer and persistent species due to plant genotype	(Sánchez-Gallén and Guadarrama, 2000)
Cedrela odorata, Tabebuia donell-smithii.	Consortium of *Glomus* spp. (MTZ-1[a,b]) from a tropical region in Veracruz	Differential enhancement of plant growth and plant biomass production (higher for *Tabebuia donell-smithii*)	(Zulueta et al., 2000)
Manihotis esculenta	*Glomus macrocarpum*	Efficient use of rock phosphate in acidic soils	(Sánchez et al., 1986)

[a]Consortium integrated by three *Glomus* species: *Gl. claroideum, Gl. diaphanum,* and *Gl. albidum* (Chamizo, Ferrera-Cerrato, and Varela, 1998).

[b]Fungal species not identified.

rehabilitation through introduction of the grass *B. decumbens*. The importance of AM fungi to the restoration of such lands is supported by the authors' finding that, of the native plants reestablished by the various treatments, 81 percent were mycorrhizal.

Phytoremediation of Petroleum-Contaminated Soils

It appears that AM fungal populations could also play an important role in the phytoremediation of petroleum- and polycyclic aromatic hydrocarbon (PAH)-contaminated soils, which are a severe problem in some tropical areas. In certain cases, increases in mycorrhizal root colonization have been reported when the contaminant is dispersed from the rhizosphere by the action of N_2-fixing free-living hydrocarbonoclastic bacteria (García et al., 2000; Hernández-Acosta et al., 1998, 2000). It has been shown that, in some cases, AM fungi can increase the survival and tolerance of plants when PAHs are present in the rhizosphere (Binet, Portal, and Leyval, 2000; Leyval and Binet, 1998), but in others, AM fungal colonization can be drastically affected by the presence of PAHs (Joner and Leyval, 2001). It is primarily oxidative enzymes that are responsible for initiating degradation and ring fission in the case of PAHs present in soil. AM fungi seem to be able to increase the release of 2,7-diaminoflourene-peroxidases, and the activity of these enzymes may contribute to the degradation of anthracene (Criquet et al., 2000).

A number of studies have been conducted in Mexico to clarify the role of AM fungi in tropical areas where plants are established in substrates contaminated with petroleum hydrocarbons (Figure 6.1C, D). It has been found that a strain of *Gi. margarita,* isolated from petroleum-contaminated soil from Veracruz, Mexico, is able to germinate at 100 $\mu g\ g^{-1}$ of benzo[a]pyrene (BaP) (Alarcón, Davies, and Ferrera-Cerrato, 2003a). On the other hand, the AM symbiosis between this fungus and *Echinochloa polystachya* is not affected by the presence of BAP. In this particular case, the mycorrhizal condition of *E. polystachya* is a significant factor in dehydrogenase activity, but not in the activity of polyphenol oxidase or the dispersion of this PAH from the rhizosphere (Alarcón, Davies, and Ferrera-Cerrato, 2003b). The precise physiological mechanism by which *E. polystachya* affects BaP-degradation has not been elucidated. There is evidence that AM fungi show tolerance to PAH-contaminated soil, which is likely related to the adaptation, survival, establishment, and fitness of plants under conditions of soil contamination. Joner et al. (2001) propose the following mechanisms by which AM fungi may contribute to the dispersion/degradation of organic contaminants: (1) mycorrhizal modification of plant and microbial metabolism, (2) enhanced root peroxidase activity, and (3) modification of the microbial composition of the rhizosphere as a consequence of the establishment of AM symbiosis. The latter mechanism is related both qualitatively and quantitatively to the modification of root biomass and exudation patterns; it is also related to the synthesis of simple phenolic compounds in roots that could act as inducers of PAH-degradation, and to the formation of abundant extraradical mycelium that allows the exudation of organic compounds such as glomalin and other non-characterized compounds (Wright and Upadhyaya, 1999). These hyphae-derived compounds may enhance specific bacterial activity, which can drive cometabolic degradation of persistent and recalcitrant organic contaminants.

FUTURE RESEARCH

Investigation of AM fungi in tropical agriculture remains incomplete. The need for experimental research is particularly evident in certain areas, including the following: (1) studies of AM fungal mycelial networks under field conditions; (2) taxonomic work on AM fungi based not only on morphotypes, but also on direct determinations in host plant roots using molecular tools; (3) ecophysiological studies of the influence of AM fungal populations on the functioning of tropical agroecosystems, particularly those with low inputs; (4) in situ ecological studies of the interrelationships between AM fungi and other organisms, mainly plant-growth-promoting rhizobacteria and meso- and microfaunal components; and (5) development of large-scale AM fungal inoculum production techniques adapted to tropical conditions.

Determination of the structure and functioning of AM fungal populations in tropical ecosystems is an enormous challenge. The difficulties involved in identifying field-collected spores and detecting non-sporulating members of populations (Douds and Miller, 1999) add more complexity to the problem. Experimental research on AM fungal mycelial networks associated with tropical plants is in its infancy. Research on these networks could be of paramount importance to an understanding of the functioning of certain tropical agroecosystems, particularly those with low inputs, because the networks provide interconnections among plant roots, extend far beyond the conventional "rhizosphere," and provide important pathways for nutrient and C movement (Leake et al., 2004). Because AM fungal hyphae are involved in the mobilization of carbohydrates from plants into the soil and the release of enzymes, exudates, and dead cells, they may differentially influence the structure and functioning of other components of the soil micro- and mesobiota, such as bacteria, fungi, and animals. In general, in situ biotic interrelationships among AM fungi and other soil organisms in tropical areas are very poorly understood. A number of studies, however, may indicate that certain microbial groups, such as N_2-fixing bacteria (Azcón, 2000; Barea, Azcón, and Azcón-Aguilar, 1992; Sánchez-Colín and Ramírez, 2000) and mycophagous nematodes (Hussey and Roncardi, 1981; Salawu and Esty, 1979; Smith, Harnett, and Rice, 2000), could significantly affect AM fungi functioning in agroecosystems. At the present time, large-scale use of AM fungi in tropical areas is limited to a few situations in which plants are transplanted or produced in nursery beds or greenhouses (Pérez-Moreno, Alvarado-López, and Ferrera-Cerrato, 2002). One reason for this limited use of AM fungi in tropical crop production is the high cost and bulk of inoculants. The development, preferably in situ, of cheaper methods of producing less bulky AM fungal inoculum, adapted to tropical conditions, is currently an urgent need.

REFERENCES

Abbott, L.K. and A.D. Robson (1981). Infectivity and effectiveness of five endomycorrhizal fungi: Competition with indigenous fungi in field soils. *Australian Jounal of Agricultural Research* 32:621-630.

Abbott, L.K. and A.D. Robson (1982). The role of vesicular arbuscular mycorrhizal fungi in agriculture and the selection of fungi for inoculation. *Australian Jounal of Agricultural Research* 33:389-408.

Abbott, L.K., A.D. Robson, and I.R. Hall (1983). Introduction of vesicular arbuscular mycorrhizal fungi into agricultural soils. Australian *Jounal of Agricultural Research* 34:741-749.

Abbott, L.K., A.D. Robson, and G. De Boer (1984). The effect of phosphorus on the formation of hyphae in soil by the vesicular-arbuscular mycorrhizal fungus, *Glomus fasciculatum*. *New Phytologist* 97:437-446.

Aguilar, S., R. Flores-Bello, J.L. Jiménez–Hernández, and E. Soriano-Richards (2000). Crecimiento y producción de fríjol de condiciones de trópico seco después de la colonización micorrízica-arbuscular. In *Ecología, fisiología y biotecnología de la micorriza arbuscular,* A. Alarcón and R. Ferrera-Cerrato (eds.). Mexico: Mundi Prensa, pp. 149-155.

Alarcón, A. and R. Ferrera-Cerrato (1996). Dinámica de colonización y efecto de hongos endomicorrízicos sobre el crecimiento de *Casuarina equisetifolia* L. In *Nuevos horizontes en agricultura: Agroecología y Desarrollo Sostenible,* J. Pérez-Moreno and R. Ferrera-Cerrato (eds.). Montecillo, Mexico: Colegio de Postgraduados, pp. 298-302.

Alarcón, A. and R. Ferrera-Cerrato (1999). Manejo de la micorriza arbuscular en los sistemas de propagación de plantas frutícolas. *Terra* 17:179-192.

Alarcón, A., F.T. Davies Jr., and R. Ferrera-Cerrato (2003a). Influence of two polycyclic aromatic hydrocarbons on spore germination of *Gigaspora margarita*. In *Proceedings of the 4th International Conference on Mycorrhiza*. Montreal, Canadá. p. 164.

Alarcón, A., F.T. Davies Jr., and R. Ferrera-Cerrato (2003b). Phytoremediation potential and tolerance of *Echinocloa polystachya-Gigaspora margarita* symbiosis to benzo[a]pyrene. In *Proceedings of the 4th International Conference on Mycorrhiza*. Montreal, Canadá. p. 165.

Alarcón, A., M.C. González-Chávez, and R. Ferrera-Cerrato (2003). Growth and physiology of *Citrus volkameriana* Tan et Pasq in arbuscular mycorrhizal symbiosis. *Terra* 21:503-511.

Alarcón, A., R. Ferrera-Cerrato, M.C. González-Chávez, and A. Villegas-Monter (2000). Hongos micorrízicos arbusculares en la dinámica de aparición de estolones y nutrición de plantas de fresa cv. fern obtenidas por cultivo *in vitro*. *Terra* 18:211-218.

Alarcón, A., F.T. Davies Jr., J.N. Egilla, T.C. Fox, A.A. Estrada-Luna, and R. Ferrera-Cerrato (2002). Short term effects of *Glomus claroideum* and *Azospirillum brasilense* on growth and root acid phosphatase activity of *Carica papaya* L. under phosphorus stress. *Revista Latino-Americana de Microbiologia* 44:31-37.

Alexander, I. (1989). Mycorrhizas in tropical forests. *Special Publication of the British Ecological Society* 89:169-188.

Altieri, M.A. (1987). *Agroecology, the scientific basis of alternative agriculture*. Boulder, CO: Westview Press, 227 p.

Álvarez-Solís, J.D. and N.S. León-Martínez (1997). Fertilidad del suelo y sistemas simbióticos. In *Los Altos de Chiapas: Agricultura y crisis rural,* V.M. Parra and B. Díaz. (eds.). San Cristóbal de Las Casas, Chiapas, Mexico: El Colegio de la Frontera Sur, pp. 43-64.

Álvarez-Solís, J.D. and M. de J. Anzueto-Martínez (2004). Actividad microbiana del suelo bajo diferentes sistemas de producción de maíz en los altos de Chiapas, México. *Agrociencia* 38:13-22.

Anderson, E.L., P.D. Milner, and H.M. Kunishi (1987). Maize root length density and mycorrhizal infection as influenced by tillage and soil phosphorus. *Jounal of Plant Nutrition* 10:1349-1356.

Augé, R.M. (2001). Water relations, drought and vesicular-arbuscular mycorrhizal symbiosis. *Mycorrhiza* 11:3-42.

Azcón, R. (2000). Papel de la simbiosis micorrízica y su interacción con otros microorganismos rizosféricos en el crecimiento vegetal y sostenibilidad agrícola. In *Ecología, fisiología y biotecnología de la micorriza arbuscular,* A. Alarcón and R. Ferrera-Cerrato (eds.). Mexico: Mundi Prensa, pp. 1-15.

Azcón, R. and R.M. Tobar (1998). Activity of nitrate reductase and glutamine synthetase in shoot and root of mycorrhizal *Allium cepa,* effect of drought stress. *Plant Science* 133:1-8.

Azizah-Chulan, H. and P. Ragu (1986) Growth response of *Theobroma cacao* L. seedlings to inoculation with vesicular-arbuscular mycorrhizal fungi. *Plant and Soil* 96:279-285.

Bååth, E. and J. Spokes (1989). The effect of added nitrogen and phosphorus on mycorrhizal growth response and infection in *Allium schoenoprasum. Canadian Journal of Botany* 67:3227-3232.

Bago, B., C. Azcón-Aguilar, A. Goulet, and Y. Piché (1998). Branched absorbing structures (BAS): a feature of the extraradical mycelium of symbiotic arbuscular mycorrhizal fungi. *New Phytologist* 139:375-388.

Bagyaraj, D.J. (1992). Vesicular-arbuscular mycorrhiza: Application in agriculture. In *Methods in Microbiology,* J.R. Norris, D.J. Read, and A.K. Varma (eds.). London: Academic Press Inc., pp. 819-833.

Barea, J.M., R. Azcón, and C. Azcón-Aguilar (1992). Vesicular arbuscular mycorrhizal fungi in nitrogen-fixing systems. In *Methods in Microbiology,* J.R. Norris, D.J. Read, and A. Varma (eds.). London: Academic Press Inc., pp. 391-416.

Bécard, G. and A. Fortin (1988). Early events of vesicular-arbuscular mycorrhiza formation on Ri T-DNA transformed roots. *New Phytologist* 108:211-218.

Benhamou, N., J.A. Fortin, C. Hamel, M. St-Arnaud, and A. Shatilla (1994). Resistance response of mycorrhizal RiT-DNA-transformed carrot roots to infection by *Fusarium oxysporum* f.sp. *chrysanthemi. Phytopathology* 84:958-968.

Berbee, M.L. and J.W. Taylor (1993). Dating the evolutionary radiations of the true fungi. *Canadian Jounal of Botany* 71:1114-1127.

Bethlenfalvay, G.J. (1992). Mycorrhizae and crop productivity. In *Mycorrhizae in Sustainable Agriculture,* G.J. Bethlenfalvay and R.G. Linderman (eds.). Madison, WI: American Society of Agronomy, Special Publication No. 54, pp. 1-27.

Bethlenfalvay, G.J. (1993). The mycorrhizal plant-soil system in sustainable agriculture. In *Agroecologia, sostenibilidad y educación,* R. Ferrera-Cerrato and R. Quintero-Lizaola (eds.). Texcoco, Mexico: Colegio de Postgraduados, pp. 127-137.

Bethlenfalvay, G.J. and R.G. Linderman (1992). *Mycorrhizae in sustainable agriculture.* American society of agronomy, special publication No. 54. Madison, WI, 124 pp.

Binet, P., J.M. Portal, and C. Leyval (2000). Fate of polycyclic aromatic hydrocarbons (PAH) in the rhizosphere and mycorrhizosphere of ryegrass. *Plant and Soil* 227:207-213.

Boddington, C.L. and J.C. Dodd (1998). A comparison of the development and metabolic activity of mycorrhizas formed by arbuscular mycorrhizal fungi from different genera on two tropical forage legumes. *Mycorrhiza* 8:149-157.

Brundrett, M.C. (2002). Coevolution of roots and mycorrhizas of land plants. *New Phytologist* 154:275-304.

Brundrett, M.C., D.A. Jasper, and N. Ashwath (1999). Glomalean mycorrhiza fungi from tropical Australia II: The effect of nutrient levels and host species on the isolation of fungi. *Mycorrhiza* 8:315-321.

Brundrett, M.C., N. Bougher, B. Dell, T. Grove, and N. Malajczuk (1996). *Working with mycorrhizas in forestry and agriculture*. Australian Center for International Agricultural Research. Canberra, Australia. 374 p.

Cabello, M.N. (2001). Mycorrhizas and hydrocarbons. In *Fungi in Bioremediation*. G.M. Gadd (ed). Cambridge: British Mycological Society. Cambridge University Press, pp. 456-471.

Cairney, J.W.G. and A.A. Meharg (1999). Influences and anthropogenic pollution on mycorrhizal fungal communities. *Environmental Pollution* 106:169-182.

Calvet, C., J. Pera, V. Estaún, and C. Camprubi (1989). Vesicular-arbuscular mycorrhizae on kiwifruit in an agricultural soil: Inoculation of seedlings and hardwood cuttings with *Glomus mosseae*. *Agronomie* 9:181-185.

Cantrelli, C.I. and R.G. Linderman (2001). Preinoculation of lettuce and onion with VA mycorrhizal fungi reduces deleterious effects of soil salinity. *Plant and Soil* 233:269-281.

Carpenter, F.L., S. Palacios, E. Gonzalez-Q, and M. Schroeder (2001). Land-use and erosion of a Costa Rican ultisol affect soil chemistry, mycorrhizal fungi and early regeneration. *Forest Ecology and Management* 144:1-17.

Chacón, A.M. and G. Cuenca (1998). Efecto de las micorrizas arbusculares y de la fertilizacion con fósforo, sobre el crecimiento de la guayaba en condiciones de vivero. *Agronomía Tropical* 48:425-440.

Chamizo, A., R. Ferrera-Cerrato, and L. Varela (1998). Identificación de especies de un consorcio del género *Glomus*. *Revista Mexicana de Micologia* 14:37-40.

Chen, Y.L., M.C. Brundrett, and B. Dell (2000). Effects of ectomycorrhizas and vesicular-arbuscular mycorrhizas, alone or in competition, on root colonization and growth of *Eucalyptus globulus* and *E. urophylla*. *New Phytologist* 146:545-556.

Clark, R.B. and S.K. Zeto (2000). Mineral acquisition by arbuscular mycorrhizal plants. *Journal of Plant Nutrition* 23:867-902.

Copeman, R.H., C.A. Martin, and J.C. Stutz (1996). Tomato growth in response to salinity and mycorrhizal fungi from saline or nonsaline soils. *HortScience* 31:341-344.

Criquet, S., E. Joner, P. Leglize, and C. Leyval (2000). Anthracene and mycorrhiza affect the activity of oxidoreductases in the roots and the rhizosphere of lucerne (*Medicago sativa* L.). *Biotechnology Letters* 22:1733-1737.

Cuenca, G. and E. Meneses (1996). Diversity patterns of arbuscular mycorrhizal fungi associated with cacao in Venezuela. *Plant and Soil* 183:315-322.

Cuenca, G., R. Herrera, and E. Meneses (1990). Effects of VA mycorrhiza on the growth of cacao seedlings under nursery conditions in Venezuela. *Plant and Soil* 126:71-78.

Cuenca, G., Z. Andrade, and G. Escalante (1998). Arbuscular mycorrhizae in the rehabilitation of fragile degraded tropical lands. *Biology and Fertility of Soils* 26:107-111.

Dalal, R.C. and R.J. Henry (1988). Cultivation effects on carbohydrate contents of soil and soil fractions. *Soil Science Society of America Journal* 52:1361-1365.

Daniell, T.J., R. Husband, A.H. Fitter, and J.P.M. Young (2001). Molecular diversity of arbuscular mycorrhizal fungi colonizing arable crops. *FEMS Microbiology Ecology* 36:203-209.

Dann, P.R., J.W. Derrick, D.C. Dumaresq, and H.M. Ryan (1996). The response of organic and conventionally grown wheat to superphosphate and reactive rock phosphate. *Australian Journal of Experimental Agriculture* 36:71-78.

Davies, F.T. Jr., J.D. Puryear, R.J. Newton, J.N. Egilla, and J.A. Saraiva (2001). Mycorrhizal fungi enhance accumulation and tolerance of chromium in sunflower *(Helianthus annuus). Journal of Plant Physiology* 158:777-786.

Davies, F.T. Jr., V. Olalde-Portugal, L. Aguilera-Gómez, M.J. Alvarado, R. Ferrera-Cerrato, and T.W. Boutton (2002). Alleviation of drought stress of chile ancho pepper (*Capsicum annuun* L. cv. San Luis) with arbuscular mycorrhiza indigenous to Mexico. *Scientia Horticulture* 92:347-359.

Declerck, S., C. Plenchette, and D.G. Strullu (1995). Mycorrhizal dependency of banana *(Musa acuminata)* cultivars. *Plant and Soil* 176:183-187.

Declerck, S., J.M. Risede, G. Rufyikiri, and B. Delvaux (2002). Effects of arbuscular mycorrhizal fungi on severity of root rot of bananas caused by *Cylindrocladium spathiphylli. Plant Pathology* 51:109-115.

Declerck, S., D.G. Strullu, C. Plenchette, T. Guillemette (1996). Entrapment of *in vitro* produced spores of *Glomus versiforme* in alginate beads: *In vitro* and *in vivo* inoculum potentials. *Journal of Biotechnology* 48:51-57.

Dehne, H.W. and F. Schoenbeck (1979). The influence of endotrophic mycorrhiza on plant diseases. II. Phenol metabolism and lignification. *Phytopathology Zeitschrift* 95:210-216.

Dew, E.A., R.S. Murray, S.E. Smith, and I. Jakobsen (2003). Beyond the rhizosphere: Growth and function of arbuscular mycorrhizal external hyphae in sands of varying pore soils. *Plant and Soil* 251:105-114.

Dodd, J.C., I. Arias, I. Koomen, and D.S. Hayman (1990a). The management of populations of vesicular-arbuscular mycorrhizal fungi in acid-infertile soils of a savanna ecosystem: I. The effect of pre-cropping and inoculation with VAM-fungi and plant growth and nutrition in the field. *Plant and Soil* 122:229-240.

Dodd, J.C., I. Arias, I., Koomen, and D.S. Hayman (1990b). The management of populations of vesicular-arbuscular mycorrhizal fungi in acid-infertile soils of a savanna ecosystem: II. The effect of pre-cropping on the spore populations of native and introduced VAM-fungi. *Plant and Soil* 122:241-247.

Douds, D.D. Jr. and P.D. Miller (1999). Biodiversity of arbuscular mycorrhizal fungi in agroecosystems. *Agriculture Ecosystems and Environment* 74:77-93.

Douds, D.D., R.R. Janke, and S.E. Peters (1993). VAM fungus spore populations and colonization of roots of maize and soybean under conventional and low-input sustainable agriculture. *Agriculture Ecosystems and Environment* 43:325-335.

Duponnois, R., C. Plenchette, J. Thioulouse, and P. Cadet (2001). Mycorrhizal soil infectivity and AM fungal spore communities of different aged fallows in Senegal. *Applied Soil Ecology* 17:239-251.

Duponnois, R., E. Diédhiou, J.L. Chotte, and J.L. Ouroy Sy (2003). Relative importance of the endomycorrhizal and (or) ectomycorrhizal associations in *Allocasuarina* and *Casuarina genera. Canadian Journal of Microbiology* 49:281-287.

Eason, W.R., J. Scullion, and E.P. Scott (1999). Soil parameters and plant responses associated with arbuscular mycorrhizas from contrasting grassland management regimes. *Agriculture Ecosystems and Environment* 73:245-255.

Eissenstat, D.M., J.H. Graham, J.P. Syvertsen, and D.L. Drouillard (1993). Carbon economy of sour orange in relation to mycorrhizal colonization and phosphorus status. *Annals of Botany* 71:1-10.

Elsen, A., S. Declerck, and D. De Waele (2001). Effects of *Glomus intraradices* on the reproduction of the burrowing nematode *(Radophus similis)* in deixenic culture. *Mycorrhiza* 11:49-51.

Elsen, A., S. Declerck, and D. De Waele (2002). Efecto de tres hongos micorriza arbusculares sobre la infección de *Musa* con el nematode nodulador de las raíces (*Meloidogyne* spp.). *Infomusa* 11:21-23.

El-Tohamy, W., W.H. Schnitzler, U. El Behairy, and M.S. El Behairy (1999). Effect of VA mycorrhizas on improving drought and chilling tolerance of bean plants. *Journal of Applied Botany* 73:178-183.

Escalona, M., D. Trejo, A., J. Díaz R., L. Lara C., and A. Rivera (2000). Efecto de la endomicorriza arbuscular y diferentes fechas de fertilización sobre el crecimiento de papaya en campo. In *Ecología, fisiología y biotecnología de la micorriza arbuscular,* A. Alarcón and R. Ferrera-Cerrato (eds.). Mexico: Mundi Prensa, pp. 194-205.

Estrada-Luna, A.A., F.T. Davies Jr., and J.N. Egilla (2000). Mycorrhizal fungi enhancement of growth and gas exchange of micropropagated guava plantlets (*Psidium guajava* L.) during *ex vitro* acclimatization and plant establishment. *Mycorrhiza* 10:1-8.

Evans, D.G. and M.H. Miller (1988). Vesicular-arbuscular mycorrhizas and the soil-disturbance induced reduction of nutrient absorption in maize: I. Causal relations. *New Phytologist* 110:67-74.

Ezeta, F.N. and O.M. Santos (1981). Importancia de endomicorriza na nutricao mineral do cacaueiro. *Revista Brasileira do la Ciencia do Solo* 5:22-27.

Feldman, F. and E. Idczak (1992). Inoculum production of vesicular-arbuscular mycorrhizal fungi for use in tropical nurseries. In *Methods in Microbiology,* J.R. Norris, D.J. Read, and A.K. Varma (eds.). London: Academic Press Inc., pp. 799-817.

Fernández, F.E., G.G. Cañizares, R. Rivera, and R. Ferrera (1992). Efectividad de tres hongos micorrízico vesiculo-arbusculares (MVA) y una cepa de bacterias solubilizadoras de fósforo (BSF) sobre el crecimiento de café *(Coffea arabica). Cultivos tropicales* 13:23-27.

Ferrera-Cerrato, R. (1995). Importancia de las asociaciones microbianas en el cultivo de frijol. In *Diversidad genética y patología del frijol,* J. Pérez-Moreno, R. Ferrera-Cerrato, and R. García-Espinosa (eds.). Texcoco, Mexico: Colegio de Postgraduados, pp. 122-131.

Ferrera-Cerrato, R. and D. Jaen C. (1989). Respuesta del cedro blanco (*Cupressus lindleyi* K.) a la inoculación con hongos formadores de endomicorriza vesículo-arbuscular (V-A). In *Proceedings of the XXII Mexican Congress of Soil Science,* M.A. Vergara G. Alcántar, and A. Aguilar S. (eds.). Montecillo, Mexico: Mexican Soil Science Society, p. 152.

Founoune, H., R. Duponnois, A.M. Ba, and F. El-Bouami (2002). Influence of the dual arbuscular endomycorrhizal/ectomycorrhizal symbiosis on the growth of *Acacia holosericea* (A. Cunn. ex G. Don) in glasshouse conditions. *Annals of Forest Science* 59:93-98.

Francl, L.J. (1993). *Interaction of nematodes with mycorrhizae and mycorrhizal fungi.* In *Nematode Interactions.* M. Wajid Khan (ed.). London: Chaplan and Hall, pp. 203-216.

Franco-Ramírez, A., M.J. Manjarrez-Martínez, A. Alarcón, and R. Ferrera-Cerrato (2004). Crecimiento de anonas inoculadas con hongos micorrízico arbusculares aislados de la rizosfera de anonáceas. In *Proceedings of the IV Mexican Symposium of Mycorrhizal Symbiosis,* M.C. González-Chávez, J. Pérez-Moreno, R. Ferrera-Cerrato, M.P. Ortega-Larrocea, V. Carreón-Abud, and E. Valencia-Cantero (eds.). Morelia, Mexico: Colegio de Postgraduados, p. 28.

Frank, A.B. (1885). Ueber die auf Wurzelsymbiose beruhende Ernährung gewisser Baume durch unterirdische Pilze (On the root-symbiosis depending nutrition through hypogeous fungi of certain trees). *Ber. Deut. Bot. Gesell.* 3:128-145 (translated by J.M.Trappe, *Proceedings of the 6th North America Conference on Mycorrhiza*, pp. 18-25).

Friese, C.F. and M.F. Allen (1991). The spread of VA mycorrhizal fungal hyphae in the soil: Inoculum types and external hyphal architecture. *Mycologia* 83:409-418.

Galvez, L., D.D. Dounds, L.E. Drinkwater, and P. Wagoner (2001). Effect of tillage and farming system upon VAM fungus populations and mycorrhizas and nutrient uptake of maize. *Plant and Soil* 228:299-308.

García, G.E., R. Ferrera-Cerrato, J.J. Almaraz Suárez, and R. Rodríguez (2000). Colonización micorrízica arbuscular en gramíneas creciendo en un suelo contaminado con hidrocarburos. In *Ecología, fisiología y biotecnología de la micorriza arbuscular,* A. Alarcón and R. Ferrera-Cerrato (eds.). Mexico: Mundi Prensa, pp. 236-243.

García-Rodríguez, L., J. Pérez-Moreno, A. Aldrete, and V.M. Cetina-Alcala (2005). Caracterización y cultivo del hongo silvestre ectomicorrízico *Pisolithus tinctorius* procedente de plantaciones tropicales de pinos y eucaliptos en México. *Agrociencia* (In press).

Gardezi, A.K. and R. Ferrera-Cerrato (1989). The effect of four levels of phosphorus on mycorrhizal colonization, dry root weight, and nitrogen and phosphorus contents of *Acacia saligna* inoculated with *Rhizobium* sp. and endomycorrhiza in a Mexican andisol. *Nitrogen Fixing Tree Research Report* 7:43-45.

Gardezi, A.K. and R. Ferrera-Cerrato (1992). Mycorrhizal inoculation of *Caesalpinia cacalaco. Nitrogen Fixing Tree Research Reports* 10:116-118.

Gardezi, A.K., R. Ferrera-Cerrato, and V. Lara (1988). Effect of the double inoculation of *Rhizobium* sp. and V-A endomycorrhizae on *Acacia cyanophylla* in an andosol in Mexico. *Nitrogen Fixing Tree Research Reports* 6:31-33.

Gardezi, A.K., R.A. Guzmán-Plazola, and R. Ferrera-Cerrato (1991). Growth response of *Pithecellobium dulce* to mycorrhizal inoculation. *Nitrogen Fixing Tree Research Reports* 9:111-113.

Gardezi, A.K., D. Jean, R.A. Guzmán-Plazola, and R. Ferrera-Cerrato (1990). Growth of *Acacia farnesiana* associated with mycorrizal fungi in three types of Mexican soils. *Nitrogen Fixing Tree Research Reports* 8:99-102.

Gardezi, A.K., R. García E., R. Ferrera-Cerrato, and C.A. Pérez (1995). Endomycorrhiza, rock phosphate, and organic matter effects on growth of *Erythrina americana. Nitrogen Fixing Tree Research Reports* 13:48-50.

Gardezi, A.K., E. Zavaleta-Mejia, R. García-Espinosa, R. Ferrera-Cerrato, M.A. Musalem, and C.A. Pérez-Mercado (1996). Efecto de endomicorriza, roca fosfórica y materia orgánica en el desarrollo de *Erythrina americana* y sus usos en México. In *Nuevos horizontes en agricultura: Agroecología y Desarrollo Sostenible,* J. Pérez-Moreno and R. Ferrera-Cerrato (eds.). Montecillo, Mexico: Colegio de Postgraduados, pp. 355-360.

Gianinazzi, S. and M. Vosátka (2004). Inoculum of arbuscular mycorrhizal fungi for production systems: Science meets business. *Canadian Journal of Botany* 82:1264-1270.

González-Chávez, M.C. and R. Ferrera-Cerrato (1993). Influencia de la endomicorriza vesiculo-arbuscular en cuatro variedades de café. In *Avances de Investigación,* J. Pérez-Moreno and R. Ferrera-Cerrato (eds.). Montecillo, Mexico: Área de Microbiología de suelos. IRENAT-Colegio de Postgraduados, pp. 100-112.

González-Chávez, M.C. and R. Ferrera-Cerrato (1994). Interacción de la micorriza V-A y la fertilización fosfatada en diferentes portainjertos de cítricos. *Terra* 12:338-344.

González-Chávez, M.C. and R. Ferrera-Cerrato (1996). Influencia de la endomicorriza vesículo-arbuscular en cuatro variedades de café. In *Avances de Investigación,* J. Pérez-Moreno and R. Ferrera-Cerrato (eds.). Texcoco, Mexico: Microbiología de Suelos. Colegio de Posgraduados, pp. 100-112.

González-Chávez, M.C., R. Ferrera-Cerrato, A. Villegas-Monter, and J.L. Oropeza (2000). Selección de sustratos de crecimiento en microplántulas de cítricos inoculadas con *Glomus* sp. Zac-19. *Terra* 18:369-380.

González-Chávez, M.C., R. Carrillo-González, S.E. Wright, and K.A. Nichols (2004). The role of glomalin, a protein produced by arbusuclar mycorrhizal fungi, in sequestering potentially toxic elements. *Environmental Pollution* 130: 317-323.

Graham, J.H. and D.M. Eissenstat (1998). Field evidence for the carbon cost of citrus mycorrhizas. *New Phytologist* 140:103-110.

Griffioen, W.A.J. and W.H.O. Ernst (1989). The role of VA mycorrhiza in the heavy metal tolerance of *Agrostis capillaris* L. *Agriculture Ecosystems and Environment* 29:173-177.

Guillemin, J.P., S. Gianinazzi, and A. Trouvelot (1992). Screening of arbuscular endomycorrhizal fungi for establishment of micropropagated pineapple plants. *Agronomie* 12:831-836.

Guzmán-Plazola, R.A., R. Ferrera-Cerrato, and J.D. Etchevers (1987). Reducción de la defoliación de *Leucaena leucocephala* debida al transplante cuando es inoculada con micorriza V-A. In *Abstracts of the XX Mexican Congress of Soil Science,* A. Aguilar, S. and J. Baus P. (eds.). Zacatecas, Mexico: Mexican Soil Science Society, p. 172.

Guzmán-Plazola, R.A., Ferrera-Cerrato, R., and J.D. Etchevers (1988). *Leucaena leucocephala,* a plant of high mycorrhizal dependence in acid soils. *Leucaena Research Reports* 9:69-73.

Guzmán-Plazola, R.A., R. Ferrera-Cerrato, and G.J. Bethlenfalvay (1992). Papel de la endomicorriza VA en la transferencia de exudados radicales entre el frijol y maíz sembrados en asociación bajo condiciones de campo. *Terra* 10:236-248.

Guzmán-Plazola, R.A., R. Ferrera-Cerrato, J.D. Etchevers, and T. Corona (1984). The symbiosis *Rhizobium-Glomus* in *Leucaena leucocephala. Proceedings of the 6th North American Conference on Mycorrhizae.* Bend, OR, p. 237.

Harinikumar, K.M. and D.J. Bagyaraj (1988). Effect of crop rotation on native vesicular arbuscular mycorrhizal propagules in soil. *Plant and Soil* 110:77-80.

Harrison, M.J. and M.L. Van Buuren (1995). A phosphate transporter from the mycorrhizal fungus *Glomus versiforme. Nature* 378:626-629.

Heizemann, J., E. Sieverding, and C. Diederichs (1992). Native populations of the Glomales influenced by terracing and fertilization under cultivated potato in the tropical Highlands of Africa. In *Mycorrhizas in Ecosystems,* D.J. Read, D.H. Lewis, A.H. Fitter, and I.J. Alexander (eds.). Wallingford, Oxon: C.A.B. International, p. 382.

Helgason, T., T.J. Daniell, R. Husband, A.H. Fitter, and J.P.W. Young (1998). Ploughing up the wood-wide web? *Nature* 394:431.

Henderson, J.C. and F.T. Davies Jr. (1990). Drought acclimation and the morphology of mycorrhizal *Rosa hybrida* L. cv. "Ferdy" is independent of the leaf elemental content. *New Phytologist* 115:503-510.

Hernández-Acosta, E., R. Ferrera-Cerrato, J.J. Almaraz-Suárez, and R. Rodríguez-Vasquez. (1998). Efecto de un complejo de hidrocarburos sobre la simbiosis endomicorrízica en frijol de invernadero. In *Avances de la investigación micorrízica en México,* R. Zulueta Rodriguez, M.A. Escalona Aguilar, and D. Trejo Aguilar (eds.). Xalapa, Mexico: Universidad Veracruzana, pp. 253-258.

Hernández-Acosta, E., R. Ferrera-Cerrato, L. Fernández-Linares, R. Rodríguez-Vásquez (2000). Ocurrencia de la micorriza arbuscular y bacterias fijadoras de N atmosférico en un suelo contaminado por hidrocarburos. In *Ecología, fisiología y biotecnología de la micorriza arbuscular,* A. Alarcón and R. Ferrera-Cerrato (eds.). Mexico: Mundi Prensa, pp. 227-235.

Hernández-Meza, V., M.C. González-Chávez, R. Ferrera-Cerrato, and A. Villegas-Monter (1998). Efecto de la inoculación endomicorrízica en el transplante de portainjertos de cítricos micropropagados tolerantes al virus de la tristeza. In *Avances de la investigación micorrízica en México,* R. Zulueta R., M.A. Escalona A., and D. Trejo A. (eds.). Xalapa, Mexico: Universidad Veracruzana, pp. 241-252.

Hooker, J.E., M. Jaizme-Vega, and D. Atkinson (1994). Biocontrol of plant pathogens using arbuscular mycorrhizal fungi. In *Impact of Arbuscular Mycorrhizas on Sustainable Agriculture and Natural Ecosystems,* S. Gianinazzi and H. Schüepp (eds.). Basel, Switzerland: Birkhäuser Verlag, pp. 191-200.

Howeler, R.H. and E. Sieverding (1983). Potentials and limitations of mycorrhizal inoculation illustrated by experiments with field grown cassava. *Plant and Soil* 75:245-261.

Howeler, R.H., E. Sieverding, and S. Saif (1987). Practical aspects of mycorrhizal technology in some tropical crops and pastures. *Plant and Soil* 100:249-283.

Hussey, R.S. and R.W. Roncadori (1981). Influence of *Aphelenchus avenae* on vesicular-arbuscular endomycorrhizal growth response in cotton. *Journal of Nematology* 2:48-52.

Ingham, R.E. (1988). Interactions between nematodes and vesicular arbuscular mycorrhizae. *Agriculture, Ecosystems and Environment* 24:169-182.

Jaizme-Vega, M.C., B. Sosa-Hernández, and J.M. Hernández-Hernández (1998). Interaction of arbuscular mycorrhizal fungi and the soil pathogen *Fusarium oxysporum* f.sp. *cubense* on the first stages of micropropagated Grande Naine banana. *Acta Horticulturae* 490:285-295.

Jaizme-Vega, M.C., M. Esquivel-Dalamos, P. Tenoury-Domínguez, and A.J. Rodríguez-Romero (2002). Effectos de la micorrización sobre el desarrollo de dos cultivares de platanera micropropagada. *Infomusa* 11:25-28.

Jaizme-Vega, M.C., P. Tenoury, J. Pinochet, and M. Jaumot (1997). Interactions between the root-knot nematode *Meloidogyne incognita* and *Glomus mosseae* in banana. *Plant and Soil* 196:27-35.

Jakobsen, I., S.E. Smith, and E.A. Smith (2002). Function and diversity of arbuscular mycorrhizae in carbon and mineral nutrition. In *Mycorrhizal Ecology,* M.G.A. van der Heijden and I.R. Sanders (eds.). The Netherlands: Springer, Ecological studies 157, pp. 75-92.

Janos, D.P. (1980). Vesicular arbuscular mycorrhizae affect lowland tropical rain forest plant growth. *Ecology* 61:151-162.

Janse, J.M. (1896). Les endophytes radicaux de quelques plantes Javanaises. *Annales du Jardin Botanique Buitenzoro* 14:53-212.

Jeffries, P., S. Gianinazzi, S. Peroto, K. Turnau, and J.M. Barea (2003). The contribution of arbuscular mycorrhizal fungi in sustainable maintenance of plant health and soil fertility. *Biology and Fertility of Soils* 37:1-16.

Johnson, N.C. (1993). Can fertilization of soil select less mutualistic mycorrhizae? *Ecological Applications* 3:749-757.

Johnson, N.C. and F.L. Pfleger (1992). Vesicular-arbuscular mycorrhizae and cultural stresses. In *Mycorrhizae in Sustainable Agriculture,* G.J. Bethlenfalvay and

R.G. Linderman (eds.). Madison, WI: American Society of Agronomy Special Publication No. 54, pp. 71-99.

Johnson, N.C. and D.A. Wedin (1997). Soil carbon, nutrients, and mycorrhizae during conversion of dry tropical forest to grassland. *Ecological Applications* 7:171-182.

Johnson, N.C., J.H. Graham, and F.A. Smith (1997). Functioning of mycorrhizal associations along the mutualism-parasitism continuum. *New Phytologist* 135: 575-586.

Johnson, N.C., D.R. Zak, D. Tilman, and F.L. Pfleger (1991). Dynamics of vesicular-arbuscular mycorrhizae during old-field succession. *Oecologia* 86:349-358.

Johnson, N.C., F.L. Pfleger, R.K. Crookston, S.R. Simmons, and P.J. Copeland (1991). Vesicular-arbuscular mycorrhizas respond to corn and soybean cropping history. *New Phytologist* 177:657-663.

Johnston, A. (1949). Vesicular-arbuscular mycorrhiza in Sea Island cotton and other tropical plants. *Tropical Agriculture Trin* 26:118-121.

Joner, E.J. and C. Leyval (2001). Influence of arbuscular mycorrhiza on clover and ryegrass grown together in a soil spiked with polycyclic aromatic hydrocarbons. *Mycorrhiza* 10:155-159.

Joner, E.J. and C. Leyval (2003). Rhizosphere gradients of polycyclic aromatic hydrocarbon (PAH) dissipation in two industrial soils and the impact of arbuscular mycorrhiza. *Environmental Science and Technology* 37:2371-2375.

Joner, E.J., A. Johansen, A.P. Loibner, M.A. De la Cruz, O.H.J. Solar, J.M. Portal, and C. Leyval (2001). Rhizosphere effects on microbial community structure and dissipation and toxicity of polycyclic aromatic hydrocarbons (PAHs) in spiked soil. *Environmental Science and Technology* 35:2771-2777.

Kabir, Z. and R.T. Koide (2002). Effect of autumn and winter mycorrhizal cover crops on soil properties, nutrient uptake and yield of sweet corn in Pennsylvania, USA. *Plant and Soil* 238:205-215.

Kabir, Z., I.P. O'Halloran, J.W. Fyles, and C. Hamel (1998). Dynamics of the mycorrhizal symbiosis of corn (*Zea mays L.*): Effects of host physiology, tillage practice and fertilization on spatial distribution of extra-radical mycorrhizal hyphae in the field. *Agriculture Ecosystems and Environment* 68:151-163.

Kabir, Z., I.P. O'Halloran, P. Widden, and C. Hamel (1998). Vertical distribution of arbuscular mycorrhizal fungi under corn (*Zea mays* L.) in no-till and conventional tillage systems. *Mycorrhiza* 8:53-55.

Karasawa, T., Y. Karahara, and M. Takebe (2002). Differences in growth responses of maize to preceding cropping caused by fluctuation in the population of indigenous arbuscular mycorrhizal fungi. *Soil Biology and Biochemistry* 34:851-857.

Khan, A.G. (1972). The effect of vesicular-arbuscular mycorrhizal associations on growth of cereals. I. Effects on maize growth. *New Phytologist* 71:613-619.

Kleinschmidt, G.D. and J.W. Gerdemann (1972). Stunting of citrus seedlings in fumigated nursery soils related to the absence of endomycorrhizae. *Phytopathology* 62:1447-1453.

Kormanik, P.P., W.C. Bryan, and R.C. Schultz (1980). Increasing endomycorrhizal fungus inoculum in forest nursery soil with cover crops. *Southern Journal of Applied Forestry* 4:151-153.

Kurle, J.E. and F.L. Pfleger (1994). Arbuscular mycorrhizal fungus spore populations respond to conversions between low-input and conventional management practices in a corn-soybean rotation. *Agronomy Journal* 86:467-475.

Kyllo, D.A., V. Velez, and M.T. Tyree (2003). Combined effects of arbuscular mycorrhizas and light on water uptake of the neotropical understory shrubs, *Piper* and *Psychotria. New Phytologist* 160:443-454.

Lara-Fernández, V. and R. Ferrera-Cerrato (1986). Study of vesicular arbuscular endomycorrhizal-*Leucaena leucocephala* symbiosis. *Leucaena Research Reports* 7:94-96.

Laycock, D.H. (1945). Preliminary investigations into the function of the endotrophic mycorrhiza of *Theobroma cacao* L. *Tropical Agriculture Trin* 22:77-80.

Leake, J., D. Johnson, D. Donnelly, G. Muckle, L. Boddy, and D. Read (2004). Networks of power and influence: The role of mycorrhizal mycelium in controlling plant communities and agroecosystem functioning. *Canadian Journal of Botany* 82:1016-1045.

Leyval, C. and P. Binet (1998). Effect of polyaromatic hydrocarbons in soil on arbuscular mycorrhizal plants. *Journal of Environmental Quality* 27:402-407.

Linderman, R.G. (1988). Mycorrhizal interactions with the rhizosphere microflora: The mycorrhizosphere effect. *Phytopathology* 78:366-371.

Linderman, R.G. (1992). Vesicular arbuscular mycorrhizae and soil microbial interactions. In *Mycorrhizae in sustainable agriculture,* G.J. Bethlenfalvay and R.G. Linderman (eds.). Madison, WI: American Society of Agronomy. Special publication No. 54, pp. 45-70.

Linderman, R.G. (1993). Effects of microbial interactions in the mycorrhizosphere on plant growth and health. In *Agroecología, sostenibilidad y educación,* R. Ferrera-Cerrato and R. Quintero-Lizaola (eds.). Texcoco, Mexico: Colegio de Postgraduados, pp. 138-152.

Lopes, F.S., F. Oliveira, A.M.L. Neptuno, and F.R.P. Morais (1983). Efeito da inoculacao do cafeeto com diferentes especies de fongos micorrizicos vesicular-arbusculares. *Revista Brasileira Ciencia Solo* 7:137-141.

Louis, L. and G. Lim (1987). Spore density and root colonization of vesicular-arbuscular mycorrhizas in tropical soil. *Transactions of the British Mycological Society* 88:207-212.

Lovato, P.E., J.P. Guillemin, and S. Gianinazzi (1992). Application of commercial arbuscular fungal inoculants to the establishment of micropropagated grapevine rootstock and pineapple plants. *Agronomie* 12:873-880.

Lovato, P.E., V. Gianinazzi-Pearson, A. Trouvelot, and S. Gianinazzi (1996). The state of art of mycorrhizas and micropropagation. *Advances in Horticutural Sciences* 10:46-52.

Lovelock, J. (1995a). *Gaia, a new look at life on earth.* Oxford: Oxford University Press.

Lovelock, J. (1995b). *The ages of Gaia.* 2nd edition. Oxford: Oxford University Press.

Lovelock, C.E., K. Andersen, and J.B. Morton (2003). Influence of host tree species and environmental variables on arbuscular mycorrhizal communities in tropical forests. *Oecologia* 135:268-279.

Lovelock, C.E., S.F. Wright, A. Deborah, D.E. Clark, and R.W. Ruess (2003). Soil stocks of glomalin produced by arbuscular mycorrhizal fungi across a tropical rain forest landscape. *Journal of Ecology* 92:278-287.

Mäder, P., S. Endenhofer, T. Boller, A. Wiemken, and U. Niggli (2000). Arbuscular mycorrhizae in a long-term field trial comparing low-input (organic, biological) and high-input (conventional) farming systems in a crop rotation. *Biology and Fertility of Soils* 31:150-156.

Maldonado-Mendoza, I.E., G.R. Dewbre, and M.J. Harrison (2001). A phosphate transporter gene from the extra-radical mycelium of an arbuscular mycorrhizal fungus *Glomus intraradices* is regulated in response to phosphate in the environment. *Molecular Plant-Microbe Interactions* 14:1140-1148.

Malencon, G. (1938). Les truffes européennes. *Revue Mycologie* 3:1-92.

Malloch, D.W., K.A. Pirozynski, and P.H. Raven (1980). Ecological and evolutionary significance of mycorrhizal symbioses in vascular plants (a review). *Proceedings of the National Academy of Sciences of the United States* 77:2113-2118.

Martin-Laurent, F., S.K. Lee, F.Y. Tham, H. Jie, and H.G. Diem (1999). Aeroponic production of *Acacia mangium* saplings inoculated with AM fungi for reforestation in the tropics. *Forest Ecology and Management* 122:199-207.

Martínez-Reyes, M., J. Pérez-Moreno, A. Aldrete, and V.M. Cetina-Alcalá (2005). Recolección de hongos silvestres comestibles ectomicorrízicos, producción de inoculo a base de esporas e inoculación en pinos en invernadero. *Agrociencia* (In press).

Marx, D.H. (1980). Ectomycorrhizal fungus inoculations: A tool for improving forestation practices. In *Tropical Mycorrhizal Research*. P. Mikola (ed.). Oxford: Oxford University Press, pp. 13-71.

Marx, D.H. and W.C. Bryan (1975). Growth and ectomycorrhizal development of loblolly pine seedlings in fumigated soil infested with the fungal symbiont *Pisolithus tinctorius*. *Forest Science* 21:245-254.

Matías-Crisostomo, S. and R. Ferrera-Cerrato (1993). Efecto de microorganismos y adición de materia orgánica en la colonización micorrízica en la recuperación de tepetates. In *Avances de Investigación*, J. Pérez-Moreno and R. Ferrera-Cerrato (eds.). Montecillo, Mexico: Sección de Microbiologia de Suelos, Colegio de Postgraduados, pp. 52-61.

McGonigle, T.P. and M.H. Miller (2000). The inconsistent effect of soil disturbance on colonization of roots by arbuscular mycorrhizal fungi: A test of the inoculum density hypothesis. *Applied Soil Ecology* 14:147-155.

Meharg, A.A. (2001). The potential for utilizing mycorrhizal associations in soil bioremediation. In *Fungi in Bioremediation*. G.M. Gadd (ed.). Cambridge: British Mycological Society. Cambridge University Press, pp. 445-455.

Meharg, A.A. and J.W.G. Cairney (2000). Co-evolution of mycorrhizal symbionts and their hosts to metal-contaminated environments. *Advances in Ecological Research* 30:69-112.

Menge J.A., D. Steirle, D.J. Bagyaraj, E.L.V. Johnson, and R.T. Leonard (1978). Phosphorus concentrations in plants responsible for inhibition of mycorrhizal infection. *New Phytologist* 80:575-578.

Michelsen, A. (1993). Growth improvement of Ethiopian Acacias by addition of vesicular-arbuscular mycorrhizal fungi or roots of native plants to non-sterile nursery soil. *Forest Ecology and Management* 59:193-206.

Mikola, P. (1970). Mycorrhizal inoculation in afforestation. *International Review of Forestry Research* 3:123-196.

Mikola, P. (1980). *Tropical mycorrhizal research.* Oxford, UK: Oxford University Press.

Miller, R.M. and J.D. Jastrow (1992a). The role of mycorrhizal fungi in soil conservation. In *Mycorrhizae in Sustainable Agriculture,* G.J. Bethlenfalvay and R.G. Linderman (eds.). Madison, WI: American Society of Agronomy. Special publication No. 54, pp. 29-44.

Miller, R.M. and J.D. Jastrow (1992b). The application of VA mycorrhizae to ecosystem restoration and reclamation. In *Mycorrhizal Functioning.* M.F. Allen (ed.). New York: Rutledge, Chapman and Hall, pp. 438-467.

Miranda, J.C.C. and P.J. Harris (1994). Effects of soil phosphorus on spore germination and hyphal growth of arbuscular mycorrhizal fungi. *New Phytologist.* 128:103-108.

Mohammad, A. and A.G. Khan (2002). Monoxenic *in vitro* production and colonization potential of AM fungus *Glomus intraradices. Indian Journal of Experimental Biology* 40:1087-1091.

Morandi, D. (1996). Occurrence of phytoalexins and phenolic compounds in endomycorrhizal interactions, and their potential role in biological control. *Plant and Soil* 185:241-251.

Mosse, B. (1973). Plant growth responses to vesicular-arbuscular mycorrhiza: IV. In soil given additional phosphate. *New Phytologist* 72:127-136.

Mosse, B. and D.S. Hayman (1980). Mycorrhiza in agricultural plants. In *Tropical Mycorrhizal Research,* P. Mikola (ed.). Oxford: Oxford University Press, pp. 213-230.

Mulligan, M.F., A.J.M. Smucker, and G.F. Safir (1985). Tillage modification of dry edible bean root colonization by VAM fungi. *Agronomy Journal* 77:140-144.

Muthukumar, T. and K. Udaiyan (2002). Growth and yield of cowpea as influenced by changes in arbuscular mycorrhiza in response to organic manuring. *Journal of Agronomy and Crop Science* 188:123-132.

Muthukumar, T., K. Udaiyan, and S. Manian (1996). Vesicular-arbuscular mycorrhizae in tropical sedges of southern India. *Biology and Fertility of Soils* 22:96-100.

Nadian, H., S.E. Smith, A.M. Alston, and R.S. Murray (1997). Effects of soil compactation on plant growth, phosphorus uptake and morphological characteristics of vesicular-arbuscular mycorrhizal colonization of *Trifolium subterraneum. New Phytologist* 135:303-311.

Nagahashi, G., D.D. Douds Jr., and G.D. Abney (1996). Phosphorus amendment inhibits hyphal branching of the VAM fungus *Gigaspora margarita* directly and indirectly through its effect on root exudation. *Mycorrhiza* 6:403-408.

Oehl, F., E. Sieverding, K. Ineichen, P. Mader, T. Boller, and A. Wiemken (2003). Impact of land use intensity on the species diversity of arbuscular mycorrhizal fungi in agroecosystems of Central Europe. *Applied Environmental Microbiology* 69:2816-2824.

O'Halloran, I.P., M.H. Miller, and G. Arnold (1986). Absorption of P by corn (*Zea mays* L.) as influenced by soil disturbance. *Canadian Journal of Soil Science* 66:287-302.

Olsson, P.A., I. Thingstrup, I. Jakobsen, and E. Baath (1999) Estimation of the biomass of arbuscular mycorrhizal fungi in a linseed field. *Soil Biology and Biochemistry* 31:1879-1887.

Onguene, N.A. and T.W. Kuyper (2002). Importance of the ectomycorrhizal network for seedling survival and ectomycorrhiza formation in rain forests of south Cameroon. *Mycorrhiza* 12:13-17.

Peng, S., D.M. Eissensat, and J.H. Graham (1991). Mycorrhizal citrus on soil of high phosphorus: Carbon cost and growth depression. *Bulletin of the Ecological Society of America* 72:217.

Pérez-Moreno, J. and R. Ferrera-Cerrato (1993). *Avances de Investigación*. Área de Microbiología de suelos. IRENAT-Colegio de Postgraduados. Montecillo, Mexico. 200 p.

Pérez-Moreno, J. and R. Ferrera-Cerrato (1996). *New horizons in agriculture: Aagroecology and sustainable development.* Colegio de Postgraduados, Montecillo, Mexico 435 p.

Pérez-Moreno, J. and R. Ferrera-Cerrato (1997). Mycorrhizal interactions with plants and soil organisms in sustainable agroecosystems. In *Soil Ecology in Sustainable Agricultural Systems,* L. Brussaard and R. Ferrera-Cerrato (eds.). Boca Raton, FL: CRC Lewis, pp. 91-112.

Pérez-Moreno, J., J. Alvarado-López, and R. Ferrrera-Cerrato (2002). *Producción y control de calidad de inoculantes agrícolas y forestales.* Colegio de Postgraduados. Montecillo, Mexico. 167 pp.

Perotto, S. and E. Martino (2001). Molecular and cellular mechanisms of heavy metal tolerance in mycorrhizal fungi: What perspectives for bioremediation? *Minerva Biotechnologica* 13:55-63.

Perry, D.A., M.P. Amaranthus, S.L. Borchers, and R.E. Brainerd (1989). Bootstrapping ecosystems. *Bioscience* 39:230-237.

Picone, C.M. (2000). Diversidad y abundancia de esporas de micorrizas vesiculo-arbusculares en bosques y pastos tropicales. *Biotropica* 32:734-750.

Plenchette, C. (2000). Receptiveness of some tropical soils from banana fields of Martinique to the arbuscular fungus *Glomus intraradices. Applied Soil Ecology* 15:253-260.

Pyke, E.E. (1935). Mycorrhiza in cacao. *Reports of Cacao Research Trin.* (1934). pp. 41-48.

Quiñones-Aguilar, E.E., D. Trejo, T. Aguas R., R. Ferrera-Cerrato, and M.C. González-Chávez (1998). Hongos endomicorrízicos arbusculares y diferentes sustratos en el crecimiento de plantas de papaya (*Carica papaya* L.). In *Avances de la investigación micorrízica en México,* R. Zulueta Rodríguez, M.A. Escalona Aguilar, and D. Trejo Aguilar (eds.). Xalapa, Mexico: Universidad Veracruzana, pp. 127-140.

Quiroga-Madrigal, R. (1996). Uso de leguninosas en la producción de maíz: una experiencia regional en Chiapas, México. In *Nuevos horizontes en agricultura: Agroecología y desarrollo sostenible,* J. Pérez-Moreno and R. Ferrera-Cerrato (eds.). Texcoco, Mexico: Colegio de Postgraduados, pp. 35-57.

Rashid, A., T. Ahmed, N. Ayub, and A.G. Khan (1997). Effect of forest fire on number, viability and post-fire re-establishment of arbuscular mycorrhizae. *Mycorrhiza* 7:217-220.

Redecker, D., R. Kodner, and L.E. Graham (2000). Glomalean fungi from the Ordovician. *Science* 289:1920-1921.

Redhead, J.F. (1980). Mycorrhiza in natural tropical forests. In *Tropical Mycorrhiza Research.* P. Mikola (ed.). Oxford: Clarendon Press, pp. 127-142.

Rey, C.J.S., B. Gaiska A., M. Camacho F., D. Jean C., and R. Ferrera-Cerrato (1992). Desarrollo de plantas endomicorrizadas en áreas erosionadas. In *Proceedings of the XXV Mexican Congress of Soil Science,* J.L. Tovar S. and R. Quintero L. (eds.). Acapulco, Mexico: Mexican Soil Science Society, p. 229.

Reyes, A.J.C. (2000). *Micorriza arbuscular, bacterias y vermicomposta en el desarrollo y fisiología de un portainjerto de aguacate, raza mexicana (*Persea americana *Mill.) en un sustrato alternativo de vivero.* Tesis de Maestría. Colegio de Postgraduados. Montecillo, México.

Rillig, M.C. and M.F. Allen (1999). What is the role of arbuscular mycorrhizal fungi in plant-to-ecosystem responses to elevated atmospheric CO_2? *Mycorrhiza* 9:1-8.

Rillig, M.C., S.F. Wright, and M.S. Torn (2001). Unusually large contribution of arbuscular mycorrhizal fungi to soil organic matter pools in tropical forest soils. *Plant and Soil* 233:167-177.

Rillig, M.C., S.F. Wright, and V.T. Evanir (2002). The role of arbuscular mycorrhizal fungi and glomalin in soil aggregation: Comparing effects of five plant species. *Plant and Soil* 238:325-333.

Rios-Garay, S. (1994). *Manejo de la endomicorriza V-A en plantas arbóreas para la rehabilitación de tepetate.* Thesis of Bachelor degree. Universidad Autónoma Chapingo, Chapingo, Mexico.

Rose, S.L. (1988). Above and belowground community development in a marine sand dune ecosystem. *Plant and Soil* 109:215-226.

Rufyikiri, G., S. Declerck, B. Delvaux, and J.E. Dufey (2000). Arbuscular mycorrhizal fungi might alleviate aluminium toxicity in banana plants. *New Phytologist* 148:343-352.

Ruiz-Lozano, J.M., R. Azcón, and M. Gómez (1996). Alleviation of salt stress by arbuscular mycorrhizal *Glomus* species in *Lactuca sativa* plants. *Physiologia Plantarum* 98:767-772.

Ryan, M.H. and J.H. Graham (2002). Is there a role for arbuscular mycorrhizal fungi in production agriculture? *Plant and Soil* 244:263-271.

Salawu, E.O. and R.H. Esty (1979). Observations on the relationships between a vesicular-arbuscular fungus, a fungivorous nematode, and the growth of soybeans. *Phytoprotection* 60:99-102.

Salazar-García, S. (2002). *Nutrición del aguacate, principios y aplicaciones.* INIFAP, INFOPOS, Texcoco, Mexico, pp. 36-67.

Sánchez, M.J.H., R. Ferrera-Cerrato, R., Nuñez E., and J.D. Etchevers (1986). Efecto de la micorriza VA *(Glomus macrocarpum)* sobre la eficiencia de la roca fosfórica en yuca desarrollada en un suelo ácido. *Proceedings of the XIX Mexican congress of soil science.* Manzanillo, Mexico, p. 105.

Sánchez-Colín, M.J. and V.N. Ramírez (2000). Micorriza arbuscular y *Rhizobium* presentes en leguminosas establecidas en suelo andosol. In *Ecología, fisiología y biotecnología de la micorriza arbuscular.* A. Alarcón and R. Ferrera-Cerrato (eds). Mexico: Mundi Prensa, pp. 46-55.

Sánchez-Espindola, M.E., M.C. González-Chávez, R. Ferrera-Cerrato, and D. Teliz-Ortiz (1996). Inducción del vigor en plántulas de *Carica papaya* L. bajo el efecto de la micorriza vesiculo-arbuscular *Glomus* sp. como factor de desarrollo. In *Avances de Investigación,* J. Pérez-Moreno and R. Ferrera-Cerrato (eds.). Texcoco, Mexico: Microbiología de suelos. 2nd. Ed. Colegio de Posgraduados, pp. 124-133.

Sánchez-Gallén, I. and P. Guadarrama (2000). Influencia de las micorrizas arbusculares en el crecimiento de plántulas de la selva húmeda de Los Tuxtlas, Veracruz. In *Ecología, fisiología y biotecnología de la micorriza arbuscular,* A. Alarcón and R. Ferrera-Cerrato (eds). Mexico: Mundi Prensa, pp. 70-78.

Sanii, O. (1976). Vesicular arbuscular mycorrhiza in some Nigerian soils. The effect of *Gigaspora gigantea* on the growth of rice. *New Phytologist* 77:673-674.

Schereiner, R.P., K.L. Mihara, H. McDaniell, and G.J. Bethlenfalvay (1997). Mycorrhizal fungi influence plant and soil functions and interactions. *Plant and Soil* 188:199-209.

Schimel, D.S. (1986). Carbon and nitrogen turnover in adjacent grassland and cropland ecosystem. *Biogeochemistry* 2:345-357.

Schmidt, S. and G.R. Stewart (2003). δ^{15}N values of tropical savanna and monsoon forest species reflect root specialisations and soil nitrogen status. *Oecologia* 134:569-577.

Schübler, A., D. Schwarzott, and C. Walker (2001). A new fungal phylum, the Glomeromycota phylogeny and evolution. *Mycological Research* 105:1413-1421.

Schweiger, P.F. and I. Jakobsen (2000). Laboratory and field methods for measurement of hyphal uptake of nutrients in soil. *Plant and Soil* 226:237-244.

Schwob, I., M. Ducher, and A. Coudret (1999). Effects of climatic factors on native arbuscular mycorrhizae and *Meloidogyne exigua* in a Brazilian rubber tree *(Hevea brasiliensis)* plantation. *Plant Pathology* 48:19-25.

Sieverding, E. (1985). Influence of method of VA mycorrhizal inoculum placement on the spread of root infection in field grown cassava. *Journal of Agronomy and Crop Science* 154:161-170.

Sieverding, E. (1990). Ecology of VAM fungi in tropical systems. *Agriculture Ecosystems and Environment* 29:369-390.

Sieverding, E. (1991). *Vesicular-arbuscular mycorrhiza management in tropical agroecosystems.* Deutsche Gesellschaft fur Technische Zusammenabeit. GTZ. Bremer, Alemania.

Sieverding, E. and R.H. Howeler (1985). Influence of species of VA mycorrhizal fungi on cassava yield response to phosphorus fertilization. *Plant and Soil* 88:213-222.

Sieverding, E. and D.E. Leihner (1984). Effect of herbicides on population dynamics of VA-mycorrhiza with cassava. *Angewandte Bot anik* 58:283-294.

Siqueira, J.O., J. Colozzi-Filho, and E. Oliveira (1989). Ocurrencia de mycorrizas vehículo-arbusculares em agro e ecossistemas naturais do estado de minas gerais. *Pesquisa Agropecuaria Brasileira* 24:1499-1506.

Smith, S.E. and D.J. Read (1997). *Mycorrhizal symbiosis.* 2nd Edition. Academic Press, London

Smith, M.R., I. Charvat, and R.L. Jacobson (1998). Arbuscular mycorrhizae promote establishment of prairie species in a tallgrass prairie restoration. *Canadian Journal of Botany* 76:1947-1954.

Smith, M.D., D.C. Hartnett, and C.W. Rice (2000). Effects of long-term fungicide applications on microbial properties in tallgrass prairie soil. *Soil Biology and Biochemistry* 32:935-946.

St-Arnaud, M., C. Hamel, M. Caron, and J.A. Fortin (1994). Inhibition of *Pythium ultimum in* roots and growth substrate of mycorrhizal *Tagetes patula* colonised with *Glomus intraradices. Canadian Journal of Plant Pathology* 16:187-194.

St John, T.V. (1980). Root size, root hairs and mycorrhizal infection: A reexamination of Baylis's hypothesis with tropical trees. *New Phytologist* 84:483-487.

Strickland, T.C., P. Sollins, D.S. Schimel, and E.A. Kerle (1988). Aggregation and aggregate stability in forest and range soils. *Soil Science Society of America Journal* 52:829-833.

Suresh, C.K., D.J. Bagyaraj, and D.D.R. Reddy (1985). Effect of vesicular arbuscular mycorrhiza on survival, penetration and development of root-knot nematode in tomato. *Plant and Soil* 87:305-308.

Sylvia, D.M. and S.E. Williams (1992). Vesicular-arbuscular mycorrhizae and environmental stress. In *Mycorrhizae in Sustainable Agriculture,* G.J. Bethlenfalvay and R.G. Linderman (eds.). Madison, WI: American Society of Agronomy. Special publication No. 54, pp. 101-124.

Taylor, T.N., W. Remy, H. Hass, and H. Kerp (1995). Fossil arbuscular mycorrhizae from the Early Devonian. *Mycologia* 87:560-573.

Theodorou, C. and G.D. Bowen (1970). Mycorrhizal responses of radiata pine in experiments with different fungi. *Australian Forestry* 34:183-191.

Thompson, J.P. and G.B. Wildermuth (1988). Colonization of crop and pasture species with vesicular-arbuscular mycorrhizal fungi and a negative correlation with root infection by *Bipolaris sorokiniana. Canadian Journal of Botany* 69:687-693.

Thomas, R.S., S. Dakessian, R.N. Ames, M.S. Brown, and G.J. Bethlenfalvay (1986). Aggregation of a silty clay loam by mycorrhizal onion roots. *Soil Science Society of America Journal* 50:1494-1498.

Thomas, R.S., R.L. Franson, and G.J. Bethlenfalvay (1993). Separation of vesicular-arbuscular mycorrhizal fungus and root effects on soil aggregation. *Soil Science Society of America Journal* 57:77-81.

Tisdall, J.M. (1991). Fungal hyphae and structural stability of soil. *Australian Journal of Soil Research* 29:729-743.

Tisdall, J.M. and J.M. Oades (1982). Organic matter and water stable aggregates in soils. *Journal of Soil Science* 33:141-163.
Trappe, J.M. (1977). Selection of fungi for ectomycorrhizal inoculation in nurseries. *Annual Review of Phytopathology* 15:203-222.
Trejo, D., G. Solis, M. Escalona, and R. Ferrera-Cerrato (2000). Estudio preliminar del efecto de la asociación micorrízica en café en condiciones de campo. In *Ecología, fisiología y biotecnología de la micorriza arbuscular,* A. Alarcón and R. Ferrera-Cerrato (eds.) Mexico: Mundi Prensa, pp. 206-212.
Trejo-Aguilar, D. (1997). *Ecología y comportamiento de la endomicoriza arbuscular en el cultivo de café (*Coffea arabica *L.).* Tesis de Maestría en Ciencias. Universidad Nacional Autónoma de México. Mexico City, Mexico. 130 p.
Treub, M. (1885). Onderzoekingen over Sereh-zick Suikerriet. *Meded. Pl. Tuin. Batavia II.*
Van Duin, W.E., W.A.J. Griffioen, and J.H. Ietswaart (1991). Occurrence and function of mycorrhiza in environmentally stressed soils. In *Ecological Responses to Environmental Stresses,* J. Rozema and J.A.C. Verkleij (eds.) The Netherlands: Kluwer Academic Publishers, pp. 114-123.
Vejsadová, H. (1992). The influence of organic and inorganic fertilization on development of indigenous VA fungi in roots of red clover. In *Mycorrhizas in Ecosystems,* D.J. Read, D.H. Lewis, A.H. Fitter, and I.J. Alexander (eds.). Wallingford, Oxon: C.A.B. International, pp. 406-407.
Vivekanandan, M. and P.E. Fixen (1991). Cropping systems effects on mycorrhizal colonization, early growth, and phosphorus uptake of corn. *Soil Science Society of America Journal* 55:136-140.
Vozzo, J.A. and E. Hacskaylo (1971). Inoculation of *Pinus caribaea* with ectomycorrhizal fungi in Puerto Rico. *Forest Science* 17:239-245.
Waceke, J.W., S.W. Waudo, and R. Sikora (2001). Response of *Meloidogyne hapla* to mycorrhiza fungi inoculation on pyrethrum. *African Journal of Science and Technology* (AJST) Science and Engineering Series 2:63-70.
Waidyanatha, U.P.S. (1980). Mycorrhizae of *Hevea* and leguminous ground covers in rubber plantations. In *Tropical Mycorrhiza Research,* P. Mikola (ed.). Oxford: Oxford University Press, pp. 238-241.
White, R.E. (1985). The influence of macropores on the transport of dissolved and suspended matter through soil. *Advances in Soil Science* 3:95-120.
Wright, S.F. (2000). A fluorescent antibody assay for hyphae and glomalin from arbuscular mycorrhizal fungi. *Plant and Soil* 226:171-177.
Wright, S. (2004). Role of glomalin in soils. In *Proceedings of the IV Mexican Syposium of Mycorrhizal Symbiosis,* M.C. González-Chávez, J. Pérez-Moreno, R. Ferrera-Cerrato, M.P. Ortega-Larrocea, V. Carreón-Abud and E. Valencia-Cantero (eds.). Morelia, Mexico: Colegio de Postgraduados, p. 97.
Wright, S.F. and A. Upadhyaya (1996). Extraction of an abundant and unusual protein from soil and comparison with hyphal protein of arbuscular mycorrhizal fungi. *Soil Science* 161:575-586.
Wright, S.F. and A. Upadhyaya (1998). A survey of soils for aggregate stability and glomalin, a glycoprotein produced by hyphae of arbuscular mycorrhizal fungi. *Plant and Soil* 198:97-107.

Wright, S.F. and A. Upadhyaya (1999). Quantification of arbuscular mycorrhizal fungi activity by the glomalin concentration on hyphal traps. *Mycorrhiza* 8: 283-285.

Wright, S.F. and R.L. Anderson (2000). Aggregate stability and glomalin in alternative crop rotation for the central Great Plains. *Biology and Fertility of Soils* 31:249-253.

Wright, S.F., A. Upadhyaya, and J.S. Buyer (1998). Comparison of n-linked oligosaccharides of glomalin from arbuscular mycorrhizal fungi and soils by capillary electrophoresis. *Soil Biology and Biochemistry* 30:1853-1857.

Wright, S.F., J.L. Starr, and I.C. Paltineanu. (1999). Changes in aggregate stability and concentration of glomalin during tillage management transition. *Soil Science Society of America Journal* 63:1852-1859.

Wright, S.F., M. Franke-Snyder, J.B. Morton, and A. Upadhyaya (1996). Time course study and partial characterization of a protein on hyphae of arbuscular mycorrhizal fungi during active colonization of roots. *Plant and Soil* 181:193-203.

Yang, W.Q. and B.L. Goulart (2000). Mycorrhizal infection reduces short-term aluminum uptake and increases root cation exchange capacity of highbush blueberry plants. *HortScience* 35:1083-1086.

Yano-Melo, A.M., O.J. Saggin Jr., J.M. Lima-Filho, N.F. Melo, and L.C. Maia (1999). Effect of arbuscular mycorrhizal fungi on the acclimatization of micropropagated banana plantlets. *Mycorrhiza* 9:119-123.

Zulueta, R., M. Alejandro, M. Escalona, D. Trejo, and L. Lara (2000). Respuesta de dos especies forestales tropicales a la inoculación micorrízica. In *Ecología, fisiología y biotecnología de la micorriza arbuscular,* A. Alarcón and R. Ferrera-Cerrato (eds.). Mexico: Mundi Prensa, pp. 184-193.

Chapter 7

Mycorrhizae in Indian Agriculture

Seema Sharma
Deepak Pant
Sujan Singh
Radha R. Sinha
Alok Adholeya

Arbuscular mycorrhizal (AM) fungi by virtue of their ubiquity and their key role as biofertilizers, bioregulators, and bioprotectors are essential components of the soil biota (Dodds, 2000). The production of many agricultural, horticultural, and fruit crops in soil are dependent on the formation of AM fungi, making this symbiosis an essential factor in low-input sustainable agriculture. AM fungi are commonly described as "the universal plant symbiont," and are found in practically every taxonomic group of plants. About 80 percent of all terrestrial plant species form this type of symbiosis. Widespread distribution in habitats and host species, symbiotic existence, enhancement of host growth and protection, obligate nature and specificity for hosts, and positive and negative interaction with other rhizosphere microbes are a few characteristics of AM fungi that lead to the study of these microorganisms. Arbuscular mycorrhizal fungi are widely distributed and usually abundant in soil (Mukerji and Rani, 1989). Their population varies greatly in size and composition according to habitat, and type of soil and vegetation. Because of their beneficial effect on plant growth, AM fungi are important for natural and managed ecosystems (Brundrett et al., 1996).

Mycorrhizae in Crop Production
© 2007 by The Haworth Press, Inc. All rights reserved.
doi:10.1300/5425_07

With the development of isolation techniques, mass production methods, and inoculations, AM fungi have been a boon for agriculture, forestry, and the restoration of disturbed ecosystems (Quilambo, 2003). The diversity in soil and climate and the vast cultivable land in India provide tremendous scope for conducting a range of research focused on mycorrhiza. Scientists in India extensively studied the AM fungi association in a great variety of plant species grown and cultivated in India, including cereals, pulses, fruits, vegetables, and other beneficial crops.

This chapter aims to summarize the work done by Indian researchers in identifying the benefits of AM fungi to Indian agriculture and the role of these fungi in environmental protection. The viability and efficacy of AM fungi and their role in Indian agriculture has been studied taking into account various factors. This information is presented within the following six broad topics: fungal diversity; inocula and inoculation; crop management; environmental influences; effects of pollution; and microbial interactions, including symbiotic nitrogen fixers and pathogens.

MYCORRHIZAL DIVERSITY AND EFFICACY IN INDIAN AGRICULTURE

Arbuscular mycorrhizal associations have been found in the majority of agricultural crops in India. Arbuscular mycorrhizal fungi are present in the soil as extraradical hyphae and resting spores, and as intraradical mycelium in colonized roots. Extensive surveys have been carried out to explore the AM fungi diversity associated with the main agricultural crops. Out of ten known genera, *Glomus* is the most frequently encountered genus. The large number of species isolated are included in the genus *Glomus*. Among these, *G. fasciculatum* and *G. macrocarpum* are the dominant species, with a frequency of occurrence of 46 and 40 percent, respectively (Mukerji and Rani, 1989).

Geographically, India can be divided into six ecological zones in which some 288 AM fungal species have been reported. The highest number of mycorrhizal species diversity is found in the forests, followed by agricultural soils. *Glomus* is the genus most commonly occurring in forest ecosystems. *Casuarina equisetifolia,* a species of tree, was associated with 35 species of mycorrhizal fungi. The indigenous *Dalbergia sissoo* hosted all the genera of mycorrhizal fungi (Verma et al., 1998). Generally, when a particular plant species/variety/cultivar/genotype produces greater biomass in association with specific AM fungi than in its absence, under a given set of conditions, it is considered to be mycorrhiza dependent (Singh, 2001) (Table 7.1). Furthermore, to achieve the best yield results, it is also essential to know the soil phosphorus (P) level at which maximum mycorrhizae-derived benefit may be provided to a plant. Generally, the efficiency of mycorrhizal infection in a plant is inversely proportional to the level of phosphorus in the soil. This section deals with the diversity of AM fungi and the selection of efficient mycorrhizal fungi for a variety of agricultural plants.

Throughout India, various agricultural crops–AM fungal species combinations have been tested in field and greenhouse experiments. The agricultural crops tested include cereals, pulses, vegetables, and oil seed crops, all of which, with mycorrhizal partners, showed increases in yield and decreases in phosphate fertilizer requirement. Some crops showing enhanced production after mycorrhizal inoculation are shown in Figure 7.1.

Cereal Crops

Alfalfa (Medicago sativa). The biomass of alfalfa increased by 20 percent after inoculation with a native AM fungi consortium in a low phosphorus sandy loam field soil (Subhashini, Rana, and Potty, 1988; Gaur and Adholeya, 2002).

Berseem (Trifolium alexandrium). Berseem showed a mycorrhizal dependency of 72 percent in a low phosphorus sandy loam field soil. Inoculation with a native AM fungi consortium increased its biomass by 57 percent, (Subhashini, Rana, and Potty, 1998; Gaur and Adholeya, 2002).

Finger millet (Eleusine coracana). Inoculation of finger millet with *G. caledonium* increased its root colonization and plant biomass compared to those of control plants in the greenhouse. Mycorrhizal inoculation also increased plant shoot mineral content and uptake of phosphate, nitrogen, zinc, and copper (Tewari, Johri, and Tandon, 1993; Krishna et al., 1985).

TABLE 7.1. List of efficient AM fungi for different agricultural crops.

Plant species	Efficient mycorrhizal fungus	Mycorrhizal fungi tested	Parameters evaluated	References
Allium cepa	*Gigaspora margarita*	*G. margarita, G. calospora*	Leaf number, dry matter, P-uptake, bulb yield	Ramana and Babu, 1999
	Scutellospora calospora	*G. margarita, S. calospora*	Root colonization	Ramana and Babu, 1999
Amorphophallus paeonufolius		*Glomus mosseae, G. aggregatum, Gigaspora albida and Pisolithus tinctorius*	Tuber yield/plant	Ganeshan and Mahadevan, 1994
Arachis hypogeae	*Glomus fasciculatum*	*G. fasciculatum* and seven other VAM fungi	Root-shoot length, dry weight, P-content and root colonization	
Capsicum	*Glomus intraradices* AM004	*G. intraradices*, indigenous mixed culture, commercial inoculum (Mycorise & C)	Fruit yield	Gaur et al., 1998
Colocasia esculenta	*Glomus mosseae* + *G. aggregatum*	*G. mosseae, G. aggregatum, Gigaspora albida*	Tuber yield/plant	Ganeshan and Mahadevan, 1994
Cyamopsis tetragonoloba	*Glomus fasciculatum*	*G. fasciculatum, G. mosseae, Acaulospora morrowae, G. constrictum, Glomus rubiforme, Scutellospora calospora*	Protein and sugar contents in pods	Mathur and Vyas, 1995

Eleusine coracana	*Glomus caledonium*	*Glomus caledonium*, *G. fasciculatum*, *G. mosseae*, *G. epigaeum* (*G. versiforme*), *Gigaspora calospora*, *G. margarita*	Mycorrhizal efficacy and root colonization	Tewari et al., 1993
Manihot esculenta	*Glomus fasciculatum*	*G. fasciculatum*, *G. mosseae*, *G. constrictum*, *G. etunicatum*, *Acaulospora morrowae*	Root colonization, plant weight and shoot and root dry weight	Prasad et al., 1990
Manihot esculenta	*G. mossae* + *G. aggregatum*	*G. mosseae*, *G. aggregatum*, and *Gigaspora albida*	Tuber yield/plant	Ganeshan and Mahadevan, 1994
Oryza sativa cv. Prakash	*Glomus intraradices*	*Glomus intraradices*, *G. fasciculatum*	Grain yield	Secilia and Bagyaraj, 1994
Oryza sativa (*upland rice*)	*Acaulospora spinosa*	*Acaulospora spinosa*, *A. scrobiculata*	Plant biomass, grain yield and root colonization	Ammani and Rao, 1996
Polianthes tuberose	*Glomus intraradices* AM004	*G. intraradices*, indigenous mixed cultures, commercial inoculum (*Mycorise*)	Fruit yield and spike length	Gaur et al., 1998
Vigna mungo	*Glomus epigaeum* (=*G. versiforme*)	*G. epigaeum*, *Acaulospora spinosa*, *A. morrowae*	Root colonization, shoot dry weight, N and P concentrations	Rao and Rao, 1996
Vigna radiata	*Glomus epigaeum* (=*G. versiforme*)	*Glomus epigaeum*, *Acaulospora spinosa*, *A. morroweae*	Root colonization, shoot dry weight, N and P concentrations	Rao and Rao, 1996

FIGURE 7.1. Yield enhancement achieved by (a) 30 percent in *Glycine max* (b) 15 percent in *Cajanus cajan* and (c) wheat grown on raised beds.

Maize (Zea mays). Inoculation with an AM fungi consortium increased maize shoot biomass by 28 percent in a non-sterile, P-deficient, sandy loam field soil amended with organic matter. Root colonization reached 54, 50, and 58 percent following inoculation with *G. margarita, G. fasciculatum,* and the AM fungi consortium, respectively (Singh, Singh, and Johri, 2002).

Oat (Avena sativa). Oat increased by 50 percent in biomass upon inoculation with a native AM fungi consortium in a low phosphorus sandy loam field soil (Subhashini, Rana, and Potty, 1998; Gaur and Adholeya, 2002).

Rice (Oryza sativa). Sondergaard and Lindegaard (1977) first reported the presence of mycorrhizae in aquatic plants. Since then mycorrhizae have been found in aquatic macrophytes colonizing lakes and streams (Beck-Nilsen, and Madsen, 2001). At the Ganges river, mangroves and other plants known as non-mycotrophic under natural conditions were found to be mycorrhizal (Sengupta and Chaudhuri, 2002). Most of the studies in India on mycorrhizal association in rice have been conducted on upland rice. Inoculation of cv. Prakash with *G. fasciculatum* and *G. intraradices* caused 8 and 11 percent increase respectively in yield under field conditions compared to the uninoculated plants. Further, the quantity of phosphate fertilizer usually applied to the rice crop was reduced by 50 percent, without affecting the grain yield. The low phosphorus tolerant variety RCPL (101) responded to mycorrhiza up to a soil phosphorus level of 20 kg ha^{-1}, whereas the susceptible variety RCPL (104) responded to a level of up to 80 kg P ha^{-1}. The variety requiring high phosphorus levels depended on AM fungi at higher soil phosphorus levels. In upland rice, *Glomus* species are more efficient in colonizing roots, forming vesicles and spores, and increase plant biomass and grain yield when compared to *Acaulospora* species. Within the *Acaulospora, A. spinosa* was more effective than *A. scrobiculata*. AM fungal association has also been reported in the wetland rice "Kranti," which yielded more with mycorrhizal association (Gupta and Ali, 1993; Secilia and Bagyaraj, 1994; Singh and Mishra, 1995).

Sorghum (Sorghum vulgare). In a low phosphorus sandy loam field soil, inoculation with a native AM fungi consortium increased the biomass of sorghum by 6 percent; sorghum mycorrhizal dependency was 5.7 percent (Subhashini, Rana, and Potty, 1998; Gaur and Adholeya, 2002). Root AM colonization varied from 25 percent to 95 percent in various cultivars (PD-3-1-11, PVR-10, M-148, and RS-71).

Wheat (Triticum aestivum). Wheat cultivars and varieties have a differential response to mycorrhizal infection. Singh and Adholeya (2002a,b,c; 2004) studied the symbiotic performance of modern, high yielding wheat cultivars; ancient landraces; varieties including the six modern hexaploid wheat varieties of *Triticum aestivum* (UP2338, HW2004, HD2687, PBW343, HW2045, and Kundan), two tetraploid wheats *(T. dicoccoides* and *T. turgidum),* and five diploid wheats *(Aegilops squarrosa, A. speltoides, A. sharonensis, T. uratu,* and *T. monococcum).* These were tested against *G. caledonium* (AM WG 08), *G. etunicatum* (AM WG 14), *G. intraradices* (AM WG 19), *G. mosseae* (AM WG 23), and *Scutellospora calospora* (AM WS 26).

They reported that the old hexaploid wheat landraces and tetraploid cultivars showed higher growth response toward mycorrhizal symbiosis than did modern cultivars.

Oilseed Crops

Coconut (Cocos nucifera). Seventeen tall and dwarf cultivars and four hybrids of coconut showed 56.8 to 95.2 percent root colonization respectively; higher root colonization was generally concurrent with abundant sporulation (Thomas and Ghai, 1987).

Groundnut (Arachis hypogeae). In greenhouse-grown groundnut crops *(Arachis hypogeae), G. fasciculatum* lead to the greatest root and shoot length, dry weight, phosphorus content, and percentage of root colonization compared to eight indigenous AM fungal species in nonphosphorus fertilized soil (Vijaykumar and Bhiravamurthy, 1999).

Groundnut, soybean, coconut, and sunflower are four important oilseed crops in India. These crops responded well to mycorrhizal inoculation and confirmed that the use of AM fungi as biofertilizer as the potential to increase oilseed production (Manoharachary, Sulochana, and Ramaro, 1990).

Soybean (Glycine max). Inoculation of pot-grown soybean with *G. margarita* in a phosphorus deficient soil to which the equivalent of 80 kg of phosphorus fertilizer ha^{-1} was added resulted in a higher seed yield than a crop without phosphorus fertilization, with or without *G. margarita* (Lingaraju, Srinivasa, and Bablad, 1994). This result shows that AM symbiosis requires enough soil-available P for optimum performance.

Sunflower (Helianthus annuus). *G. fasciculatum* was the most efficient AM species on sunflower crops. Inoculation with this fungus produced root colonization, in the range of 74.6 percent to 88 percent. Greenhouse-grown sunflower plants inoculated with *G. fasciculatum* and given 38 kg of phosphorus fertilizer ha^{-1} produced a yield similar to uninoculated plants supplied with 75 kg of phosphorus fertilizer ha^{-1} (Chandrashekars, Patil, and Sreenivasa, 1995).

Pulses

Pulses are important crops in India. They rely on symbiotic rhizobia for their nitrogen nutrition, but their phosphorus and moisture requirements are fulfilled partially by mycorrhizal fungi (Gautam and Mahmood, 2002a) (Figure 7.1b).

Chickpea (Cicer arietinum). Studies on chickpea crops ("Avrodhi" variety) inoculated with various AM fungi *(G. mosseae, G. aggregatum, G. constrictum, G. fasciculatum, Gigaspora gigantea,* and *Acaulospora scrobiculata)* in greenhouses showed that AM inoculation improved not only mycorrhizal colonization of roots but also nodulation, plant biomass, and nutrient status (Adholeya, 1988). *G. fasciculatum* increased plant nitrogen, phosphorus, and potassium most effectively, while *G. mosseae* was the species with the second highest degree of performance (Gautam and Mahmood, 2002a,b).

Black gram (Vigna mungo) *and green gram* (Vigna radiata). Rao and Rao (1996) inoculated black gram and green gram crops with *Acaulospora spinosa, A. morroweae,* and *G. versiforme.* They reported root colonization levels ranging from 68 to 89 percent for black gram, and from 69 to 93 percent for green gram. Colonization in both species was lowest in pots inoculated with *A. morroweae* + superphosphate and highest with *G. versiforme* without phosphorus.

Vegetable Crops

Various studies in India have revealed that many vegetables are highly responsive to mycorrhizal infection.

Amorphophallus (Amorphophallus baionifolius). The tuber yield of Amorphophallus was increased per plant by inoculation with a mix of *G. mosseae* and *G. aggregatum* (Ganesan and Mahadevan, 1994).

Brinjal (Solanum writti). Brinjal genotypes PS-8, Dorli, Pragati, and Borgaon 1 were more responsive to inoculation with *G. fasciculatum* than Vaishali, Manjari, Gota, Annamali, Krishnakathi, and PP long, as tested by Indi, Konde, and Sonar (1990) in pot culture.

Carrot (Daucus carrota). *G. mosseae* inoculated on two carrot varieties, a local carrot and the SKG cultivar, increased shoot and root dry and fresh weight. The yield of the local variety was 270 percent greater than that of uninoculated control plants and the yield of the SKG cultivar was 42 percent greater (Reddy and Gudige, 1999).

Cassava (Manihot esculenta). A *G. mosseae* and *G. aggregatum* inoculum mix increased tuber yield per cassava plant (Ganesan and

Mahadevan, 1994). In alluvial soils, cassava inoculated with AM and fertilized with 50 kg P fertilizer ha^{-1} had similar dry matter production than uninoculated cassava fertilized with 75 kg P fertilizer ha^{-1}. The phosphorus content of cassava recorded for 100 kg P fertilizer/ha with and without AM fungi was similar to 75 kg P fertilizer/ha with AM fungi (Tholkappian, Sivasaravanan, and Sundaram, 2000).

Coco yam (Colocasia esculenta). Coco yam crops inoculated with a mix of *G. mosseae* and *G. aggregatum* had higher tuber yield per plant than those without inoculation (Ganesan and Mahadevan, 1994). Coco yam inoculated with *G. fasciculatum* had higher shoot and root dry weight, plant height, and AM fungi infection than coco yam inoculated with other AM fungi such as *G. mosseae, G. constrictum, G. etunicatum,* or *A. morroweae* (Tholkappian, Sivasaravanan, and Sundaram, 2000).

Garlic (Allium sativum). Kunwar, Reddy, and Manoharachary (1999) defined the AM fungal population associated with garlic. Thirty-five AM fungal species were isolated from rhizosphere soils and identified. There were 11 species of *Acaulospora,* 1 species of *Entrophospora,* 4 species of *Gigaspora,* 16 species of *Glomus,* and 3 species of *Scutellospora,* reflecting the mycorrhizal dependency of the plant. Four garlic varieties, namely Pusa white flat, Pusa white round, Early grano, and Pusa madhvi, were inoculated with a mix of indigenous AM species or left uninoculated, and grown in phosphorus deficient alfisol at two phosphorus levels (25 kg P ha^{-1} and 50 kg P ha^{-1}) (Sharma and Adholeya, 2000). At harvest, all the inoculated garlic varieties had larger bulb diameter, shoot dry weight, shoot phosphorus content and bulb yield than did uninoculated plants. Early grano and Pusa white flat plants exhibited maximum mycorrhizal dependency at 50 kg P ha^{-1} whereas Pusa madhvi and Pusa white round plants exhibited maximum mycorrhizal dependency at 25 kg P ha^{-1}.

Pepper (Capsicum annuum). Studies conducted with six pepper cultivars reported a good response to inoculation with *G. macrocarpum* (Sreenivasa and Gaddagimath, 1993). The Byadagi cultivar had the highest percentages of root colonization and spore counts while the G-3 cultivar had the lowest level of colonization. Similar trends were observed on shoot phosphorus concentration, plant dry mass, and yield. Gaur, Adholeya, and Mukerji (1998) assessed various inoculum formulations (soil based, soil bead, sheared root, and fly ash beads) of mixed indigenous species and *G. intraradices* on pepper and tuberose *(Polianthes tuberosa).* The soil based and soil beads inocula produced the highest response, and *G. intraradices* produced the best yield in both crops (tuberose showed a 5 percent increase in spike length and pepper showed a 112 percent increase in fruit yield).

Tomato (Lycopersicon esculentum). Gaur, Sharma, and Adholeya. (2003) tested the theory that multiple-strain inocula may be more effective than single-strain inocula on tomato crops. Indeed, inoculation with a mix of *G. constrictum, G. fasciculatum,* and *G. mosseae* produced the best growth stimulation in greenhouse-grown tomatoes. A mix of *G. mosseae* and *G. constrictum* constituted the best two-species inoculum, followed by a mix of *G. constrictum* and *G. fasciculatum,* and a mix of *G. mosseae* and *G. fasciculatum. G. mosseae* produced the largest plant growth stimulation and spore count of single-strain inocula, followed by *G. constrictum* and *G. fasciculatum* (Iqbal and Mahmood, 1998).

Micropropagated Plants

Mycorrhizal associations enhance the rooting and promote the growth of edible-fruit plants raised from cuttings or by tissue culture.

Banana (Musa Xparadisiaca) and Alocasia. Tissue culture plantlets of banana and *Alocasia* species inoculated with native *G. fasciculatum* and *G. etunicatum* had better growth, vigor, and biomass production during acclimatization. These fungi improved the establishment of *Alocasia* from 73.6 to 100 percent and of banana from 68.5 to 92 percent, and considerably increased plant phosphorus and zinc content (Thaper and Fasrai, 2002).The shoot length of mycorrhizal plantlets increased by more than 83 percent and shoot weight by 85 percent over those of non-mycorrhizal plants. Root weight increased by 115 percent and root length by 17.14 percent. The leaves of mycorrhizal plants were also longer by 53 percent and wider by 63 percent, had a 159 percent larger area, and a chlorophyll content increase of 42 percent compared to those of control plants. The benefits of AM fungi in field-grown banana are well-known: 10,000 producers used a commercial *G. intraradices* inoculant.

Dracaena sp. Inoculation of micropropagated *Dracaena* with the multiple AM strain formulation had a beneficial influence at the acclimatization and weaning stages, and saved almost five days in the *Dracaena* hardening process (Gaur and Adholeya, 2000a).

Lilium sp. The effect of three AM inocula with respective available phosphorus levels increased the growth, flowering, phosphorus uptake, and root colonization of micropropagated lily bulblets (Varshney et al., 2002).

Nephthytis (Syngonium podophyllum) *and* Dracaena *sp.* Inoculation with a multiple AM strain formulation had a significantly favorable effect on the acclimatization of micropropagated plantlets at the weaning stage, reducing the total hardening process by almost 15 days (Gaur and Adholeya, 2000a).

Strawberries (Fragaria × ananassa). Inoculation of strawberry plants with a mixture of indigenous AM fungi, mainly *Glomus, Gigaspora,* and *Scutellospora* species, resulted in higher fruit yield, fruit mass, number of runners, shoot dry matter, and shoot phosphorus content compared to those of uninoculated plants (Sharma and Adholeya, 2004). Inoculation relieved the negative effects of alfisols in semiarid India, where low soil organic carbon and phosphorus levels combined with high fixation power drastically affects the availability of phosphorus (Adholeya and Cheema, 1990).

Medicinal Plants

There is a dearth of reports on mycorrhizal colonization and its role in essential medicinal plants in India (Selvaraj, Murugan, and Bhaskaran, 2001).

Acacia nilotica. Mycorrhizal dependency was negatively correlated with soil phosphorus levels (Sharma, Bhatia, and Adholeya, 2001).

Albizzia lebbeck. Mycorrhizal dependency was negatively correlated with soil phosphorus levels (Sharma, Bhatia, and Adholeya, 2001).

Chanca piedra (Phyllanthus fraternus *Webster synonym:* Phyllanthus niruri *L.*). This is an annual herbaceous plant with very high medicinal values in the Ayurvedic system. Earlier reports had shown that the species of *Phyllanthus* such as *Phyllanthus madraspatensis, Phyllanthus simplex,* and *Phyllanthus niruri* were non-mycorrhizal plants, but recent studies on *P. fraternus* (Mulani, Prabhu, and Dinkaran, 2002) showed 75 to 80 percent mycorrhizal colonization of roots.

Indian borage (Plectranthus amboinicus). This is an important medicinal plant used for the treatment of urinary diseases, epilepsy, chronic asthma, cough, bronchitis, and malarial fever. It also acts as a powerful aromatic carminative. Hemalatha and Selvaraj (2003) screened eight AM fungi on Indian borage. Inoculation with *G. aggregatum* produced larger plant biomass, percentage of root colonization, number of spores, and enhanced plant uptake of phosphorus and potassium more so than the other AM fungi tested.

Kashini (Cichorium entybus) is an important medicinal plant commercially cultivated in the Vellore district of Tamil Nadu, India. Kashini leaves and roots are used in the treatment of stomach ailments, as a diuretic, and to treat fever, vomiting and joint pain. Kashini also helps in the enrichment and purification of blood. Murugan and Selvaraj (2003) screened eight native AM fungi on Kashini for symbiotic response. They reported that *G. margarita* best stimulated plant growth, followed by *G. aggregatum, G. mosseae,* and *G. macrocarpum.*

Other Plants

Ornamentals. Petunia *(Petunia hybrida),* China aster *(Callistephus chinensis),* and balsam *(Impatiens balsamina)* were inoculated with a mix of indigenous AM fungi and grown in marginal wasteland amended with organic matter. Mycorrhizal inoculation improved flower number and dry matter markedly in all three plants. Petunia plants had three times more flowers when mycorrhizal as when not, while balsam and China aster had twice as many (Gaur and Adholeya, 2000d).

Sugarcane *(Saccharum officinarum)* CO 419 variety was inoculated with one of four AM fungal species *(G. fasciculatum, G. mosseae, G. margarita,* or *Acaulospora laevis). G. margarita* and *A. laevis* improved plant root and shoot biomass, plant height, and leaf area more effectively than the other AM fungal inocula. The intensity of mycorrhizal root colonization was almost 100 percent with all the inoculants (Reddy et al., 2004).

INOCULUM PRODUCTION AND APPLICATION

An inoculum is a material that carries infective propagules in a usable form, that is, packaged AM fungal spores and hyphae, and colonized root fragments. Inocula can be produced using various methods such as soil based inoculum production (Gaur and Adholeya, 2000c), aeroponic culture, the nutrient film technique, and in vitro culture (Adholeya, Verma, and Bhatia, 1997; Tiwari and Adholeya, 2001). In India, a soil-based inoculum is commonly produced directly in the nurseries or fields of the crops that are to be inoculated. The fact that a soil-based inoculum is bulky and heavy, which may cause problems with transport and commercial distribution, makes in vitro culture a very desirable production method.

An inoculum or inoculant product gives the best results where indigenous soil populations of AM fungi are low or where native AM fungi are no longer effective. It is particularly important to inoculate plants when out-planting to nonirrigated sites, wastelands, or disturbed soils where the establishment of plants' root systems is impaired. The efficiency of inoculum or inoculant products varies with products, environmental conditions, delivery methods (broadcasting, seedcoating, root dipping, and in-furrow application), and a number of other variables. The amount of inoculum to apply, that is, the number of AM fungal propagules per plant or per area, is calculated based on various parameters such as the following: (1) the weight or volume of the material; (2) the concentration of AM fungal propagules in the material; (3) where the inoculum is applied, that is, to the seed or to the soil; (4) the extent of adhesion of the product to the seed; and (5) the seeding rate (Adholeya, Tiwari, and Singh, 2005).

The application of AM fungi in Indian agriculture is limited by the availability of bulk quantities of AM fungi inoculum. The mass production of AM fungal inoculum through the in vitro culture technique by The Energy and Resources Institute (TERI) overcame the major part of this problem. Efforts for the conservation and multiplication of strains started with a culture depository set up in 1993 at TERI, with the support of the Department of Biotechnology of the Government of India (DBT, 1993-1997). It now maintains Asia's largest germplasm bank, the Center for Mycorrhizal Culture Collection (CMCC), that contains over 450 isolates of mycorrhizal fungi from 12 out of 20 of the agroecological zones of India (Gadkar and Adholeya, 2000; Tiwari and Adholeya, 2001, 2003). TERI has transferred this technology to four companies for the mass production of inoculum for large-scale application by producers, rehabilitation agencies, and other agro industries. Two are agro-based companies (KCP Sugar and Industries Limited, Andhra Pradesh, and Majestic Agronomics, Himachal Pradesh), and two are pharmaceutical companies (Cadilla Pharmaceuticals Limited, Gujrat, and Cosme Pharma, Goa).

In India, the centralized production of AM fungal inocula is feasible. At this time, however, the business is at the concept selling stage, and product penetration at the regional level and demonstrations must be done. Once the product is tested in various agro-climatic zones and regions of the country and its potential proven, consolidation may occur.

EFFECT OF MANAGEMENT PRACTICES ON AM FUNGI

AM fungi play an important role in ecosystems and they could be managed for economic or environmental purposes. In spite of considerable research on the large-scale production of AM fungi, field-scale inoculation has been limited. Currently, the commercial use of AM fungi is limited to special locations where the natural population has been destroyed. Though it is possible to manipulate the indigenous populations of AM fungi by means of appropriate crop and soil management practices, little systematic work has been done in this area in India. The following section deals with the effects of management practices such as tillage, cropping patterns, organic amendments, and the application of fertilizers and pesticides on the mycorrhizal association of crop plants. Thereafter, management practices that affect mycorrhizal associations and crop production are identified.

Tillage Practices

Tillage operations disturb the soil and change its physical, chemical, and biological properties, thereby affecting AM fungal colonization (Table 7.2). In a study by Maiti, Variar, and Singh (1996) on the effect of three tillage practices on rice crops, namely, conventional tillage (CT, soil disturbed up to 5 to 10 cm depth twice consecutively prior to seeding), deep tillage (DT, soil disturbed up to 20 to 25 cm depth), and no tillage (NT), the highest population of infective propagule and percentage of colonization was obtained in CT. In DT, the drastic reduction in the population and efficacy was attributed to the disruption of the mycelial network in the soil, as evidenced by reduced percentage of root colonization. In NT, the mycelial network established with the roots of predominant weed species was affected by the destruction of these hosts with herbicide. The build-up phase of the AM fungal population was adversely affected by the low density of roots for colonization and multiplication of the AM fungi. Given the minimum soil disturbance of shallow ploughing depth in CT, however, there was minimum disruption of the established mycelial network. The minimum disruption resulted in a comparatively more steady buildup of the native AM fungal population in the soil over a period of time (five years in this study), which in turn increased the percentage of root colonization (Maiti et al., 1997).

One of the newly emerging issues related to increasing mechanization in the rice-wheat areas of India is the management of straw after combine harvesting of rice. Singh and Adholeya (2004) studied the impact of rice straw mulches and burnt straw on the AM fungal population in the NT fields. They reported a higher microbial and AM fungal spore count in the straw mulched fields than in the burned fields. No significant difference in pH and electrical conductivity was ob-

TABLE 7.2. Effect of different tillage practices on AM fungi and plant growth.

Management practices	Crops used	Effect on crops	Effect on mycorrhizal fungi	References
Conventional tillage	Rice	+ve	+ve	Maiti et al., 2000
Deep tillage	Rice	−ve	−ve	Maiti et al., 2000
No tillage	Rice	−ve	−ve	Maiti et al., 1996
Mulching	Wheat	+ve	+ve	Singh and Adholeya, 2004
Residue burned	Wheat	−ve	−ve	Singh and Adholeya, 2004

served between fields. However, organic carbon, nitrogen, phosphorus, and potassium contents were significantly higher in the straw mulched fields. Thus, this study suggests that straw mulched fields are better for soil microbes and soil fertility.

Cropping Pattern

Crop management can affect indigenous AM fungal populations positively or negatively, as these fungi are obligate symbionts. Crop species differ in their ability to form AM associations and the inclusion of appropriate plant species in crop rotation, or intercropping, can increase the native population of AM fungi (Table 7.3), which often is the objective of inoculation Harinikumar and Bagyaraj (1989). Kale and mustard, two nonhost species, introduced in a crop rotation reduced the AM colonization of barley, the succeeding crop, to the same extent as did the following (Bagyaraj, 1992). After the harvest of a crop of mustard, spore density was very low (180 spores/100 g of soil) compared to the density found after a potato crop (*Solanum tuberosum* L.) (430 spores/100 g of soil). Wheat after mustard had only 40 percent colonization whereas after potato, colonization reached 85 percent (Ghosh, Bhattacharya, and Verma, 2004).

TABLE 7.3. Effect of cropping pattern on AM fungi and plant growth.

Cropping pattern	Subsequent crop	Effect on crops	Effect on mycorrhiza	References
Castor + groundnut	Sunflower	+ve	+ve	Kumar and Bagyaraj, 1989
Cowpea + Finger millet	Sunflower	−ve	−ve	Kumar and Bagyaraj, 1989
Rice (variety: Vandana) + Tomato	Rice	+ve	+ve	Harinikumar and Bagyaraj, 1989
Rice + maize	Rice	−ve	−ve	Maiti et al., 2000
Rice + okra	Rice	−ve	−ve	Maiti et al., 2000
Rice + Red gram	Rice	−ve	−ve	Maiti et al., 2000
Rice (variety: Brown gora) + Red gram (variety: Laxmi)	Rice	+ve	+ve	Maiti et al., 1997
Mustard	Barley	−ve	−ve	Bagyaraj, 1994
Potato	Wheat	+ve	+ve	Ghosh et al., 2004

In another study, on the effects on the population dynamics of native AM fungi of intercropping rice (local variety, Brown Gora) with red gram (*Cajanus cajan* L., local variety, Laxmi), it was demonstrated that rice-red gram intercropping leads to a substantial increase in the native AM fungal population, which declined during the off-season (Maiti, Variar, and Singh, 1996), but remained higher than under onocultured rice. In the first season (1993), only a marginal yield increase was measured, but in the subsequent season (1994), the increment was statistically significant. When upland rice (variety, Vandana) was intercropped with maize, okra (*Abelmoschuc esculentus* L.) and red gram, in a rice intercrop row ratio of 4:1, rice had lesser affinity for AM fungal colonization than the intercrops. When used as monocultures, all the intercrop plant species stimulated AM fungal spore production compared to when they were intercropped with rice, and after one and two growing seasons, there were more spores in soil under monoculture than intercropped. Little difference in spore production was found among the intercrops. These results were attributed to the lower affinity between rice and AM fungi than between the intercrops tested and AM fungi (Maiti, Singh, and Saha, 2000).

Precropping with plant species with various mycotrophy levels can also modify the AM fungal potential of soils and the performance of any subsequent mycorrhiza-dependent plants. The net growth benefit due to precropping with maize, paspalum *(Paspalum vaginatum),* millet, soybean, onion, tomato, mustard, and ginger or weeds varied from 0 to 50 percent, depending on the mycorrhizal root mass, AM fungi spore number, and infective inoculum density remaining in the soil following the preceding crops and weeds (Panja and Chaudhary, 1998). This rotation effect can be taken into consideration in the design of more effective crop rotation sequences.

Organic Amendments

Organic fertilization enhances the biological activity of soil. In a study by Harinikumar and Bagyaraj (1989) it was reported that fertilization with 7.5 tonnes ha^{-1} of farmyard manure significantly increased the numbers of AM fungi in the soil and AM root colonization in the second and third season of application. When onion *(Allium cepa),* potato, and garlic were inoculated with AM fungi in nutrient deficient soil amended with composted leaves of albizzia *(Albizzia lebek),* Poplar *(Populus deltoids),* and Leucaena *(Leucaena leucocophala),* colonization in onion was approximately 85 percent, and that of garlic and potato was 65 percent. The best yield increase due to inoculation was in onion (70 percent), while garlic and potato yield increased by 30 and 48 percent (Gaur and Adholeya, 2000c). Four organic amendments (leaf compost, vegetable compost, poultry compost, and sewage sludge) applied at four rates (40, 80, 100, and 120 tonnes ha^{-1}) were evaluated for their effect on three varieties of citronella oil *(Cymbopogon winterianus)* (varieties Manjusa, Mandakini, and Bio-13). Poultry compost, applied at 100 tonnes ha^{-1}, most effectively increased herbage yield, essential oil content, and dry matter yield, and was followed by sewage sludge. Bio-13 performed better and produced the highest herbage, essential oil, content and dry matter yield. The highest number of AM fungal propagules was recorded in the leaf compost amended plots in all three varieties. Among the varieties, the highest native mycorrhizal inoculum was recorded in Bio-13 (DBT, 1998-2002a; Tanu, Prakash, and Adholeya, 2004).

Pesticides and Fungicides

Pesticides may affect the growth of AM fungi (Parvati, Venkateswarlu, and Rao, 1985). In India, a few studies done concluded that, generally, systemic fungicides adversely affect AM fungi and such effects were specific to a particular combination of the host and the AM fungal species (Table 7.4). The mode of fungicide application also governs the severity of the fungicide effect on AM fungal infection. It has been observed that the application of fungicides as seed treatment had no effect on mycorrhizal infection (in maize), whereas fungicides applied as soil drenches inhibited mycorrhizal infection (Vyas and Singh, 1992). In another study on maize, the growth of maize plants was reduced by benomyl (fungicide) at a low but not at a high soil phosphorus supply, demonstrating that benomyl has no detrimental effect on maize growth, but affects the phosphorus supply of the plant (Amitava et al., 2002).

TABLE 7.4. Effect of fungicides on establishment of AM fungi in host plants.

Fungicides	Crop	AM fungi	Effect on AM fungi	References
Thiram	Wheat	*G. mosseae*	−ve	Jalali, 1979
Benomyl	Onion	*G. mosseae*	−ve	Manjunath and Bagyaraj, 1984
Benomyl	Onion	*G. intraradices*	−ve	Parvathi et al., 1985
Fosetyl-Al	Soybean	*G. mosseae*	+ve	Vyas et al., 1990
Metalaxyl	Soybean	*G. mosseae*	+ve	Vyas et al., 1990

EFFECT OF EDAPHO-CLIMATIC FACTORS ON MYCORRHIZAL COLONIZATION AND PLANT GROWTH

The occurrence of AM fungal species in agricultural soils varies with edaphic and climatic factors such as soil moisture, texture, temperature, pH, and light, which affect the efficacy and diversity of mycorrhizal fungi. AM fungi may increase the resistance of plants to drought, waterlogged conditions, soil salinity, and extreme pH conditions by a number of mechanisms such as increased root hydraulic conductivity, stomatal regulation, hyphal water uptake, osmotic adjustment, nutrient transportation, phytohormone production, and through augmented root surface. Studies have been carried out on the effect of various edaphic and climatic factors on AM colonization and plant growth. The following section highlights the work done regarding the effect of edapho-climatic factors on the development of AM fungi in various agricultural crops.

Soil Moisture

Mycorrhizal association is known to help plants under drought conditions. In moong *(Phaseolus mungo),* water stress affects not only the microbial spectrum of the rhizosphere but also the intensity of root infection and the population of AM spores in the root region (Kehri and Chandra, 1995). *G. fasciculatum* showed maximum root colonization and sporulation in severe water stressed conditions (watering filled to capacity once in five days) in cowpea grown in pots; the number of spores and infective propagules decreased when the pots were maintained at 50 percent of field capacity (Pai, Bagyaraj, and Prasad, 1990). In water stressed maize, grown on sterilized and unsterilized soil and inoculated and not inoculated with AM fungus (*G. caledonium* or indigenous AM fungal population), Ramakrishnan, Johri, and Gupta (1990) reported that the leaf water potential at various levels of water stress was nearly of the same order in both mycorrhizal and non-mycorrhizal plants. Photosynthetic and photorespiratory activities decreased with increasing osmotic stress but the chlorophyll content remained the same. There was no significant increase in photosynthesis under various conditions in unstressed plants, but it increased significantly in *G. caledonium* and indigenous AM fungi + *G. caledonium* treated plants than it did in the other treatments. Some increase in chlorophyll a and chlorophyll b content was observed in *G. caledonium,* indigenous AM fungi + *G. caledonium* and phosphorus amended unstressed plants. At osmotic potentials of 0 bars, −2 bars, −5 bars, and −10 bars (leaf water potentials of −3 bars, −5 bars, −8 bars, and −12 bars, respectively), free proline content increased markedly with decrease in leaf water potential. This increase was greatest in phosphorus amended non-mycorrhizal plants followed by *G. caledonium* and then indigenous AM fungi + *G. caledonium* inoculated plants. Further, Ramakrishnan, Johri, and Gupta (1990) worked on maize plants that were five weeks old (non-AM fungal and AM fungal plants inoculated with *G. caledonium*) and exposed to osmotic potentials of 0 bars, −2 bars, −5 bars, and −10 bars. After eight hours of exposure, the leaf water potentials were about −3 bars, −5 bars, −8 bars, and −12 bars in both AM fungal and non-AM fungal plants. Net photosynthesis did not increase significantly at 0 bars, −2 bars, and −5 bars osmotic potentials, but at −10 bars osmotic potential, there was a significant increase in the assimilation of carbon dioxide by AM plants (10 percent to 36 percent). On lowering the water potentials, the rate of decrease in the assimilation of carbon dioxide in AM fungal plants was lower compared to the rate of decrease in non-AM fungal plants. However, among the AM and non-AM fungal plants, the differences in the total chlorophyll content were not significant at various low osmotic potentials.

Soil Texture and pH

Soil type and physicochemical properties have a great impact on the occurrence and efficiency of AM fungi. Most AM fungi adapt to a broad spectrum of edaphic conditions, but few of them are specially adapted to the extremes of soil pH conditions. In *Ipomoea batatas* and *Manihot esculenta,* a higher soil pH and phosphorus retarded the growth and establishment of AM fungi. However, artificial inoculation with spores and colonized roots in an inert carrier (lignite) was effective. Arbuscular mycorrhizal root colonization levels of 80 percent to 91 percent in *Ipomoea batata* and of 80 to 85 percent in *Manihot esculenta* were measured within 15 to 25 days of inoculation. In soybean grown in acidic soil (pH 5.1), AM inoculation increased plant dry matter production and phosphorus uptake. Shoot dry matter increased with the rate of phosphorus application while the root dry matter was highest with the lowest rate of phosphorus application. Shoot phosphorus uptake increased up to 20 g P g^{-1} soil, and root phosphorus uptake increased up to 40 g P kg^{-1} soil (Singh and Reddy, 1994). Mycorrhizal presence in soil is also influenced by soil texture. A compact soil texture affects the partial atmospheric pressure and moisture levels and inhibits the germination and hyphal growth of AM fungi (Bhatia, Sundari, and Adholeya, 1996).

Soil Salinity

Indian research studies on the effects of mycorrhizal inoculation on plant metabolism in a saline environment have been scarce. A few plants, such as moong and mulberry, have been studied; they showed a positive response to AM colonization. Mycorrhizal inoculation had a beneficial effect on moong (cv. ML-131) metabolism under NaCl salinity with an increase in chlorophyll content and proline accumulation. The concentration of potassium, zinc, iron, and manganese was also improved in mycorrhizal plants under salt stress (12.5 mM NaCl) (Jindal et al., 1992; Theoder and Vivekanandan, 1994).

Simiyon, Theoder, and Vivekanandan (1994) studied the occurrence of salt tolerant AM fungi in 11 mulberry varieties planted in the coastal region of Tamil Nadu. Mycorrhizal infection was highest in S-36, BC2-59, and MR-2 (81 to 86 percent), moderate in S-30, ACC-235, S-41, M-5, and Tr-10 (70 to 78 percent), and was lowest in Tr-4, Tr-8, and C-1 (60 to 70 percent). Hyphal density was positively correlated with the level of root colonization and *Glomus* spore density. Phenol and phosphate content were concurrently higher, supporting favorable AM fungal symbiotic associations.

Abdi and Dube (2003) reported that *Suaeda mudiflora*, a Chenopodiaceae, could be colonized by AM fungi (*Glomus* sp. 18 RA, *G. microaggregatum*, G. sp. 12 RA, and G. sp 3 RA). This plant could survive and tolerate soil salinity up to 75,000 mg g^{-1}. The perennial plant along with *Glomus* spp. colonizes saltpan habitats with a salt concentration that no other plant can tolerate.

Seasonal Variation

Studies on the seasonal fluctuations of AM fungi on some commonly cultivated crops at various locations in Aligarh (India) revealed that the predominant AM fungal species were *G. mosseae*, *G. constrictum*, and *G. fasciculatum* (Iqbal and Mahmood, 1998). Species of *G. aggregatum*, *G. gigantia*, and *A. scrobiculata* rarely occurred and their abundance differed from one location to another. The intensity of AM fungal colonization and spore formation varied with locations and seasons. At all locations, the level of AM colonization was very high from February to the first week of May, and from the end of July to September, being highest in August and September, occasionally reaching 90 percent. Arbuscule density in the roots showed similar fluctuations and was also highest in August and September, reaching 57 percent in some locations. Arbuscule mycorrhizal fungal spore abundance was negatively related with root colonization. In all locations, plant roots were extensively colonized by AM fungi throughout the growing season, when AM fungal spore density in soil was very low. Khade and Rodrigues (2004) studied seasonal variations in the mycorrhizal colonization of banana. They found the largest spore densities of AM fungi premonsoon season (426 spores per 100 g of soil), average densities during monsoon season (384 spores per 100 g of soil), and the lowest densities postmonsoon season (250 spores per 100 g of soil).

EFFECT OF POLLUTION AND METAL TOXICITY ON MYCORRHIZAL COLONIZATION AND PLANT GROWTH

The discharge of industrial effluents into the soil, mining operations, and deposits of acids, sulfur, or ammonia on the soil surface or the addition of heavy metals to the soil cause pollution and metal toxicity. These pollutants affect the mycorrhizal colonization and diversity as well as plant growth. The soils of disturbed sites are frequently low in available nutrients and lack the nitrogen-fixing bacteria and mycorrhizal fungi usually associated with plant roots (Cooke and Lefor, 1990). Land restoration in semiarid areas, therefore, faces a number of constraints related to soil degradation and water shortage (Vallejo et al., 2000). Degraded soils, the common targets of revegetation efforts in the tropics, often exhibit low densities of AM fungi (Michelsen and Rosendahl, 1990). These low densities may limit the degree of mycorrhizal development of transplanted seedlings and consequently hamper their plant establishment and growth in those areas.

Mining Operations

Mining operations and surface subsidence due to underground mining operations, have a serious impact on the growth of the vegetation and the general landscape of an area. Microorganisms benefit plant growth in coal wastes, and AM fungi are especially important (Reeves et al., 1979). In limestone mine spoils, soil inoculation with *G. mosseae* has significantly enhanced plant growth and biomass production (Rao and Tak, 2002), as it has in coal mine spoils. Coal wastes are generally deficient in nutrients and the plant cover is often difficult to establish. A study conducted in the coal wastes of Kothagudam showed that the pH and electrical conductivity of coal wastes may fluctuate from 5.2 to 5.6 and from 0.1 to 0.2, respectively (Andhra Pradesh, South India). There were 5.25 mg K 100 g^{-1} coal waste, 1.6 μg Zn g^{-1}, 17.2 μg Fe g^{-1}, 3.75 μg Mn g^{-1}, and 0.43 μg Co g^{-1} (Ganeshan and Mahadevan, 1994). Eighteen plant species from 11 families were examined for AM fungal association; 12 species had mycorrhizal infection. Root colonization ranged from 1 to 86 percent. Spores of *Acaulospora foveata, Entrophosphora colombiana, G. ambisporum, G. aggregatum, G. botryoides, G. fasciculatum, G. claroideum, G. heterosporum, G. hoi, G. mosseae, G. Microcarpus,* and *S. rubiformis* were isolated. Spores were generally dark brown, globose to subglobose, and filled with oil droplets and gas.

Industrial Effluents

Industrial waste, sewage sludge, and automobile effluents severely affect plant and soil microorganisms. Research on AM fungal strains tolerant to industrial effluents has provided evidence for their rapid adaptation to contaminated soils. A study on crops such as maize, sorghum, sugarcane, and coco yam cultivated in petro-effluent-irrigated fields (Joseph and Kothari, 1993) showed that all four crops had AM fungal association. The highest level of colonization was seen in coco yam (98 percent), followed by sorghum (90 percent), maize (73 percent), and sugarcane (73 percent). *Glomus* and *Gigaspora* species dominated the rhizosphere of the plants growing in the petro-effluent-irrigated fields. Selvaraj and Bhaskaran (1996) reported that among the 20 species of AM fungi belonging to five genera found in soils contaminated by paper mill effluents, *G. intraradices* was the dominant AM species. In *Vetiver zizanioides,* inoculation with *G. fasciculatum* increased the percentage of root infection and number of AM fungal spores compared to inoculation with *G. mosseae* or non-inoculated control plants. This study indicates that petro-effluents do not necessarily inhibit mycorrhizal development in *V. zizanioides.*

Arul and Vivekanandan (1994) studied the occurrence of AM fungal colonization in legumes grown in a soil polluted with dust from the cement kiln exhausts of the Tamil Nadu Cements Corporation Limited. The population of beneficial microbes was not affected by a soil pH of 8.7, and AM fungal infection was observed in all legume crops examined, that is, *Vigna mungo, Vigna catjung, Cajanus cajan, Arachis hypogea,* and *Glycine max.* Root infection ranged between 50 and 62 percent. *Glomus* spores were abundant in the polluted soil. Generally, spores were globose to subglobose in shape with two to three layered walls. Among 49 plants grown in soil polluted with industrial waste, sewage sludge, and automobile effluents, all the examined plants were colonized by AM fungi, except *Amaranthus viridis, Portulaca oleraceae,* and *P. quadrifides* (Lakshman, 2000). The stabilization of soils polluted by industries and automobiles depended on the choice of plant community and AM fungi.

Metal Toxicity

Soil pollution caused by heavy metals severely affects mycorrhizal colonization and plant growth (Gaur and Adholeya, 2004a,b). A study conducted at Chennai University, in India, showed that ten *Glomus,* two *Gigaspora,* two *Acaulospora,* and one *Sclerocystis* spp. isolated from soils polluted by heavy metals had a variable level of tolerance to heavy metal pollution (Singh, 1996). A survey of soil polluted by heavy metals (Cu, Pb, Ni, and Cd, but particularly Zn) in Tamil Nadu by Sambandan, Kannan, and Raman (1992) revealed that 16 of 18 plant species examined were mycorrhizal, and 10 of 15 AM fungi observed were *Glomus* spp., two were *Gigaspora* or *Acaulospora,* and one was a *Sclerocystis* spp. Mycorrhizal colonization may enhance heavy metal tolerance in *Sorghum bicolor* (L.) Moeneh. AM fungi could protect plants against heavy metal toxicity only up to the "LD-50" dose, namely 50 μM (Senthil and Arockiasamy, 1994). In another experiment, Sharma et al. (1992) reported that the absorption of Zn by mycorrhizal maize was greater at low concentrations (less than 4 mM^{-3}) and decreased at higher levels.

Effects of Fire

Forest fires may drastically change the population of soil microbes. The effects of fires on AM fungi are not well documented in India. Deka, Mishra, and Sharma (1990) studied the effects of slash and burn cultivation on endomycorrhizal fungi and their influence on the early colonizing plant species. The study was carried out at Burnihat (Meghalaya), which is dominated by bamboo trees *(Dendrocalamus hamiltonii).* The AM fungi associated with a few early colonizers *(Eupatorium odoratum, Bothriochola intermedia, Desmodium trifolium, Dendrocalamus hamiltonii, Imperata cylindrical, Osbeckia crinata, Panicum maximum, Setaria palmiflora,* and *Xanthium straumarium)* were extracted from burnt soil and studied to evaluate the effects of fire. The intensity of AM colonization and the density of AM spores were drastically reduced after the fire. After the emergence of seedlings in the burnt area, root colonization with *G. fasciculatum* expanded as the plants grew older.

Fly Ash Pollution and Reclamation: A Success Story

New coal-based power plants have been established to meet the increasing demands for electricity. However, these power plants annually generate up to 150 million tonnes of fly ash that is usually dumped in wastelands. It destroys the vegetation in addition to being a source of groundwater contamination. Fly ash contains toxic metals such as Pb, Cr, Co, and Cu.

Researchers at TERI worked on the reclamation of the fly ash pond of Badarpur (Figure 7.2), New Delhi. This work led to the development of a technology that uses the association of AM fungi with flowering plants and trees and enables producers and plantation companies to grow plants of economic value on fly ash. A variety of plants that could grow on fly ash with the help of a suitable mycorrhizal fungus were screened. Five species of herbaceous plants, four trees (eucalyptus, poplar, acacia, and prosopis), and a mycorrhizal fungus (*G. intraradices* AM1004) were identified. Thorough evaluation of the biological material was first done under greenhouse or nursery conditions. Strains with high nutrient uptake ability, that is, the

FIGURE 7.2. Mycorrhiza application in reclamation of fly ash overburdens: Fugitive dust emission and restoration of green cover.

ability to increase the yield of plants, and a good shelf life in fly ash were selected. The application doses of organic amendments and mycorrhizal biofertilizer were optimized. Seedlings of suitable flowering plants were grown in a nursery. The plant roots were inoculated with various rates of in vitro cultured AM fungi inoculum. Compost and manure were added at various rates (3 kg and 5 kg) in the root zone to enhance the capacity of fly ash to retain moisture and support plant growth. Other than supplying nutrients, organic matter has another benefit: as it decomposes, it produces acids that combine with the heavy metals to form compounds that are less mobile, therefore less likely to pollute surface and groundwater. Seedlings were transplanted in six treatment blocks. The study showed that, with the help of a suitable AM fungi and plants, fly ash can be transformed into a substrate fit for plant production, without the use of costly and wasteful chemicals (Gaur and Adholeya, 2004a,b). This technology is expected to turn waste into wealth (*The Tribune,* 2004).

In another study on the reclamation of fly ash ponds at Korba, Chattisgarh (NTPC, 2001-2004), the same technology was used and AM and ectomycorrhizal (EM) fungi isolates were screened for their adaptability and efficacy, under pot and field conditions (Ray et al., 2005). Drawing on their large collection of mycorrhizal fungi cultures, TERI researchers identified EM and AM fungal species that seemed most likely to help bind the fine ash particles and sequester heavy metals in the ash pond. Various amounts of organic matter were added near the roots to enhance the capacity of the fly ash to retain moisture and support plant growth (Tables 7.5 and 7.6).

Proper management of the site improved the physical, chemical, and biological properties of the substrate. The porosity and water-holding capacity increased over time. The availability of phosphorus, nitrogen, and potassium also increased. Soil dehydrogenase activity and total microbial population were measured two years after planting. The heavy metal toxicity of fly ash also decreased after two years of planting. The technology developed at TERI allows the growth of plants on polluted sites, and also helps prevent leaching of heavy metal pollutants in groundwater. It was found that the heavy metal content of the lechates decreased drastically as a result of reclamation activities (Fly ash mission, TIFAC project report, 1997-2000, 1997-2001, 1999-2004; NTPC project report, 2001-2004).

TABLE 7.5. Comparison of fly ash properties before and after applying mycorrhizal technology at Badarpur and Korba sites.

Properties of fly ash	Badarpur		Korba	
	Zero time	After two years of mycorrhizal technology application	Zero time	After two years of mycorrhizal technology application
pH	7.4	6.6	8.7	8.1
Available phosphorus ($\mu g\ g^{-1}$)	11.7	20.5	7.1	21.1
Total nitrogen (%)	0.01	0.3	0.1	0.6
Organic carbon (%)	0.6	1.8	0.2	0.9

Source: National thermal power corporation limited (NTPC, 2001), Korba and Fly ash mission; Technology information forecasting and assessment council (TIFAC) funded project report (1997a, 1997b, and 1999).

TABLE 7.6. Percentage of reduction in heavy metal profile after two years of applying mycorrhizal technology at Badarpur and Korba sites.

Heavy metal	Badarpur	Korba
Aluminum	–	62
Chromium	36	90
Manganese	ns	44
Iron	ns	27
Cobalt	ns	50
Nickel	ns	64
Copper	75	44
Zinc	33	32
Arsenic	ns	43
Lead	59	53
Cadmium	–	97

Source: NTPC Korba and Fly ash mission, technology information forecasting and assessment council funded project report (2000 and 2004).

Note: ns, non-significant.

Another study on fly ash demonstrated the importance of the root-associated AM fungus *G. mosseae* and nodulating *Rhizobia* on the establishment, growth, and yield of the black gram variety T9 in the enriched and abandoned ash ponds of the Neyveli Thermal Power Plant. The best effects came from inoculation with the AM fungus

G. mosseae and the *Rhizobium* sp. PAB-1, which produced the highest level of mycorrhizal colonization (88 percent), largest plant height, dry weight, grain nitrogen, and phosphorus content, and pod yield (Sheela and Sundaram, 2003).

INTERACTION OF MYCORRHIZAE WITH SOIL MICROORGANISMS

The soil microflora consists of free-living microorganisms of a ubiquitous nature and specific microbial communities in the rhizosphere and rhizoplane regions. All the microorganisms, including fungi, actinomycetes, bacteria, rhizobacteria promoting plant growth, chitin-decomposing bacteria, acid-producing bacteria, soil-inhabiting pseudomonads, nitrogen-fixing bacteria, and phosphorus-solubilizing bacteria, affect mycorrhizal growth. These organisms may either promote or inhibit mycorrhizal development in plant roots and their survival in soil. These interactions, overall, have a positive effect on plant growth. The most important interactions involve AM fungi, and nitrogen fixers (including bacteria and actinomycetes) or phosphorus-solubilizing bacteria and fungus, as these interactions promote plant growth in many ways.

Nodule-Forming Nitrogen Fixers

Studies conducted in various research laboratories in India and worldwide have shown that the dual inoculation of plants with AM fungi and nitrogen-fixing nodule-forming organisms improves plant growth parameters such as plant biomass, dry matter, leaf biomass, leaf number, seed yield, and the nitrogen and phosphorus contents, in comparison to inoculating plants with the AM fungi alone, to inoculating them with nitrogen-fixing nodule-forming organism alone, or not to inoculating them at all. The yield of 60-day-old moong bean plants dually inoculated with *Glomus caledonium* and *Rhizobium* (strain KM-[1]) was 22 g, 14.2 g in plants inoculated only with *Rhizobium,* 15.65 g in plants inoculated only with AM fungi, and 10.83 g in uninoculated control plants (Adholeya, 1988). Dual inoculation with *Rhizobium* and the AM fungi also increases the total nitrogen in plants. Tilak (1985) conducted experiments involving the combined inoculation of *Cenchrus ciliaris* and *Macroptilium atropurpureum* with AM fungi and *Rhizobium* (TAL555) and the application of ^{15}N as labeled ammonium sulfate, at a rate of 10 kg N ha^{-1}. Dual inoculation increased the total annual dry matter yield of plants over that of uninoculated plants from 19.2 t ha^{-1} to 23.9 t ha^{-1}, and the total nitrogen yield from 250 kg ha^{-1} to 300 kg ha^{-1} in the first year, reaching 297 kg ha^{-1} N in the second year. The amount of fixed nitrogen was reduced from 79 kg ha^{-1} in the first year to 39 kg ha^{-1} in the second year. Dual inoculation with AM fungi and *Rhizobium* also resulted in maximum use of phosphorus from the soil. Combined inoculation of chick pea crops *(Cicer arietinum)* with AM fungi and *Rhizobium* resulted in better use of phosphorus from mono- and di-calcium phosphate compared to the phosphorus use in plants inoculated with AM fungi alone (Chaturvedi et al., 1989).

The presence of AM fungi in nodulated legumes often increases the efficiency of *Rhizobium,* resulting in increased nodulation and nitrogenase activity. Combined inoculation of red gram, grown in phosphorus-deficient tropical soil, with *G. fasciculatum* and *Rhizobium* increased nodulation by 178 percent and nitrogenase activity by 185 percent, compared to inoculation with *Rhizobium* alone (Chaturvedi et al., 1989).

Some strains of a species of *Rhizobium* have better symbiotic capabilities than others. Bhandal, Gupta, and Pandher (1989) showed that the inoculation of pea crops *(Pisum sativum)* with *Rhizobium leguminosarum* B164 (hup^+, or hydrogenase positive) enhanced all symbiotic parameters, while inoculation with strain 23 (hup^- or hydrogenase negative) increased only nitrogenase activity and plant nitrogen content. Enhanced nitrogenase activity with the hup^- strain resulted in increased plant dry matter production and nitrogen content in the PB88 and PG3 cultivars. Inoculation with an AM fungus and the hup^+ strain resulted in significantly better symbiotic characteristics than did a single inoculation. Experiments with red gram showed that dual inoculation with the hup^+ *Rhizobium* strain and *G. fasciculatum* increased nodulation, nitrogenase activity, plant biomass, and nitrogen and phosphorus content compared to inoculation with the hup^- strain and *G. fasciculatum* or single inoculation with either of the organisms. Nodules formed by the hup^+ parent strain showed more nitrogenase activity and hup activity, and more leghemoglobin than the hup$^-$ strain.

Free-Living Nitrogen Fixers

A variety of free-living, nitrogen-fixing microorganisms can be found in agricultural soils. Such organisms are not or are only loosely associated with plant roots and thus do not form any root nodules. These free-living nitrogen fixers may be present in rhizosphere soil or may be intimately associated with mycorrhizal roots. The most common free-living nitrogen fixers associated with mycorrhizal roots are species of *Azospirillum, Azotobacter,* and *Klebsiella.*

Azospirillum species are known to fix atmospheric nitrogen and enhance plant growth. Dual inoculation of onion plants with *Azospirillum brasilense* and AM fungi (*Glomus* sp. and *Gigaspora* sp.) in phosphorus deficient soil in pot culture resulted in increased plant biomass production. Nitrogen and phosphorus in shoots and bulbs also increased following dual inoculation compared to the content in plants following inoculation with single cultures and the content in non-inoculated controls (Konde, Tambe, and Ruikar, 1988). Screening of 60 sweet potato cultivars showed that dual inoculation of sweet potato with *A. brasilense* and AM fungi *(G. fasciculatum* and *G. mosseae)* in pots significantly increased plant growth, nitrogen and phosphorus content, tuber weight, and starch content. Infection by AM fungi varied from 13.89 percent to 46.42 percent, depending on the genotype (Tilak et al., 1982). Pearl millet crops subjected to dual inoculation, with *A. brasilense* and *G. margarita* or *A. brasilense,* and *G. fasciculatum* in unsterile, phosphorus-deficient soil, had increased shoot and root biomass, and phosphorus uptake compared to the non-inoculated controls. Dual inoculation of finger millet with *A. brasilense* and *G. caledonium* improved shoot and root dry matter production and the level of mycorrhizal infection by 75 percent (Tewari, Tandon, and Johri, 1987). Dual inoculation of palmarosa (*Cymbopogon martini* var. motia) with *G. aggregatum* and *A. brasilense* increased growth, yield, and oil content more so than inoculation with AM fungi alone, *Azospirillum* alone, or no inoculation (Ratti and Janardhanan, 1996).

Inoculation with *Azospirillum* stimulates the growth of cuttings when they are planted in a soil inoculated with AM fungi. Mulberry cuttings dipped in a peat-based slurry of *A. brasilense* for 15 minutes, dried in the shade, and planted in soil inoculated with 100 g of *G. fasciculatum* (260 spores/100 g of soil), 2.5 cm below the cuttings, increased plant biomass production and leaf weight, in a pot experiment. Interestingly, inoculation with *Azospirillum* alone or AM fungi alone had little effect on the development of cuttings (Nagarajan et al., 1989).

Azotobacters are other free-living, nitrogen-fixing bacteria. Like *Azospirillum, Azotobacter* has a positive effect on plant growth when combined with AM fungi, phosphorus-solubilizing microbes, or *Rhizobium*. In experiments on linseed crops, the maximum improvement in shoot biomass production and in nitrogen and phosphorus content was obtained by triple inoculation with a P-solubilizing fungus *(Aspergillus niger)*, an AM fungus *(G. aggregatum)*, and *Azotobacter chroococcum*. Single and dual inoculations of seeds improved plant growth and nitrogen and phosphorus uptake compared to no inoculation, but inoculation with *G. aggregatum* alone produced mycorrhizal colonization levels higher than those produced with dual and triple inoculations (Kehri and Chandra, 1995). Inoculation of *Trigonella foenum-graecum* with *G. macrocarpum, Rhizobium meliloti,* and *A. chroococcum* resulted in maximum growth and yield of plants whereas dual or single inoculation or non-inoculation did not (Bhattacherjee and Mukerji, 1981). The dual inoculation of three varieties of mulberry with an AM fungus and *Azotobacter* yielded superior results than did inoculation with single organisms or no inoculation (Gowda et al., 1995).

Phosphorus-Solubilizing Microorganisms

A number of phosphorus-solubilizing microorganisms, both bacteria and fungi, enhance the effects of AM fungi. These organisms not only increase the supply of phosphorus to plants but also increase the efficiency of their associated AM fungi. The highest mean grain yield, biomass production, and phosphorus uptake in lentils *(Lens culinaris)* were from plants inoculated with *G. fasciculatum* combined with *Pseudomonas striata* (a phosphorus-solubilizing bacterium). The next highest yields were from plants inoculated with a phosphorus-solubilizing fungus *(Aspergillus awamori)* alone (Sattar and Gaur, 1985). The inoculation of wheat with *P. striata, Agrobact-*

erium radiobacter, and AM fungi *(G. fasciculatum* and *G. margarita)* improved the dry matter yield; maximum yield was derived from inoculation and phosphorus and nitrogen fertilization (Gaur and Rana, 1990). The combined inoculation of chili plants with *G. macrocarpum* and *Bacillus polymyxa* yielded better results than did individual inoculation and produced the highest fruit yield. It was possible to replace 25 percent of the phosphorus with combined inoculation compared to individual inoculation (Chandraghatgi and Srinivasa, 1995).

Phosphorus-solubilizing bacteria improved the efficiency of AM fungi. This improvement was observed in an experiment in which the inoculation of mission grass *(Pennisetum pedicillatum)* seeds with a phosphorus-solubilizing bacterium and a nitrogen-fixing bacterium increased root volume, AM fungal colonization, and number of spores in the presence of *G. macrocarpum.* However, such stimulation was more significant with the nitrogen-fixing bacterium than with the phosphorus-solubilizing bacterium (Singh, 1995).

Interaction with Other Soil Microflora

The effect of mycorrhiza on soil microflora is very pronounced. Mycorrhizal roots encourage the development of their own microflora in the mycorrhizosphere or mycorrhizoplane. *Cladosporium herbarum* was very abundant in pot cultures of *G. fasciculatum, G.margarita, A. laevis,* and *Glomus dussii.* The total bacterial population, number of nitrogen fixers and gram-negative bacteria were significantly higher in pot cultures of *G. fasciculatum, G. margarita,* and *S. dussii* than in control pots devoid of AM fungi. Spore formers decreased and urea hydrolyzers increased in all these cultures and in the pot cultures of *A. laevis.* Pot cultures of *G. fasciculatum* and *A. laevis* contained a large number of actinomycetes antagonistic to pathogens. *Fusarium solani, Raustonia solanacearum,* and pot cultures of *G. margarita* contained a large number of actinomycetes antagonistic to the pathogen *Xanthomonas compestris* var. Vignicola (Secilia and Bagyaraj, 1989). Several pathogenic fungi-like species of *Fusarium, Pythium,* and *Veticillium* occurred more frequently on the rhizosphere and rhizoplane of non-mycorrhizal plants than did the root zone fungi of mycorrhizal plants (Bansal and Mukerji, 1994).

Fluorescent *Pseudomonads* are known to promote plant growth. They interact positively with AM fungi, further enhancing plant growth. Fluorescent *pseudomonads* which are plant growth promoting rhizobacteria (PGPRs) and *Bacillus* species interacted with indigenous AM fungi synergistically increasing the growth and the economic yield of wheat plants (Kumar, Gaikwad, and Singh, 1995).

FACTORS AFFECTING EFFICIENCY OF TRIPARTITE SYSTEMS

Various factors such as fertilization, soil reaction, moisture, toxicity, and beneficial soil microflora may affect the efficiency of tripartite systems involving plants, AM fungi, and nitrogen fixers. Phosphorus fertilization increases the efficiency of the tripartite system in phosphorus deficient soils. Dual inoculation of pigeon pea plants or cowpea plants grown on phosphorus-deficient, nonsterile soil coupled with the application of 22 kg P ha^{-1} increased nodulation, the abundance of AM spores in the root zones, plant shoot dry weights, and plant nitrogen and phosphorus content (Manjunath and Bagyaraj, 1986). Similarly, in moong bean plants grown in phosphorus-deficient soil, seed inoculation with *Rhizobium* and soil inoculation with *G. fasciculatum* in combination with the application of 40 kg P ha^{-1} resulted in the highest yield (Gaur, 1985).

The application of nitrogen and phosphorus fertilizers in phosphorus deficient soils also improved the efficiency of the tripartite system. Dual inoculation of *Leucaena* with *G. fasciculatum* and *Rhizobium* strains R4 and R5 in combination with the application of phosphorus and nitrogen fertilizers improved growth, nodulation, and nitrogen fixation of the plant, in addition to increasing its nitrogen and phosphorus content, compared to single inoculation with any one of the organisms (Chaturvedi and Singh, 1989).

The application of gypsum to the soil also increased the efficiency of the tripartite system. The increase in groundnut growth, AM fungal colonization, and nutrient content was maximum with dual inoculation with *Rhizobium* and AM fungus, together with the application of 150 kg ha^{-1} of gypsum. Mycorrhizal root colonization, however, decreased with larger applications of gypsum up to 450 kg ha^{-1} (Gaur and Rana, 1990).

Certain phosphorus-solubilizing soil organisms increase the efficiency of the tripartite system when inoculated with a combination of AM fungi and *Rhizobium/Bradyrhizobium*. Inoculation of chickpea plants with *Rhizobium, G. fasciculatum,* and *Bacillus polymyxa* (a phosphorus-solubilizing microbe) resulted in significantly greater dry matter production and phosphorus uptake, compared to single or dual inoculation with any of the test organisms (Chaturvedi et al., 1989).

INTERACTION WITH PLANT PATHOGENS

Arbuscular mycorrhizae formation generally reduces the incidence of soilborne diseases in plants. Phytoprotection conferred by AM fungi varies with the nature of the host plant, mycorrhizal symbionts, and plant pathogens involved, as well as on the conditions of the soil environment. Positive research results on the suppression of soilborne pathogens by AM fungi have opened up prospects for their application in plant production. Several reports have indicated increased resistance to a number of root disease pathogens upon AM inoculation of plants. In India, studies have been carried out on the effectiveness of mycorrhizal fungi against fungal pathogens attacking cereals, pulses, vegetables, cash crops, fiber crops, and oilseed crops.

Pathogenic Fungi

Foot root rot of wheat caused by *Sclerotium (Corticium) rolfsii* was controlled by inoculation with *G. fasciculatum. S. calospora* in glass-house pot culture experiments had an inhibitory effect on the development of pigeon pea blight caused by *Phytopthora drechsleri* f. sp. Cajani; however, the presence of AM fungi had no effect on *Fusarium* wilt of pigeon peas when a high level of *Fusarium udum* was present in the soil (Bisht et al., 1985; Reddy, Rao, and Krishna, 1988). *G. etunicatum* induced cowpea tolerance against *Macrophomina* root rot. Disease incidence was 16 percent in inoculated plants compared to 33 percent in uninoculated plants (Ramraj, Shanmugan, and Dwarkanath, 1988). Dual inoculation of moong bean plants with *G. mosseae* and *Macrophomina phaseolina* reduced disease incidence from 77.9 percent in plants inoculated with only the pathogen, to 13.3 percent. Dual inoculation also resulted in an increase in total dry matter production and nitrogen, phosphorus, and potassium content of plants compared to inoculation with the pathogen (Jalali, Chabra, and Singh, 1990). Simultaneous inoculation of pigeon pea plants with *G. mosseae, G. constrictum,* or *G. monosporum,* and *Fusarium oxysporum* had no effect on the incidence of wilt in the susceptible and the resistant genotypes. However, in the wilt-tolerant genotype, AM inoculation reduced wilt severity by 25 percent (Reddy, Rao, and Krishna, 1989).

Arbuscular mycorrhizal inoculation also reduces disease incidence in vegetable, oil, and fiber cash crops. Inoculation of tomato plants with *Glomus mosseae* reduced *Fusarium* wilt from 45 percent, in non-mycorrhizal plants, to 11 percent (Ramraj et al., 1988). Colonization of cumin *(Cuminum cyminum)* plants by *Gigaspora calospora, Glomus fasciculatum, G. mosseae,* or *Acaulospora laevis* alone or in association with *Fusarium oxysporum* f. sp. Cumini enhanced nutrient uptake and reduced the severity of wilt in phosphorus-deficient sandy loam soils (Sharma and Bagyaraj, 1993). Mycorrhizae reduced disease severity caused by *Fusarium oxysporum* f. sp. vesinfectum in cotton *(Gossypium herbaceum)* and by *F. solani* in jute *(Corchorus capsularis)* (Bali and Mukerji, 1988). *G. fasciculatum* reduced the number of sclerotia produced by *Sclerotium rolfsii* in groundnuts in pot culture experiments. The mycorrhizal fungus reduced the percentage of root infection and chlamydospore production of the pathogen. Root and shoot dry weights and phosphorus content of groundnut plants were highest in AM plants non-inoculated with the pathogen and lowest in plants inoculated with the pathogen only. Further, inoculation of the mycorrhizal fungus and the pathogen reduced the severity of the disease (Krishna and Bagyaraj, 1983).

Establishment of mycorrhizal symbiosis occurs earlier and more rapidly in disease-resistant cultivars than in disease-susceptible cultivars. The potato cultivars SSC 1174 and Kafri jyoti range from highly resistant to resistant to *Phytophthora infestans;* AM fungi development occurs earlier and more quickly in these cultivars than in highly susceptible cultivars. Primary infection was observed at 19 and 12 days after emergence in highly resistant and resistant cultivars, respectively (Bhatarai and Mishra, 1984). Diseased plants have rarely been found to bear more mycorrhizal infection than healthy plants, but horsebean *(Vicia faba)* plants attacked by *F. oxysporum* had more

AM fungal infection than did healthy plants. Soil samples from the vicinity of diseased roots contained higher endomycorrhizal spores than soil samples collected from healthy root environments (Singh, Varma, and Mukerji, 1987).

The soil phosphorus level affects the efficiency of mycorrhizal fungi in imparting resistance to plants against diseases. At 0 P kg^{-1} and 40 mg P kg^{-1} soil, chickpea plants inoculated with *G. aggregatum* and the wet pathogen *F. oxysporum* f. sp. Ciceri, 80 and 20 percent of healthy chickpea plants exhibited yellowing 45 days after seeding (Singh and Singh, 1988). At 80 mg P kg^{-1} soil, however, 40 percent of plants were found to be healthy and 60 percent showed yellowing. The trend remained more or less similar after 75, 105, and 130 days of sowing. Root colonization at all three phosphorus levels was more or less similar.

Nematodes

Nematodes are widely present in the Indian soils. Numerous species of parasitic nematodes feed on living plants and cause a variety of plant diseases. The role of AM fungi in reducing the harmful effects of crop root infestation by many parasitic nematodes has been studied in India by many mycorrhiza specialists. In nature, AM fungi and plant parasitic nematodes may coexist with or without significant influence on each other. Jain and Hasan (1986, 1988), in their studies on nematodes, showed that AM spores and nematodes occurred in every sample of root and rhizosphere soil collected from 25 locations of five districts in the Bundelkhand region, which grows berseem and various forage legumes and grasses. The spiral nematodes *(Helicotylenchus dihystera)* and stunt nematodes *(Tylenchorynchus vulgaris)* were the most common plant parasitic nematodes found. *G. mosseae* was the most common AM fungus. Abundant ectoparasitic nematodes were associated with high mycorrhizal counts, indicating that these nematodes do not negatively affect AM fungi, and vice versa. However, in some crops such as gram, cow pea, and pigeon pea, a low incidence of root knot nematode in the roots with a high level of AM fungi was observed. Jain and Hasan (1986) had also observed earlier that the presence of nematodes did not adversely affect AM fungi sporulation. Soil and root samples collected from forage sorghum plantations revealed the presence of *G. fasciculatum* and *G. mosseae,* and three predominant nematodes—*Helicotylenchus dihystera, Pratylenchus zeae,* and *Tylenchorhynchus vulgaris*. The number of nematodes was lower where there was 50 percent root colonization by AM fungi. Jain and Sethi (1987) observed that *Heterodera cajini, G. fasciculatum,* and *G. epigaeus* coexisted in cowpea plants. An increase in nematode inoculum invariably resulted in reduced root infection and spore production by mycorrhizal fungi. *G. fasciculatum* had a profoundly negative effect on cyst production and nematode multiplication, while *G. epigaeus* tended to stimulate nematode activity (Mishra and Shukla, 1995).

The adverse effects of AM fungi on nematodes may either be physical or physiological in nature. Reduction in the severity of disease caused by nematodes may occur through improved plant vigor, physiological alteration of root exudates, or through direct impact of mycorrhiza on the development and reproduction of nematodes within the root tissues. Competition for occupancy may be involved, as suggested by the rapid colonization of roots by the AM fungi compared to the slow establishment of nematodes in roots. Nematodes introduced 20 days prior to AM fungi had sufficient time to penetrate and establish themselves on the root system, leaving little space for the fungus to spread mycelium within the root tissues. When AM inoculation was done 20 days before nematode inoculation, fungus establishment in the root system made it harder for the nematodes to proliferate and negatively affect plant growth (Baghel, Bhatti, and Jalali, 1990). Root extracts from plants colonized by *G. fasciculatum* caused about 50 percent mortality in the larvae of *Meloidogyne incognita* present in tomato roots, but mycorrhizal roots were not shielded against penetration by the nematode larvae. Singh, Singh, and Sitaramaiah (1990) showed that the preoccupation of tomato (variety Pusa Ruby) roots by *G. fasciculatum,* which was coupled with biochemical changes including increased lignin deposition and phenol levels, made tomato resistant to the root-knot nematode *Meloidogyne incognita*.

ADOPTION OF MYCORRHIZAE BY PRODUCERS

The fertility of most of the Indian soils is insufficient for crops to reach their full yield potential, and phosphorus is particularly limiting. This infertility is the driver behind the development and use of agricultural products that ensure a better supply of nutrients to plants. In India, the Mycorrhiza Network, headquartered in TERI, New Delhi, has been active since 1989. This network brings producers, researchers, and policy makers together in an effort to improve the Indian agricultural industry. Research throughout the country has demonstrated that AM fungi can play a major role in plant production. The results of multilocation trials conducted within an Integrated Nutrient Management network project supported by the Department of Biotechnology (DBT) also support this conclusion (DBT, 1990-2002b). It was also evident from the literature that species of the genus *Glomus* are the most abundant and functionally efficient in the various agro-climatic conditions of India. Therefore, technology for the mass production of *G. intraradices* on root organ culture was developed by TERI. The technology was then licensed to industries in India. Presently, four industries are engaged in the production of inocula through TERI licenses. This development has opened up a new era for agricultural crop production. The present production capacity and projections up to 2006 are indicated in Figure 7.3.

CONCLUSION

AM fungi are ubiquitous and have been recorded on the majority of agricultural crops. Mycorrhizae offer several benefits to the host

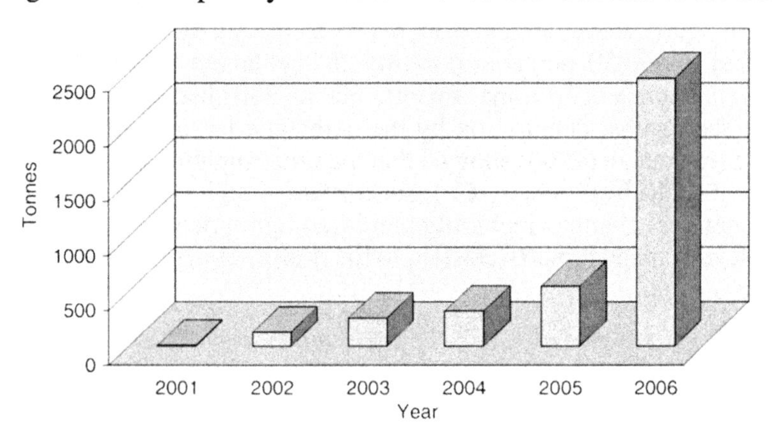

FIGURE 7.3. Industrial production of Mycorrhiza product in India based on TERI-DBT Technology.

plants, including faster growth, greater drought resistance, improved nutrition, and protection from pathogens. Plants are dependent on mycorrhiza for growth and productivity. *G. intraradices* is the most efficient at improving rice yield and *G. versiforme* enhances shoot dry weight, nitrogen, and phosphorus concentration in grams. Inoculation boosts root and shoot dry weight and plant height in plants such as colocasia, and brinjal, while for onions it results in greater bulb diameter, and dry and fresh weights. Mycorrhizal inoculation increases the mineral content of plants. Mycorrhizal association has been found to enhance rooting and promote growth of plants propagated from cuttings or by tissue. Mycorrhizal plants can do well under waterlogged conditions where AM inoculation may also increase yields. Inoculated medicinal plants are taller and more productive than non-inoculated ones. The nutrient requirements of pulses are met through *Rhizobium* while their phosphorus and moisture requirements are better fulfilled with the assistance of AM fungi. AM fungi interactions with microorganisms of the rhizosphere of host plants generally result in better plant growth, and dual or multi-organism inoculation is preferable to single inoculation. Mycorrhizal associations also protect plants against soilborne diseases caused by bacteria, fungi, and nematodes.

> A low supply of inoculum for large-scale field trials has limited the application of AM fungi in agriculture. TERI has overcome this hurdle by developing a commercially viable technology for mass production of AM inocula. (Tiwari, Prakash, and Adholeya, 2003)

TERI also maintains a vast AM fungi collection in its germplasm bank, the Center for Mycorrhizal Culture Collection. This collection of mycorrhizal fungi, the largest in Asia, was developed to enhance agricultural production in an economical and sustainable manner. Major successes have been achieved with the large-scale production of sugarcane, banana, vegetables, and cash crops, due to the geographic location of their production and their current market penetration. Slowly but gradually, more and more cropping sequences are being studied and evaluated within AM management-based cropping systems. The adoption of new practices and the confidence of researchers regarding the future success of AM technologies in India is driven by the fact that the economic strength of growers is very weak and more than 70 percent of producers are in the category of marginal producers. The ever-rising costs of widely used agrochemicals is a major problem in India, and any net input cost-reduction is most welcome for producers, a vast group in a country such as India, where approximately 25 percent of the GDP is dependent on agriculture.

The effect of tillage, cropping systems, organic amendments, and application of fertilizers and pesticides on the mycorrhizal associations of crops may have an important impact on yields. Residue-mulched fields, in contrast to residue-burned fields, are conducive to soil microbial activity and have enhanced fertility. The association of mycorrhiza with host plants is influenced by edaphic and climatic factors affecting root colonization, AM fungi sporulation, and plant uptake of minerals. Indian agriculture depends heavily on rainwater and *G. fasciculatum* shows maximum root colonization and sporulation in severe water stressed conditions. Arbuscular mycorrhizal fungi are found in high pH as well as in saline soils. They enhance host plant growth in adverse conditions, which is very helpful in the reclamation of wastelands. Mycorrhizal associations also vary seasonally, influencing crop yield. The negative impact of industrial discharges, pollution, and metal toxicity of soil can be reduced by mycorrhizal colonization. The reclamation of fly ash ponds using mycorrhizal biotechnology is convenient, as the country is rapidly developing its capacity to produce electricity.

We have seen that AM fungi benefit many agricultural crops grown in India, including cereals, coarse cereals, pulses, oilseeds, cash crops, and a wide range of fruits and vegetables. Arbuscular mycorrhizal fungi also interact with other microorganisms and, through these interactions, may have multiplier effects on plant growth and protect them from pathogens. Thus, AM associations increase plant productivity and ultimately enhance yield to meet the requirements of growing populations in an increasingly constrained environment. The future of Indian agriculture therefore depends on the success of these research efforts, which ultimately means returning to nature and reducing the dependence of agriculture on agrochemicals.

REFERENCES

Abdi, R. and H.C. Dube (2003). Distribution of AM fungi along salinity gradient in a salt pan habitat. *Mycorrhiza News* 15:20-22.

Adholeya, A. (1988). *Influence of VA fungus Glomus caledonius on nodulation and productivity in moong Vigna radiata L. Wilczek.* PhD Thesis, Microbiology Department, G.B. Pant, University of Agriculture and Technology, Pantnagar and Jiwaji University, Gwalior, India.

Adholeya, A. and G.S. Cheema (1990). Evaluation of VA Mycorrhizal inoculation in micro-propagated *Populus deltoides* Marsh clones. *Current Science* 59:1244-1247.

Adholeya, A., A. Verma, and N.P. Bhatia (1997). Influence of media gelling agents on root biomass and in vitro VA-mycorrhizal symbiosis of carrot with *Gigaspora margarita. Biotropia* 10:63-74.

Adholeya, A., P. Tiwari, and R. Singh (2005). Commercialization of Arbuscular mycorrhiza and inoculation strategies. In *Root Organ Culture of Arbuscular Mycorrhiza.* (In Press).

Amitava, R., P.B.S. Bhadoria, S. Satnam, and C. Norbert (2002). Significance of indigenous arbuscular mycorrhiza on phosphorus influx of maize grown on oxisol. In *Proceedings of the 17th WCSS,* Thailand, paper no. 248.

Ammani, K. and A.S. Rao (1996). Effect of two arbuscular mycorrhizal fungi *Acalospora spinosa* and *A. scrobiculata* on upland rice variety. *Microbiological Research* 151:235-237.

Arul, A. and M. Vivekanandan (1994). Occurrence of VAMF colonization in legumes grown in soil polluted with dust from cement kiln exhausts. *Mycorrhiza News* 5:11-12.

Baghel, P.P.S., D.S. Bhatti, and B.L. Jalali (1990). Interaction of VA mycorrhizal fungus and *Tylenchulus semipenetrans* on citrus. In *Current Trends in Mycorrhizal Research. Proceedings of the National Conference on Mycorrhiza,* B.L. Jalali and H. Chand (eds.), Hisar, India: Haryana Agricultural University. pp. 118-119.

Bagyaraj, D.J. (1991). Utilization and commercialization of mycorrhizal fungi. *Mycorrhiza News* 3:4.

Bagyaraj, D.J. (1992). Vesicular arbuscular mycorrhizae: Application in agriculture. *Methods in Microbiology* 24:359-374.

Bali, M. and K.G. Mukerji (1988). Effect of VAM fungi on fusarium wilt of cotton and jute. In *Mycorrhiza for Green Asia. Proceedings of the First Asian Conference on Mycorrhiza,* A. Mahadevan, N. Raman, and K. Natrajan (eds.), Chennai, India: University of Madras. pp. 233-234.

Bansal, M. and K.G. Mukerji (1994). Positive correlation between VAM-induced changes in root exudation and mycorrhizosphere microflora. *Mycorrhiza* 5:39-44.

Beck-Nielsen, D. and T.V. Madsen (2001). Occurrence of vesicular arbuscular mycorrhiza in aquatic macrophytes from lakes and streams. *Aquatic Botany* 71:141-148.

Bhandal, B.K., R.P. Gupta, and M.S. Pandher (1989). Synergistic effect of strain hup$^+$ of *Rhizobium leguminosarum* and vesicular arbuscular mycorrhiza on symbiotic parameters of two cultivars of *Pisum sativum. Research and Development Reporter* 6:38-45.

Bhatia, N.P., K. Sundari, and A. Adholeya (1996). Diversity and selective dominance of vesicular-arbuscular mycorrhizal fungi. In *Concepts in Mycorrhizal Research,* K.G. Mukerji (ed.), Netherlands: Kluwer Academic Publishers. pp. 133-178.

Bhattacharjee, M. and K.G. Mukerji (1981). Vesicular arbuscular mycorrhiza and plant growth—their interactions with *Rhizobium* sp. and *Azotobacter* sp. *Programme and abstracts of the fifth North American conference on mycorrhiza,* Quebec, Canada: University of Laval, p. 83.

Bhattarai, I.D. and R.R. Mishra (1984). Study on vesicular arbuscular mycorrhizae of the three cultivars of potato (*Solanum tuberosum* L.). *Plant and Soil* 79:299-303.

Bisht, V.S., K.R. Krishna, and Y.L. Nene (1985). Interaction between vesicular-arbuscular mycorrhiza and *Phytophthora drechsleri* f. sp. Cajani. *International Pigeonpea Newsletter* 4:63-64.

Brundrett, M., N. Bougher, B. Dell, T. Groove, and N. Malajczuk (1996). Working with mycorrhizas in forestry and agriculture. ACIAR Monograph 32. 374pp.

Chandraghatgi, S.P. and M.N. Sreenivasa (1995). Possible synergistic interaction between Glomus. macrocarpum and Bacillus polymyxa in chilli. In *Mycorrhizae: biofertilizers for the future. Proceedings of the Third National Conference on Mycorrhiza,* A. Adholeya and S. Singh (eds.), New Delhi, India, Tata Energy Research Institute. pp. 180-183.

Chandrashekara, C.P., V.C. Patil, and M.N. Sreenivasa (1995). Response of two sunflower (*Helianthus annuus* L.) genotypes to VA mycorrhizal inoculation levels and phosphorus levels. *Biotropia* 8:53-59.

Chaturvedi, C. and R. Singh (1989). Response of chickpea (*Cicer arietinum* L.) to inoculation with *Rhizobium* and VA Mycorrhiza. In *Proceedings of the National Academy of Sciences, India.* 59 Sec. B part II:443-446.

Chaturvedi, C., A.K. Sharma, R. Singh, and K.D. Sharma (1989). The influence of seed dressing fungicides on growth responses, mycorrhizal infection and nodulation in chickpea. *Proceedings of the National Academy of Sciences, India.* 59 Sec. B part II:232-240.

Cooke, J.C. and M.W. Lefor (1990). Comparison of veiscular-arbuscular mycorrhizae in plants from disturbed and adjacent undisturbed regions of a coastal salt marsh in Clinto, Connecticut, USA. *Environment Management* 14(1):212-337.

Deka, H.K., R.R. Mishra, and G.D. Sharma (1990). Effect of fuel burning on VA Mycorrhiza fungi and its influence on the growth of early plant colonizing species. *Acta Botanica Indica* 18:184-189.

Department of Biotechnology funded project (1993-1997). *Creation of germplasm bank and continuation of Mycorrhizal network,* New Delhi, India: TERI, TERI project report no. 1993 BM 61.

Department of Biotechnology funded project (1998-2002a). *Demonstration of integrated organic farming—organo biofertilizer package using vermicomposting and VAM fungi in aromatic plants namely cymbopogon winterianus and polianthes tuberosa,* New Delhi, India: TERI, TERI project report no. 1998 BM 61.

Department of Biotechnology funded project (1998-2002b). *Integrated nutrient management in poplar-eucalyptus based sustainable agroforestry system,* New Delhi, India: TERI, TERI project report no. 1998 BM 63.

Dodd, J.C. (2000). The role of arbuscular mycorrhizal fungi in agro- and natural ecosystems. *Outlook on Agriculture* 29:55-62.

Fly Ash Mission, Technology information forecasting and assessment council funded project (1997-2000). *Application of fly ash, organic manure and mycorrhiza biofertilizer for improvement in tree plantation under the thrust area 'Agriculture related studies and applications,'* New Delhi, India: TERI, TERI project report no. 1997 BM 63.

Fly Ash Mission, Technology Information Forecasting and Assessment Council funded project (1997-2001). *Reclaiming ash ponds by means by mycorrhizal organo-biofertilizer under the thrust area 'Agriculture Related Studies Applications,'* New Delhi, India: TERI, TERI project report no. 1997 BM 66.

Fly Ash Mission, Technology Information Forecasting and Assessment Council funded project (1999-2004). *Application of fly ash as a soil enricher in arid lands using mycorrhiza biofertilizer under the thrust area 'Agriculture Related Studies and Applications,'* New Delhi, India: TERI, TERI project report no. 1999BM 64.

Gadkar, V. and A. Adholeya (2000). Intraradical sporulation of AM *Gigaspora margarita* in long-term axenic cultivation in Ri T-DNA carrot root. *Mycological Research* 104:716-721.

Ganeshan, V. and A. Mahadevan (1994). Effect of mycorrhizal inoculation of cassava, elephant foot yam and taro. *Journal of Root Crops* 20:1-14.

Gaur, A.C. (1985). Phosphate solubilising microorganisms and their role in plant growth and crop yields. In *Proceedings of a Symposium in Soil Biology*, Hisar, India. pp. 125-138.

Gaur A. and A. Adholeya (2000a). Mycorrhizal effects on the acclimatization, survival, growth and chlorophyll of micropropagated *Syngonium* and *Dracaena* inoculated at weaning and hardening stages. *Mycorrhiza* 9:215-219.

Gaur A. and A. Adholeya (2000b). Effects of the particle size of soil-less substrates upon AM fungus inoculum production. *Mycorrhiza* 10:43-48.

Gaur, A. and A. Adholeya (2000c). On-farm production of VAM inoculum and vegetable crops in marginal soil amended with organic matter. *Tropical Agriculture* 77:1-6.

Gaur, A. and A. Adholeya (2000d). Growth and flowering in *Petunia hybrida, Callistephus chinesis* and *Impatiens balsamina* inoculated with mixed AM inocula or chemical fertilizers in a soil of low P fertility. *Scientia Horticulture* 84:151-162.

Gaur, A. and A. Adholeya (2002). Arbuscular mycorrhizal inoculation of five tropical fodder crops and inoculum production in marginal soil amended with organic matter. *Biology and Fertility of Soils* 35:214-218.

Gaur, A. and A. Adholeya (2004a). Prospects of AM fungi in phytoremediation of heavy metal contaminated soils—Mini-review. *Current Science* 86:528-534.

Gaur, A. and A. Adholeya (2004b). Prospects of arbuscular mycorrhizal fungi in phytoremediation of heavy metal contaminated soils. *Current Science* 86:4-5.

Gaur, A.C. and J.P.S. Rana (1990). Role of VA mycorrhizae phosphate solubilising bacteria and their interactions on growth and uptake of nutrients by wheat crop. In *Current Trends in Mycorrhizal Research. Proceedings of the National Conference on Mycorrhiza,* B.L. Jalali and H. Chand (eds.), Hisar, India: Haryana Agricultural University. pp. 105-106.

Gaur, A., A. Adholeya, and K.G. Mukerji (1998). A comparison of AM fungi inoculants using *Capsicum* and *Polianthes* in marginal soil amended with organic matter. *Mycorrhiza* 7:307-312.

Gaur, A., Sharma M.P., and A. Adholeya (2003). Production quality vegetables through Mycorrhizae. *Indian Horticulture* 18:19-21.

Gautam, A. and I. Mahmood (2002a). A survey of some cultivated legume crops of Aligarh district to determine the occurrence of AM fungi. *Mycorrhiza News* 14:13-15.

Gautam, A. and I. Mahmood (2002b). Comparative efficacy of different arbuscular mycorrhizal fungal species (AMF) on chickpea *(Cicer arietinum). Mycorrhiza News* 14:9-11.

Ghosh, S., D. Bhattacharya, and N.K. Verma (2004). Mustard (*Brassica campestris* L.) cultivation reduces the VA mycorrhizal advantage of successive crops. *Mycorrhiza News* 16:12-14.

Gowda, M.H., P.K. Das, R.S. Katiyar, P.S. Fatima, P.C. Choudhary, and R.K. Datta (1995). The role of synergistic effect of vesicular arbuscular mycorrhiza and azotobacter in mulberry. In *Mycorrhizae: Biofertilizers for Future. Proceedings of the Third National Conference on Mycorrhiza,* A. Adholeya and S. Singh (eds.), New Delhi, India: Tata Energy Research Institute. pp. 197-201.

Gupta, N. and S.S. Ali (1993). VAM inoculation for wetland rice. *Mycorrhiza News* 5:5-7.

Hemalatha, M. and T. Selvaraj (2003). Association of AM fungi with Indian borage (*Plectranthus amboinicus* (Lour) Spring) and its influence on growth and biomass production. *Mycorrhiza News* 15:18-20.

Indi, D.V., B.K. Konde, and K.R. Sonar (1990). Response of brinjal genotypes in terms of dry weight and phosphorus uptake as influenced by VAM inoculation. In *Current Trends in Mycorrhizal Research. Proceedings of the National Conference on Mycorrhiza,* B.L. Jalali and H. Chand (eds.), Hisar, India: Haryana Agricultural University. pp. 146-147.

Iqbal, J. and I. Mahmood (1998). Effects of single and multiple VAM inoculants on the growth parameters of tomato. *Mycorrhiza News* 10:13-15.

Jain, R.K. and N. Hasan (1986). Association of vesicular arbuscular mycorrhizal (VAM) fungi and plant parasitic nematodes with forage sorghum (*Sorghum bicolor* L.). *Sorghum Newsletter* 29:84.

Jain, R.K. and C.L. Sethi (1987). Pathogenicity of *Heterodera cajani* on cowpea as influenced by the presence of VAM fungi *Glomus fasciculatum or G. epigaeus. Indian Journal of Nematology* 18:89-93.

Jain, R.K. and N. Hasan (1988). Role of vesicular arbuscular mycorrhizal (VAM) fungi and nematode activities in forage production. *Acta Botanica Indica* 16: 84-88.

Jalali, B.L. 1979. Effects of soil fungitoxicants on the development of VA-mycorrhiza and phosphate uptake in wheat, In *Soil-Borne Plant Pathogens,* B. Schippers and W. Gams (eds.), London, UK: Academic Press. pp. 525-530.

Jalali, B.L., M.L. Chhabra, and R.P. Singh (1990). Interaction between vesicular arbuscular mycorrhizal endophyted and *Macrophomina phaseolina* in mungbean. *Indian Phytopathology* 43:527-530.

Jindal, V., A. Atwal, B.S. Sekhon, and R. Singh (1992). Effect of NaCl salinity on metabolism of vesicular arbuscular mycorrhizae inoculated moong plants. *Mycorrhiza News* 4:7.

Joseph, U. and I.L. Kothari (1993). Endomycorrhizal occurrence in some crops raised in the fields irrigated with petro-effluent. *Mycorrhiza News* 5:6-7.

Kehri, K.H. and S. Chandra (1995). Relative efficacy of single and multiple inocula of VAM fungi for chickpea. *Mycorrhiza News* 7:9-10.

Khade, S.W. and B.F. Rodrigues (2004). Populations of AM fungi associated with rhizosphere of banana (*Musa* sp.) as influenced by seasons. *Mycorrhiza News* 16:11-13.

Konde, B.K., A.D. Tambe, and S.K. Ruikar (1988). Yield of nitrogen and phosphorus uptake by onion as influenced by inoculation of VAM fungi and *Azospirillum brasilence.* In *Mycorrhiza for Green Asia Proceedings of the First Asian Conference on Mycorrhiza,* A. Mahadevan, N. Raman, and K. Natrajan Chennai (eds.), India: University of Madras. pp. 222-224.

Krishna, K.R. and D.J. Bagyaraj (1983). Interaction between *Glomus fasciculatum* and *Sclerotium rolfsii* in peanut, *Arachis hypogea. Canadian Journal of Botany* 61:2349-2351.

Krishna, K.R., K.G. Sheety, P.J. Dart, and D.J. Andrews (1985). Genotype dependent variation in mycorrhizal colonization and response to inoculation of pearl millet *Pennisetum americanum. Plant and soil* 86:113-126.

Kumar, H. and D.J. Bagayraj (1989). Effect of cropping sequence, fertilizer and farmyard manure on vesicular arbuscular mycorrhizal fungi in different crops over three consecutive seasons. *Biology and Fertility of Soils* 7:173-175.

Kumar, D., S.H. Gaikwad, and S.P. Singh (1995). Influence of plant growth promoting rhizobacteria on mycorrhizal association in wheat. In *Mycorrhizae: Biofertilizers for Future. Proceedings of the Third National Conference on Mycorrhiza,* A. Adholeya and S. Singh (eds.), New Delhi, India: Tata Energy Research Institute. pp. 206-208.

Kunwar, I.K., P.J.M. Reddy, and C. Manoharachary (1999). Occurrence and distribution of AMF associated with garlic rhizosphere soil. *Mycorrhiza News* 11:4-6.

Lakshman H.C. (2000). Occurrence and tolerance of VAM plants growing on polluted soils with sewage and industrial effluents. *Journal of Nature Conservation* 12:9-18.

Lingaraju, B.S., M.N. Srinivasa, H.B. Bablad (1994). Response of soybean to VA mycorrhizal fungi at varied levels of P. *Journal of Oilseeds Research* 11:300-302.

Maiti, D., M. Variar, and R.K. Singh (1996). Perpetuation of native VAM fungi under monocropped, rainfed upland agro-ecosystem. *Mycorrhiza News* 8:7-9.

Maiti, D., R.K. Singh, and J. Saha (2000). Influence of tillage practices and rice (*Oryza sativa* L.) based cropping systems on native VAM fungal population in rainfed upland ecosystem. *Mycorrhiza News* 12:18-20.

Maiti, D., M. Vatiar, R.K. Singh, C.V. Saha, and J. Saha (1997). Enhancing VA mycorrhizal association in rainfed upland rice (*Oryza sativa* L.) based cropping system. *Mycorrhiza News* 8:11-13.

Manjunath, A. and D.J. Bagyaraj (1984). Effect of systemic fungicides on various developmental stages in establishments of vesicular-arbuscular mycorrhizal fungi in host plants. *Plant and Soil* 80:147-185.

Manjunath, A. and D.J. Bagyaraj (1986). Response of blackgram, chickpea and mungbean to VAM inoculation in an unsterile soil. *Tropical Agriculture* 63:33-35.

Manoharachary, C., T. Sulochana, and P. Ramarao (1990). Role of Vesicular-Arbuscular Mycorrhizal fungi in oil seed production. *Mycorrhiza News* 2:2-4.

Mathur, N. and A. Vyas (1995). Production of highly nutritive pods of *Cyamopsis tetragonoloba* by VA mycorrhizae. *Indian Journal of Experimental Biology* 33:464-465.

Michelsen, A. and S. Rosendhal (1990). The effect of VA mycorrhizal fungi, phosphorus and drought stress on the growth of *Acacia nilotica* and *Leucaena leucocephala* seedlings. *Plant Soil* 133:79-83.

Mishra, A. and B.N. Shukla (1995). Studies on management of root knot *(Meloidogyne incognita)* of tomato by *Glomus fasciculatum* and some pesticides. *Mycorrhiza News* 7:8-11.

Mukerji, K.G. and R. Rani (1989). Mycorrhizal distribution in India. *Mycorrhiza News* 1:1.

Mulani, R.M., R.R. Prabhu, and M. Dinkaran (2002). Occurrence of VAM (Vesicular Arbuscular Mycorrhiza) in the roots of *Phyllanthus fraternus* Webster. *Mycorrhiza News* 14:11-14.

Murugan, R. and T. Selvaraj (2003). Reaction of Kashini (*Cichorium entybus* L.) to different native arbuscular mycorrhizal fungi. *Mycorrhiza News* 15:10-13.

Nagarajan, P., N.V. Radha, D. Kandasamy, G. Oblisami, and S. Jayaraj (1989). Effect of combined inoculation of *Azospirillum brasilence* and *Glomus fasciculatum* on mulberry. *Madras Agricultural Journal* 76:601-605.

National Thermal Power Corporation (NTPC), Korba, Chhattishgarh (2001-2004). *Reclaiming Ash Ponds and Immobilizing Heavy Metals by Mycorrhizal Organo-biofertiliser at Korba STPS*, New Delhi, India: TERI, TERI project report no. 2001CM 61.

Pai, G., D.J. Bagyaraj, and T.G. Prasad (1990). Mycorrhizal reproduction as influenced by moisture strees. In *Current Trends in Mycorrhizal Research. Proceedings of the National Conference on Mycorrhiza,* B.L. Jalali and H. Chand (eds.), Hisar, India: Haryana Agricultural University. pp. 25-26.

Panja, B.N. and S. Chaudhuri (1998). Effect of monoculture of plants on inoculum build up and quantitative distribution of VA mycorrhizal species in rhizosphere. *Mycorrhiza News* 10:13-15.

Parvati, K., K. Venkateswarlu, and A.S. Rao (1985). Effect of systemic fungicides on establishment of Vesicular-arbuscular mycorrhizal fungi in host plants. *Canadian Journal of Botany* 63:1673-1675.

Prasad, S., K.K. Sulochana, and S.K. Nair (1990). Comparative efficiency of different V A mycorrhizal fungi on cassava (*Manihot esculenta* Crantz). Journal of root crops 16:39-40.

Quilambo, O.A. 2003. The vesicular-arbuscular mycorrhizal symbiosis. *African Journal of Biotechnology* 2:539-546.

Ramakrishnan, B.R., B.N. Johri, and R.K. Gupta (1990). The response of mycorrhizal maize plants to variations in water potentials. In *Current Trends in Mycorrhizal Research. Proceedings of the National Conference on Mycorrhiza,* B.L. Jalali and H. Chand (eds.), Hisar, India: Haryana Agricultural University. pp. 61-62.

Ramana, B.V. and R.S.H. Babu (1999). Response of onion *(Allium cepa)* to inoculation of VAM fungi at different levels of phosphorus. In *Plant Growth Responses and Mycorrhizal Dependency. Proceedings of the National Conference on Mycorrhiza,* S. Singh (ed.), New Delhi, India: Tata Energy Research Institute.

Ramraj, B., N. Shanmugan, and R.K. Dwarkanath (1988). Biocontrol of Macrophomina root rot of cowpea and fusarium wilt of tomato by using VAM fungi. In *Mycorrhiza for Green Asia Proceedings of the First Asian Conference on Mycorrhiza,* A. Mahadevan, N. Raman, and K. Natrajan (eds.), Chennai, India: University of Madras. pp. 250-251.

Rao A.V. and R. Tak (2002). Growth of different tree species and their nutrient uptake in limestone mine spoil as influenced by arbuscular mycorrhizal (AM)-fungi in Indian arid zone. *Journal of Arid Environment* 51:113-119.

Rao, V.U. and A.S. Rao (1996). Response of black gram *(Phaseolus mungo)* and green gram *(P. radiatus)* to inoculation with vesicular-arbuscular mycorrhizal fungus under different sources of phosphorus fertilizer. *Indian Journal of Agricultural Science* 66:613-616.

Ratti, N. and K.K. Janardhan (1996). Response of dual inoculation with VAM and *Azospirillum* on the yield and oil content of Palmarosa (*Cymbopogon martini* var. motia). *Microbiological Research* 151:325-328.

Ray, P., R. Tiwari, U.G. Reddy, and A. Adholeya (2005). Detecting the heavy metal tolerance level in ectomycorrhizal fungi in vitro. *World Journal of Microbiology & Biotechnology* 21:309-315.

Reddy, N.C. and C.A. Gudige (1999). Growth response of carrot (*Daucus carota* L.) to a selective VA mycorrhizal inoculant in an unsterile soil. *Mycorrhiza News* 11:11-13.

Reddy, M.V., J.N. Rao, and K.R. Krishna (1988). Influence of mycorrhiza on chickpea Fusarium wilt. *International Chickpea Newsletter* 19:16.

Reddy, M.V., J.N. Rao, and K.R. Krishna (1989). Influence of mycorrhizae on fusarium wilt of pigeonpea. *International Pigeonpea Newsletter* 9:23.

Reddy, C.N., B.K. Bharti, H.G. Rajkumar, and D.N. Sunanda (2004). *Mycorrhiza News* 16:9-12.

Reeves, F., D. Brent, T.M. Wagner, and K. Jean (1979). The role of endomycorrhizae in revegetation practices in the semi-arid west. I. A comparison of incidence of mycorrhizae in severely disturbed vs. natural environments. *American Journal of Botany* 66:6-13.

Sambandan K., K. Kannan, and N. Raman (1992). Distribution of vesicular arbuscular mycorrhizal fungi in heavy metal polluted soils of Tamilnadu, India. *Journal Environmental Biology* 132:159-167.

Sattar, M.A. and A.C. Gaur (1985). Characterization of phosphate dissolving microorganisms isolated from some Bangladesh soil samples. *Bangladesh Journal of Microbiology* 2:22-28.

Secilia, J. and D.J. Bagyaraj (1989). Bacteria and actinomycetes associated with pot cultures of vesicular arbuscular mycorrhizas. *Canadian Journal of Microbiology* 33:1069-1073.

Secilia, J. and D.J. Bagyaraj (1994). Evaluation and first year field-testing of efficient vesicular arbuscular mycorrhizal fungi for inoculation of wetland rice seedlings. *World journal of Microbiology and Biotechnology* 10:381-384.

Selvaraj, T. and C. Bhaskaran (1996). Occurrence and distribution of VA-Mycorrhizal fungi in soils polluted with paper mill effluent. *Pollen Research* 15:197-300.

Selvaraj, T., R. Murugan, and C. Bhaskaran (2001). Arbuscular mycorrhizal association of Kashini (*Cichorium intybus* L.) in relation to physio-chemical characters. *Mycorrhiza News* 13:14-16.

Sengupta, A. and S. Chaudhuri (2002). Arbuscular mycorrhizal relations of mangrove plant community at the Ganges river estuary in India. *Mycorrhiza* 12:169-174.

Senthil, K.S. and D.I. Arockiasamy (1994). Role of vesicular arbuscular mycorrhiza in reducing the heavy metal toxicity in a *Sorghum bicolor* (L) Monech with reference to nitrate reduction system. *Journal Swamy Botanical Club* 11:25-27.

Sharma, M.P. and A. Adholeya (2000). Enhanced growth and productivity following Inoculation with indigenous AM fungi in four varieties of onion (*Allium cepa* L.) in an Alfisol. *Biological Agriculture and Horticulture* 18:1-14.

Sharma, M.P. and A. Adholeya (2004). Effect of arbuscular mycorrhizal fungi and phosphorus fertilization on the post vitro growth and yield of micropropagated strawberry grown in a sandy loam soil. *Canadian Journal of Botany* 82:322-328.

Sharma, M.P. and S. Bhargava (1993). Potential of vesicular arbuscular mycorrhizal fungus *Glomus fasciculatum*, against root knot nematode, *Meloidogyne incognita*, on tomato. *Mycorrhiza News* 4:4-5.

Sharma, M.P., N.P. Bhatia, and A. Adholeya (2001). Mycorrhizal dependency and growth responses of *Acacia nilotica* and *Albizzia lebbeck* to inoculation by indigenous AM fungi as influenced by available soil P in a semi-arid Alfisol wasteland. *New Forests* 21:89-104.

Sharma, A.K., P.C. Srivastava, B.N. Johri, and V.S. Rathore (1992). Kinetics of zinc uptake by mycorrhizal (VAM) and nonmycorrhizal corn *(Zea mays)* roots. *Biology and Fertility of Soils* 13:206-210.

Sheela, A.M. and M.D. Sundaram (2003). Role of VA mycorrhizal biofertilizer in establishing black gram *(Vigna mungo)* var-T9 in abandoned ash ponds of Neyveli Thermal Power Plant. *Mycorrhiza News* 15:13-16.

Simiyon, A., P. Theoder, and M. Vivekanandan (1994). Occurrence of saline-tolerant VAM in mulberry varieties planted in the coastal region of Tamil Nadu. *Mycorrhiza News* 6:5-6.

Singh, A.K. and R.R. Mishra (1995). Effect of vesicular-arbuscular mycorrhiza on growth and phosphorus uptake of phosphorus deficiency tolerant and susceptible paddy varieties (RCPL 101 and RCPL 104) under different soil phosphorus levels. In *Mycorrhizae: Biofertilizers for Future. Proceedings of the Third National Conference on Mycorrhiza,* A. Adholeya and S. Singh (eds.), New Dehli, India: Tata Energy Research Institute. pp. 314-321.

Singh, C., A.K. Singh, and B.N. Johri (2002). Variation among maize lines for colonization and responsiveness to arbuscular mycorrhizal fungi. *Mycorrhiza News* 14(3):10-13.

Singh, C.S. (1995). Impact of N_2 fixing and phosphate solubilising bacteria on root colonization and spore production. In *Mycorrhizae: Biofertilizers for Future. Proceedings of the Third National Conference on Mycorrhiza,* A. Adholeya and S. Singh (eds.), New Delhi, India: Tata Energy Research Institute. pp. 189-195.

Singh, H.P. and T.A. Singh (1988). Influence of VAM and phosphate solublizers on phosphate solubility and growth of maize (*Zea mays* L.). In *Mycorrhiza Round Table, Proceedings of National Workshop,* New Delhi, March 13-15, pp. 450-460.

Singh, K., A.K. Varma, and K.G. Mukerji (1987). Vesicular arbuscular mycorrhizal fungi in diseased and healthy plants of *Vicia faba. Acta Botanica Indica* 15: 304-310.

Singh, R. and A. Adholeya (2002a). AMF biodiversity in wheat agro-ecosystems of India. *Mycorrhiza News* 14:21-23.

Singh, R. and A. Adholeya (2002b). Biodiversity of AMF and agricultural potential II: The impact of agronomic practices. *Mycorrhiza News* 13:22-24.

Singh, R. and A. Adholeya (2002c). Biodiversity of AMF and agricultural potential III: Approaches to study the metabolic status and functional diversity of resident AMFs. *Mycorrhiza News* 14:20-22.

Singh, R. and A. Adholeya (2004). Testing the symbiotic performance of different cultivars of wheat with selected arbuscular mycorrhizal fungi. *Mycorrhiza News* 16:18-19.

Singh, R.B. and N.C. Reddy (1994). Relation between VAM association and the endogonaceous spore numbers in rhizosphere of calcareous soils. *Mycorrhiza News* 6:8-9.

Singh, S. (1996). Effect of soil pollution on mycorrhizal development. *Mycorrhiza News* 8:1-9.

Singh, S. (2001). Mycorrhizal dependency, part 1: Selection of efficient mycorrhizal fungi. *Mycorrhiza News* 13:2-16.

Singh, Y.P., R.S. Singh, and K. Sitaramaiah (1990). Mechanism of resistance of mycorrhizal tomato against root-knot nematode. In *Current Trends in Mycorrhizal Research. Proceedings of the National Conference on Mycorrhiza,* B.L. Jalali and H. Chand (eds.), Hisar, India: Haryana Agricultural University. pp. 96-97.

Sondergaard, M. and S. Lindegaard (1977). Vesicular-arbuscular mycorrhizae in some aquatic vascular plants. *Nature* 268:232-233.

Sreenivasa, M.N. and N.B. Gaddagimath (1993). Mycorrhizal dependency of chilli cultivars. *Zentralblatt fur Mikrobiologie* 148:55-59.

Subhashini, D.V., B.S. Rana, and V.P. Potty (1988). Genotype dependent variation in mycorrhizal colonization and response to phosphorus in *Sorghum bicolor.* In

Mycorrhiza for Green Asia, Proceedings of the First Asian Conference on Mycorrhiza, A. Mahadevan, N. Raman, and K. Natrajan (eds.), Chennai, India: University of Madras. pp. 255-259.

Tanu, A., A. Prakash, and A. Adholeya (2004). Effect of different organic manures/composts on the herbage and essential oil yield of *Cymbopogon winterianus* and their influence on the native AM population in a marginal alfisol. *Bioresource Technology* 92:311-319.

Tewari, L., B.N. Johri, and S.M. Tandon (1993). Host genotype dependency and growth enhancing ability of VA-mycorrhizal fungi for *Eleusine coracana* (Finger millet). *World Journal of Microbiology and Biotechnology* 9:191-195.

Tewari, L., S.M. Tandon, and B.N. Johri (1987). Effect of combined inoculation of *Azospirillum brasilense* sp. 7 and *Glomus caledonium* on yield of finger millet *(Eleusine coracana).* In *Proceedings of the Seventh North American Conference on Mycorrhizae,* Gainesville, FL. pp. 287-289.

Thaper, M.N. and Y.T. Fasrai (2002). VAM and better growth of micropropagated banana. *Mycorrhiza News* 14:16-18.

The Tribune, March 1, 2004, Chandigarh, India. <http://www.tribuneindia.com/2004/20040301/agro.htm#2> accessed January 14, 2006.

Theoder, P.A.S. and M. Vivekanandan (1994). Occurrence of saline-tolerant VAM in mulberry varieties planted in the coastal region of Tamilnadu. *Mycorrhiza News* 6:5-6.

Tholkappian, P., A. Sivasaravanan, and M.D. Sundaram (2000). Effect of phosphorus levels on the mycorrhizal colonization, growth yield and nutrient uptake of cassava (*Manihot esculenta* crantz) in alluvial soil of coastal Tamilnadu. *Mycorrhiza News* 11:15-17.

Thomas, G.V. and S.K. Ghai (1987). Genotypic dependent variation in vesicular arbuscular mycorrhizal colonization of coconut seedlings. *Proceedings of the Indian Academy of Sciences Plant Sciences* 97:289-294.

Tilak, K.V.B.R. (1985). Interaction of vesicular arbuscular mycorrhizae and nitrogen fixers. *Proceedings of National Symposiumon Current Trends in Soil Biology,* Hisar, India. pp. 219-226.

Tilak, K.V.B.R., C.S. Singh, N.K. Roy, and N.S. Subbarao (1982). *Azospirillum brasilense* and *Azotobacter chroococcum* inoculum: Effect on yield of maize *(Zea mays)* and sorghum *(Sorghum bicolor). Soil Boilogy and Biochemistry* 14:417-418.

Tiwari, P. and A. Adholeya (2001). In vitro co-culture of two AMF isolates *Gigaspora margarita* and *Glomus intraradices* on Ri T-DNA transformed roots. *FEMS Microbiology Letters* 20:639-643.

Tiwari, P. and A. Adholeya (2003). Host dependent differential spread of *Glomus intraradices* on various Ri T-DNA transformed roots in vitro. *Mycological Progress* 2:171-177.

Tiwari, P.A., Prakash, and A. Adholeya (2003). Commercialization of arbuscular mycorrhiza. In *Handbook of Fungal Biotechnology,* D.A. Arora (ed.), New York: Marcel Dekker. pp. 195-204.

Vallejo, V.R., I. Serrasloses, J. Cortina, J.P. Seva, A. Valdecantos, A. Vilagrosa (2000). Restoration for soil protection after disturbances. In *Life and Ecosystems in the Mediteranean,* L. Trabaud (ed.), Southhampton: WIT Press. pp. 301-344.

Varshney, A., M.P. Sharma, A. Adholeya, V. Dhawan, and P.S. Srivastava (2002). Enhanced growth of micropropagated *Lilium* sp. inoculated with arbuscular mycorrhizal fungi at different P fertility levels in an Alfisol. *The Journal of Horticultural Science and Biotechnology (UK)* 77:258-263.

Verma, A., Y.P. Singh, N.S. Bisht, V. Mohan, R.B. Singh, and S. Rawat (1998). Biodiversity of arbuscular mycorrhizal fungi: An Indian perspective. ICOM II Abstracts. <http://www-icom2.slu.se/ABSTRACTS/abstract.html> accessed January 4, 2006.

Vyas, S.C. and S. Singh (1992). Role of fungicides in formation of vesicular arbuscular mycorrhizain plants. *Mycorrhiza News* 4:1-6.

Vyas, S.C., A. Vyas, K.C. Mahajan, and V. N. Shroff (1990). In *Current Trends in Mycorrhizal Research,* B.L. Jalali and H. Chand (eds.), Haryana Agricultural University, Hisar, India. pp. 188-189.

Chapter 8

Mycorrhizae and Crop Production in a World of Rapid Climate Change: A Warning Call

Mayra E. Gavito

There have been many reviews of the arbuscular mycorrhizal association and global climate change in the past ten years (Staddon and Fitter, 1998; Rillig and Allen, 1999; Fitter, Heinemeyer, and Staddon, 2000; Treseder and Allen, 2000; Staddon, Heinemeyer, and Fitter, 2002; Fitter et al., 2004). In contrast, relatively few research papers have been published on the subject. Roughly 40 papers have been published on the increase in atmospheric CO_2, fewer than ten on the increase in temperature, and fewer than ten on N deposition, UV-B radiation, or ozone effects. Fewer than half of those papers have direct relevance to crop production and only a handful of those are field experiments. Almost everything that can be said at the moment has already been said in one way or another in previous reviews and little material is available for a discussion of what is specifically relevant to crop production. I must say that when I was first invited to write this chapter, I was going to decline. The main reason I accepted the job was to provide what had been missing in previous reviews, that is, to provide a clear idea of what is coming, to identify knowledge gaps, and to redefine the most pressing issues accordingly. In regard to the general biology of the arbuscular mycorrhizal association, Fitter's recently published review (Fitter et al., 2004) accomplishes that task nicely. One of the two main goals of this chapter is to attempt to focus on what is relevant to crop production.

Mycorrhizae in Crop Production
© 2007 by The Haworth Press, Inc. All rights reserved.
doi:10.1300/5425_08

The other reason for taking the job was that through this book about mycorrhizae and crop production, I thought I might be able to reach a forum that has been abandoning mycorrhizal research over the past ten years. The exodus of agronomists from mycorrhizal research has been justified, as we failed to show that the arbuscular mycorrhizal association was important to crop production in conventional, high-input agriculture. Ryan and Graham (2002) suggested that in many production-oriented agricultural systems, the C-costs derived from the establishment of arbuscular mycorrhizal associations outweighed any benefits to crop plants. The role of the arbuscular mycorrhizal association in agriculture remains, however, controversial because of other examples where mycorrhizal benefits were found (Gavito and Miller, 1998; Goicoechea et al., 2004) and because of the increasing fertilizer, tillage, and pesticide regulations pushing toward more sustainable agriculture. Sustainable agriculture is a scenario where mycorrhizae have a better chance of showing their benefits and where yields are not the main criterion of success.

At the last international conference on mycorrhizae held in Montreal, Canada, in 2003, although inoculation trials, inoculant production, and yield enhancements were the subject of a symposium and numerous papers (Janos, 2004), pest control, bioremediation, and landscape restoration had clearly taken over the field of applied mycorrhizal research. This was also justified. We have been better at showing that the arbuscular mycorrhizal association is extremely useful in fixing environmental problems and that mycorrhizal fungi are consistently useful organisms whenever plants and almost any form of stress are combined. Not surprisingly, one of the most comprehensive recent reviews of agroecosystem responses to global climate change does not include a single word about mycorrhizae (Fuhrer, 2003). Nevertheless, it is becoming increasingly clear that mycorrhizae have many important functions in ecosystem sustainability (including, of course, agroecosystems) apart from the old, plant-centered, biofertilizer function that most agronomists wished to find in them. Fortunately, with the increasing awareness of environmental damage, many countries are adopting new agricultural policies that place increasing importance on sustainability. The second main goal of this chapter is to call on agronomists and others interested in crops or applied aspects of mycorrhizal research to renew their interest in mycorrhizae because there will be a plethora of sources of crop stress in the near future. Unless we give up growing crops in open land, tackling environmental stress will be the challenge of the coming years and the arbuscular mycorrhizal association will likely regain its importance in crop production.

The review papers cited synthesize most of what has been done and what is known on the subject of mycorrhizae and global climate change and I strongly encourage readers to read them. I will attempt to focus exclusively on the information most relevant to crop production and on my two main goals.

GLOBAL CLIMATE CHANGE: WHAT IS REALLY COMING AND WHY SHOULD WE WORRY?

A few months ago, in a presidential address in the *Journal of Ecology,* John Grace (2004) described the evolution of public concern about environmental change. The story is quite recent, having started in the 1960s, and I am proud to say I belong to the generation that was born at the dawn of environmental consciousness. The most discouraging message of Grace's contribution was perhaps that the development of public awareness of environmental change has been far too slow to make us fully conscious of the magnitude of the problem.

Human activity since the beginning of the Industrial Era is undoubtedly changing the climate of our planet at an unprecedented rate (IPCC, 2001a). Atmospheric CO_2 increased from 280 ppm (years 1000-1750) to 368 (year 2000) and is predicted to reach almost 1,000 ppm by year 2100. The global mean temperature increased by 0.6°C + 0.2°C over the last century and will likely gradually increase by 1.4°C-5.8°C in this century (IPCC, 2001b). Even considering the beneficial effects of increasingly stringent environmental regulations and various compensatory activities, there is strong evidence that significant global changes relative to those observed in previous centuries will continue in the near future. Evidence continues to build that, despite the concerted efforts to reduce greenhouse gas emissions and to increase C sequestration, emissions will continue to exceed the Earth's sink capacity and global climate will continue to change for a long time. Even if we managed to stabilize atmospheric CO_2 at levels between 450 and 1,000 ppm in this century, it is likely to take several centuries to reach the equilibrium temperature rise (IPCC, 2001a).

The rate of global climate change has raised concerns about the performance and, in some cases, even the survival of some organisms that have no time to migrate or to adapt to rapidly changing environments. If the observed changes in biological and physical indicators of alterations in the Earth's atmosphere, climate, and biophysical system during the last century were not alarming enough for the septics, there are two words that comprehensively define the predictions for the world's climate during this century: extreme and variable. Briefly, we are expecting higher maximum and minimum temperatures, more hot days and heat waves, drought, fewer cold days, and fewer frost days and cold waves over nearly all land areas. Significant disruptions of ecosystems from disturbances such as fire, drought, pest infestation, invasion of species, floods, storms, and soil erosion are also expected to increase (IPCC, 2001a).

Most papers and reviews dealing with climate change and crop production in the past ten years have concentrated on the fertilizer effect of increasing atmospheric CO_2, increasing temperature, predicted yield increases, and potential yield limitations (Rogers, Runion, and Krupa, 1994; Fuhrer, 2003). This picture of the climate change scenario for this century is a definite eye-opener: the most pressing issues in crop production now have much to do with the risk of crop damage and survival.

The priority areas for mycorrhizal research and crop production in the future should include risk of damage by drought, heat, pests, wind, flooding, invasive species, and erosion. It will be of little use to continue to investigate atmospheric CO_2 or temperature increases without considering the potential accompanying events associated with climate change. Although the benefits of mycorrhizal associations have been demonstrated in crops under stress conditions in numerous experiments carried out in controlled environments, few field studies have been conducted. Field experiments manipulating temperature, water, pest injury, wind exposure, soil quality, and weed invasion are badly needed to test the potential of mycorrhizal associations to help plants and soils to overcome extreme and variable climates.

HISTORICAL DEVELOPMENT

Most of the early research involving the response of mycorrhizal fungi or mycorrhizal plants to global climate change factors was published in the 1990s and was related mainly to the increase in atmospheric CO_2. It was hypothesized that the fertilizing effect of CO_2 would result in an increased supply of C for mycorrhizal fungi, fungal growth promotion (Treseder and Allen, 2000), and consequently greater benefits for the host plant as a result of the enhancement in fungal activity (Fitter, Heinemeyer, and Staddon, 2000). However, this turned out to be a disappointment. The overall conclusions of earlier studies have been that, because increases in atmospheric CO_2 were perceived by mycorrhizal fungi only through their effects on host plants, the fungal responses were dependent on the direction and magnitude of the plant's response to atmospheric CO_2 enrichment (Fitter et al., 2004). The response was only indirect. Fungal development was always proportional to the growth response of the host plant (Figure 8.1), and the response trends of the intraradical and extraradical mycorrrhizal mycelium to fertility treatments, for example, were not altered by the CO_2 or CO_2+ warming treatments (Figure 8.2). Staddon, Graves, and Fitter (1999) tested and confirmed the proportionality between plant and fungal development at two levels of atmospheric CO_2 using ten host plant species. Fungal function, exemplified by hyphal P uptake, was not altered by exposing the host plant to increased atmospheric CO_2 (Table 8.1; Gavito, Bruhn, and Jakobsen, 2002; Gavito, Schweiger, and Jakobsen, 2003). Moreover, there was a very interesting and intriguing response in total plant P uptake. Total plant P uptake was similar in non-mycorrhizal plants, low mycorrhizal colonization plants, and high mycorrhizal colonization plants at ambient atmospheric CO_2, but increased linearly with the degree of mycorrhizal colonization when grown with atmospheric CO_2 enrichment (Figure 8.3a). This indicated that mycorrhizal colonization helped plants to increase their root P uptake capacity even though it was determined that hyphal P uptake was not directly responsible for this response. Mycorrhizal colonization may alter the capacity of roots to take up P independently of the fungal contribution to P uptake, and the lack of correlation between mycorrhizal P uptake and host growth or host total P uptake has re-

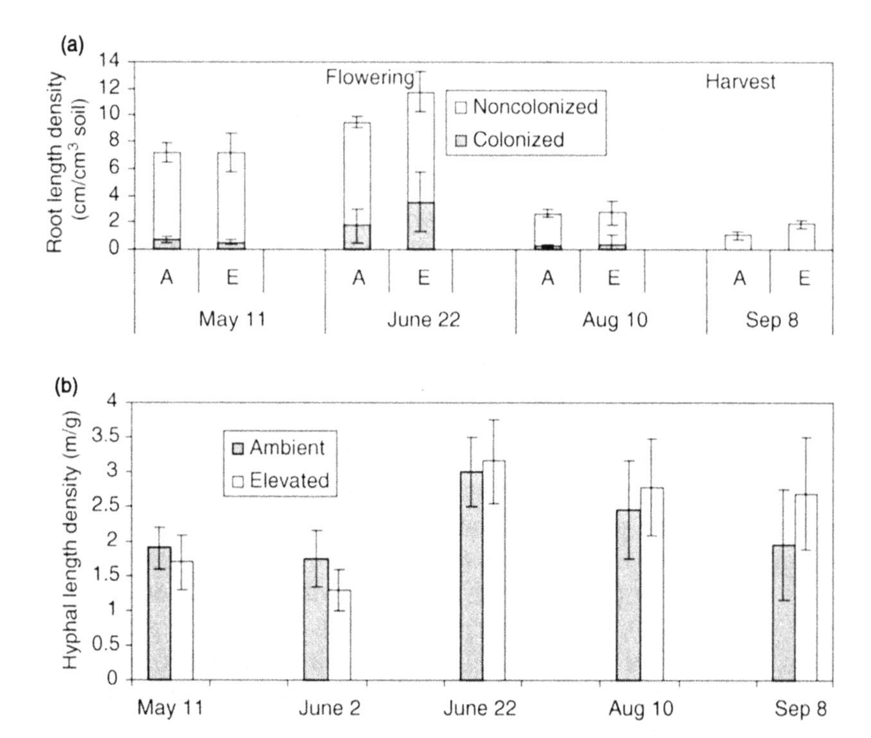

FIGURE 8.1. Root (a) and mycorrhizal hyphae (b) length density values (mean ± SE) from soil and winter wheat measured during the growing season in field plots in Foulum Denmark at ambient or elevated (500 ppm) atmospheric [CO_2]. Elevated atmospheric [CO_2] was achieved by CO_2 fumigation in FACE (Free atmospheric [CO_2] enrichment) rings. Mean minimum soil temperature was 5°C and maximum soil temperature was 17°C during the growing season.

cently been demonstrated with several host and fungal species (Smith, Smith, and Jakobsen, 2004). In addition, it was shown that neither mycorrhizal colonization nor atmospheric CO_2 enrichment had strong effects on the root dynamics of the host plant that might have contributed to this response (Gavito et al., 2001). The conclusion of those studies was that it is the host plant that sometimes is limited by C availability (as expected at least for most plants with C_3 photosynthesis) and that is why it is the host plant that responds to atmospheric CO_2 enrichment, not mycorrhizal fungi. A significant benefit of being mycorrhizal when atmospheric C availability was increased for a C-

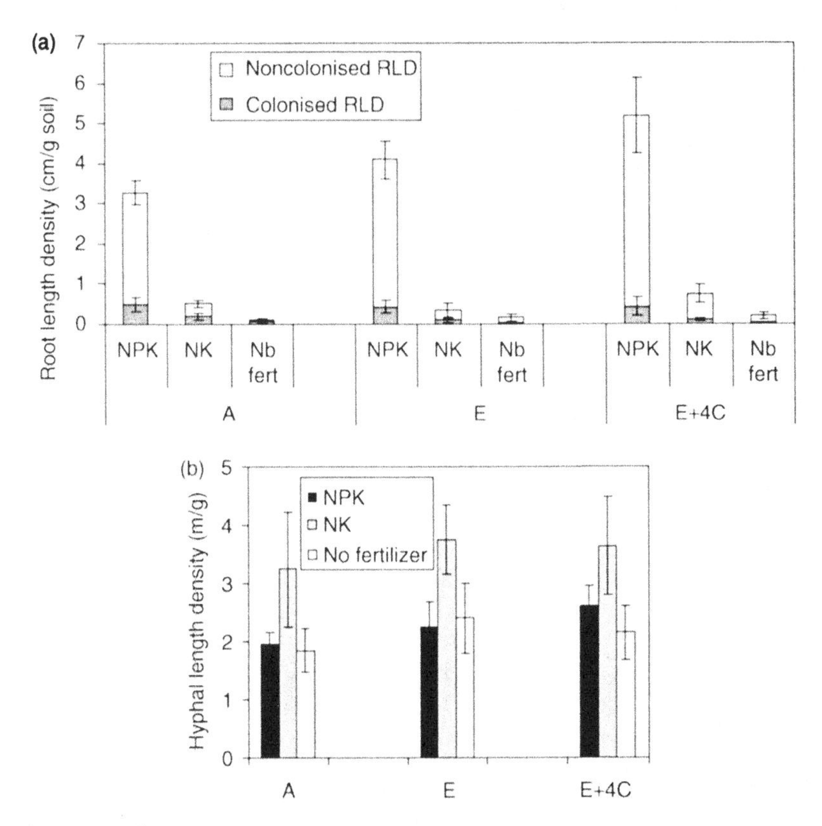

FIGURE 8.2. Root (a) and mycorrhizal hyphae (b) length density values (mean + SE) from soil and spring wheat root samples taken from 20 kg soil columns maintained at ambient atmospheric [CO_2], elevated (700 ppm) atmospheric [CO_2], or elevated atmospheric [CO_2] +4°C in soil and air temperature. Wheat plants were grown at field planting density in growth rooms simulating outdoor temperature changes during the growing period in Denmark and soil temperatures were achieved by placing the soil columns in air-cooling chambers placed inside the growth rooms. Plants were harvested and soil and roots were sampled when they reached a fixed phenological Zadok's stage in the treatment with better growth. Soils with the different fertilization treatments were collected from adjacent long-term fertilization plots in Askov, Denmark.

limited host plant resided in the possibility of maintaining its C fixation capacity and growth potential through the enhanced C sink strength exerted by the activity of the fungus (Figure 8.3b; Staddon, Fitter, and Robinson, 1999; Syvertsen and Graham, 1999; Jifon et al.,

TABLE 8.1. Total P and ^{33}P activity concentration and content, and P-use efficiency in shoots of nodulated pea plants growing at ambient or elevated atmospheric [CO_2], 10°C or 15°C soil temperature and inoculated with *Glomus caledonium* or the native AMF from field soil. All plants were harvested 74 days after emergence. ^{33}P activity originated from mycorrhizal hyphae penetrating the mesh of a ^{33}P labeled root exclusion compartment. There was basically no hyphal growth in the root exclusion compartment at 10°C Mean (SE, n = 4).

Treatment	[P] mg g^{-1}	[^{33}P activity] Bq g^{-1}	Total P mg pot^{-1}	^{33}P activity KBq pot^{-1}	P-use efficiency g d w mg^{-1} P
10°C					
Ambient CO_2					
Glomus caledonium	1.56 (0.04)	12 (7)	66 (6)	0 (0)	0.71 (0.02)
Field soil	1.53 (0.06)	19 (26)	47 (6)	0.116 (0.112)	0.74 (0.05)
Elevated CO_2					
Glomus caledonium	1.48 (0.04)	2 (4)	71 (8)	0.088 (0.173)	0.79 (0.03)
Field soil	1.55 (0.13)	32 (39)	52 (11)	0.392 (0.676)	0.78 (0.04)
15°C					
Ambient CO_2					
Glomus caledonium	1.54 (0.05)	402 (73)	116 (6)	17.88 (5.07)	0.69 (0.03)
Elevated CO_2					
Glomus caledonium	1.50 (0.09)	309 (77)	121 (2)	14.19 (4.88)	0.73 (0.05)
Field soil	1.89 (0.10)	1,692 (395)	99 (14)	45.31 (16.09)	0.56 (0.05)

Source: From Gavito, Schweiger, and Jakobsen (2003) with permission.

2002). This benefit allowed plants to achieve greater growth through sustained greater plant nutrient and water uptake under elevated atmospheric CO_2. Mycorrhizal fungi, in turn, were growing and functioning as usual (perhaps encountering a new root to colonize more often than at ambient atmospheric CO_2, allowing them to grow a little more) and had not even noticed, so to speak, that something was going on aboveground.

Few studies have dealt with mycorrhizae and global climate change factor interactions, but those that have indicate that atmospheric CO_2 will likely interact with water availability (Rillig et al., 2001; Staddon et al., 2003) and not so much with increasing temperature (Gavito,

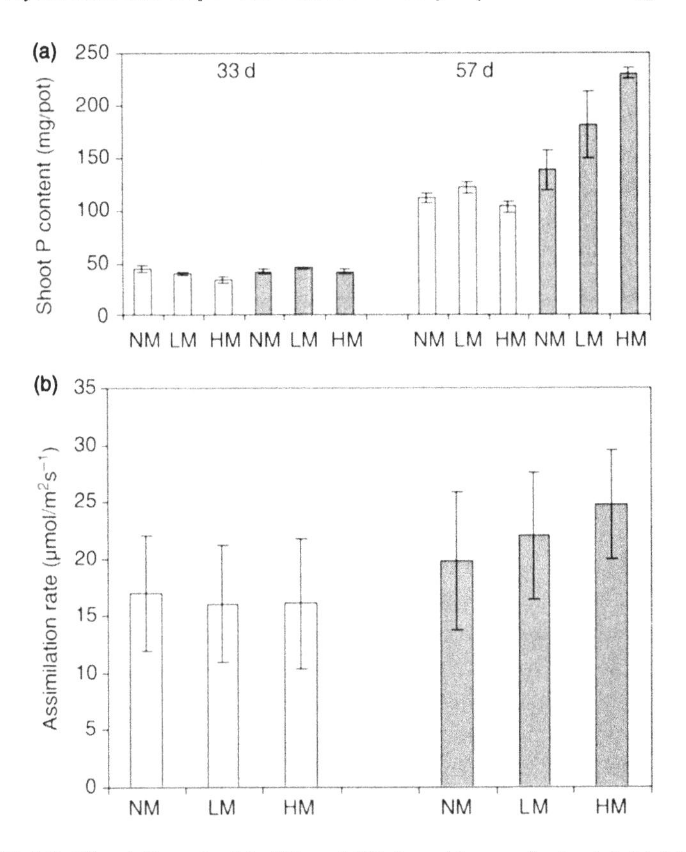

FIGURE 8.3. Shoot-P content in 33 and 57-day-old pea plants. (a). Light-saturated CO_2 assimilation rates measured at the growth [CO_2] from pea plants (b). The CO_2 assimilation rate is an average of three measuring dates (27, 31, and 44 days after emergence). Plants were grown in 20 kg soil columns at ambient (open bars) or elevated (shaded bars) atmospheric [CO_2], and inoculated with 0 percent (non-mycorrhizal, NM), 1 percent (low inoculum, LM) or 5 percent (high inoculum, HM) of soil dry weight with a mixture of three *Glomus* species. Bars denote mean ± SE, n = 5 (From Gavito, Bruhn, and Jakobsen, 2002, with permission).

Schweiger, and Jakobsen, 2003). Nutrient availability or a small increase in temperature (Figure 8.2) seemed not to interact with atmospheric CO_2 (Gamper et al., 2004). Unfortunately, most other published papers examining interactions with other factors deal only with temperate forest tree species.

After being neglected during the first ten years of climate change research, more attention has finally been given to the predicted increase in temperature. Temperature is the global climate change factor that may have the most significant effects on the arbuscular mycorrhizal association because it directly affects most living organisms as well as abiotic processes. The arbuscular mycorrhizal association is expected to respond to increasing temperature when either plants or mycorrhizal fungi are below or above their temperature optima. Plants grow typically within a 5°C to 50°C temperature range and mycorrhizal fungi are expected to have a similar temperature range. Indeed, most studies indicate that arbuscular mycorrhizal fungi develop well within 5°C and 37°C (Staddon, Heinemeyer, and Fitter, 2002), at least inside the roots, and responded positively to temperature increases when they were below their optima (Liu, Wang, and Hamel, 2004) or negatively when they were above their optima (Monz et al., 1994; Heinemeyer et al., 2003; Staddon et al., 2003). Unfortunately, very few studies have reliable, or any, measurements of fungal development in soil and it seems that intraradical and extraradical mycorrhizal development is impaired at both low (Gavito, Bruhn, and Jakobsen, 2002; Gavito, Schweiger, and Jakobsen, 2003 and unpublished; Liu, Wang, and Hamel, 2004) and high (Staddon et al., 2003) temperatures. Evidence that the intraradical and extraradical mycelium have different optimum temperatures is starting to build and further research is needed to understand this basic response and its mechanisms.

Although the first direct effects of temperature on host plants and mycorrhizal fungi were reported just a few months ago (Heinemeyer and Fitter, 2004), it appears that at least part of the response measured in the fungus is attributable to the response of the host plant (Gavito et al., unpublished). This means that there are likely both direct and indirect effects of temperature on the mycorrhizal association. The proportion of each should be investigated in the most natural conditions possible in order to obtain realistic estimates. A small increase in temperature affecting only the upper centimeters of the soil layer had little effect on growth or composition of the arbuscular mycorrhizal fungal community of a native grassland (Heinemeyer et al., 2003), but a larger temperature increase affecting a larger soil volume may produce a significant change. Mycorrhizal function, exemplified again as P uptake, translocation, and transfer to the host plant, was little affected within a 10°C to 25°C range (Wang et al., 2002; Gavito et al., unpublished) although temperatures near the freezing point prevented P movement to the plant (Wang et al., 2002). Temperature effects on fungal development did not seem to interact with P (Figure 8.4) or light availability (Heinemeyer et al., 2003) but did with water availability (Staddon et al., 2003).

There have been few studies dealing with other climate change factors, such as UV-B radiation (van den Staaij et al., 2001). Ozone and N deposition have been investigated in tree species only. Our knowledge in these areas is still too limited to draw any conclusions.

NEGLECTED AND NEW RESEARCH AREAS

The most important gaps in the general biology of the arbuscular mycorrhizal association hampering our progress toward a better understanding of the functioning of the symbiosis in current or future condi-

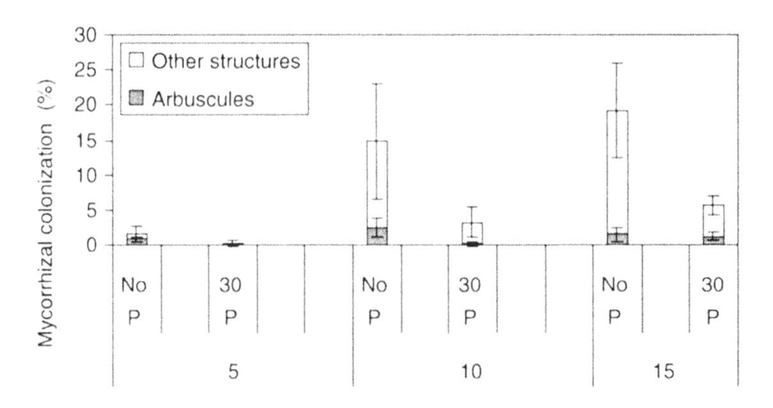

FIGURE 8.4. Intraradical colonization percentages (mean ± SE) measured in roots of spring wheat plants grown at three soil temperatures commonly experienced in the field at the beginning of the growing season in temperate areas. Plants were grown in 20 kg soil columns at field planting density and placed in growth rooms with controlled conditions. Soil temperatures were achieved by placing the soil columns in air-cooling chambers placed inside the growth rooms. Plants were harvested and soil and roots were sampled when the plants reached flowering in the treatment with better growth (15°C).

tions were addressed in a recent review (Fitter et al., 2004). Besides those gaps and the already mentioned gaps in studies investigating global climate change factor interactions, there have been too few studies on other components or side effects of global climate change, such as ozone, UV-B radiation, water availability, soil fertility, and soil protection. The latter three deserve particular attention in crop production in view of the expected climate variability.

High-performance plants growing on healthy soils will be required to face extreme conditions. Promising results have been gathered in the area of soil stabilization and protection (Rillig et al., 1999; Rillig et al., 2001; Treseder et al., 2003), but more experiments are still needed. Arbuscular mycorrhizal fungi are effective and highly resistant soil stabilizing agents and their numerous environmental and plant nutritional benefits on crop production and soil quality are just being realized. This is a very important and promising research area and merits greater attention. The potential of the arbuscular mycorrhizal association in pest control has, fortunately, already been examined in numerous studies as a result of the increasingly stringent environmental regulations on the use of agrochemicals. New research should be focused on combining these studies with realistic scenarios incorporating climate changes and their consequences.

It is also important to develop new research areas in the field of crop production. It is becoming crucial to explore crop resistance to climatic stress and to examine the potential for establishing and managing mycorrhizal associations for crop and soil protection. Field evaluations are urgently needed to evaluate:

- the susceptibility of crops, mycorrhizal fungi, and soils to climate variability and extreme conditions;
- the potential of the arbuscular mycorrhizal association to increase crop resistance to climate variability, and extreme conditions; and
- the potential of the arbuscular mycorrhizal association to reduce soil erosion and improve soil quality under highly variable and extreme climate conditions.

These evaluations should be combined with simulations of other relevant consequences of climate change, such as drought, pest infestation, invasion of species, floods, storms, soil erosion, and alterations in soil quality.

Other neglected areas in crop production are, for example, the interactions between mycorrhizae and other root symbionts whose combined activities may be synergistic or antagonistic since they improve nutrition but compete for plant photosynthate, such as nitrogen-fixing bacteria. The complexity of the response of symbiotic N_2 fixation to climate change factors, especially elevated CO_2, has been well documented (Luscher et al., 2000, and references therein). However, these kinds of interactions are extremely important to crop performance and we have to make an effort to obtain key information on the interaction of such organisms. Unfortunately, the required experiments involving several organisms and climate change factors are not only cost prohibitive, but are also so complex and difficult to design, conduct, and interpret that new approaches have to be explored to address this problem.

As suggested recently by Norby and Luo (2004), interactive modeling and experimentation targeted at improving parameterization are an affordable and powerful alternative tool for exploring responses to multiple changing factors at a larger scale. Simple experiments with one or two factors may be quite useful if they are designed specifically to strengthen parameterization and fill important gaps. Mycorrhizae have rarely been included in modeling efforts, and once again this seems justified in view of the complexity and fragmentation of the existing information (see review by Woodward and Osborne, 2000), but we clearly have to make a greater effort. Mycorrhizal research related to climate change should be incorporated into the above-mentioned parameterization and modeling efforts and we have to open a dialogue with modelers to generate the key information needed.

Perhaps the two most important outcomes of the studies conducted in this field to date are (1) that we have learned that the arbuscular mycorrhizal symbiosis is proving to be an indivisible unit and (2) that we are handicapped in making any serious predictions by the numerous gaps and methodological limitations in studying this complex indivisible unit. Both plants and mycorrhizal fungi grow and function differently when the symbiosis is established and our initial efforts to provide modelers with independent estimates of the response of plants and mycorrhizal fungi to climate change factors proved to be useless. A growing body of evidence shows that plant-fungi specificity exists and must be considered when studying mycorrhizal associations (Fitter et al., 2004). If we want to give meaningful estimates of their responses to global climate change factors, we have to use the plant-fungi combinations as established in the field and think of plants and mycorrhizal fungi as a single unit.

This would by no means diminish previous approaches or attempts to investigate each symbiotic partner independently. Those studies have been extremely useful in providing an understanding of how each partner works. However, we must bear in mind that although the processes and mechanisms operating might be the same when they live together, the numbers will not necessarily match, add or multiply in a straightforward manner. For example, attempts to expose the extraradical mycelium locally to temperature treatments may be a useful way to test for the fungus' own capacity to respond to the treatment and understand how each partner contributes to the response, but even this localized response is unavoidably linked to C supply from the host plant, which may also be influenced by the treatments. We know that arbuscular mycorrhizal fungi and host plants respond directly to temperature changes within the range tested and we know the direction of their response, but the ultimate balance of what they will do in field conditions should be tested and measured. Most important, it should be measured in directly comparable units. The development of reliable methods using molecular techniques for accurately identifying and quantifying known and unknown (uncultivable) fungi in roots and soil is impatiently awaited and the first field studies using molecular tools to trace mycorrhizal fungi in global climate change studies are now being published (Staddon et al., 2003; Heinemeyer et al., 2004). Radioactive isotopes (Gavito, Bruhn, and Jakobsen, 2002; Gavito, Schweiger, and Jakobsen, 2003) and stable isotopes (Staddon, 2004), either alone or in combination with biochemical analyses, such as signature fatty acids (Gavito and Olsson, 2003), are also useful tools for marking and especially quantifying mycorrhizal fungi or their activity in roots and soil. They do not distinguish what fungal species is responsible for the observed values measured, but this is not always relevant, and we may still provide valuable answers. Stable isotope probing (SIP), the separation of nucleic acids of different organisms according to their abilities to use substrates labeled with stable isotopes, is developing quickly and provides a very powerful research tool in microbial ecology (Anderson and Cairney, 2004). This method is very promising and has the added benefit of identifying the active microorganisms involved in each process.

CONCLUDING REMARKS

Despite the many difficulties involved in working with arbuscular mycorrhizal symbiosis and despite the large number of researchers who have abandoned this field of research, there are still many faithful enthusiasts working in this field. We have reason to be happy about that because there is indeed much material to keep us busy for many years. I hope those interested in crop production will find their way back to mycorrhizal research and I am confident that mycorrhizal plants growing on stable soils will cope better with the rapidly changing climate.

REFERENCES

Anderson, I.C. and J.W. Cairney (2004). Diversity and ecology of soil fungal communities: Increased understanding through the application of molecular techniques. *Environmental Microbiology* 6:769-779.

Fitter, A.H., A. Heinemeyer, and P.L. Staddon (2000). The impact of elevated CO_2 and global climate change on arbuscular mycorrhizae: A mycocentric approach. *New Phytologist* 147:179-187.

Fitter, A.H., A. Heinemeyer, R. Husband, E. Olsen, K.P. Ridgway, and P.L. Staddon (2004). Global environmental change and the biology of arbuscular mycorrhizae: Gaps and challenges. *Canadian Journal of Botany* 82:12-18.

Fuhrer, J. (2003). Agroecosystem responses to combinations of elevated CO_2, ozone, and global climate change. *Agriculture, Ecosystems and Environment* 97:1-20.

Gamper, H., M.J. Peter, J. Jansa, A. Luscher, U.A. Hartwig, and A. Leuchtmann (2004). Arbuscular mycorrhizal fungi benefit from seven years of free air CO_2 enrichment in well fertilized grass and legume monocultures. *Global Change Biology* 10:189-199.

Gavito, M.E. and M.H. Miller (1998). Early phosphorus nutrition, mycorrhizae development, dry matter partitioning and yield of maize. *Plant and Soil* 199: 177-186.

Gavito, M.E. and P.A. Olsson (2003). Allocation of plant carbon to foraging and storage in arbuscular mycorrhizal fungi. *FEMS Microbiology Ecology* 45:181-187.

Gavito, M.E., D. Bruhn, and I. Jakobsen (2002). Phosphorus uptake by arbuscular mycorrhizal hyphae does not increase when the host plant grows under atmospheric CO_2 enrichment. *New Phytologist* 154:751-760.

Gavito, M.E., P. Schweiger, and I. Jakobsen (2003). P uptake by arbuscular mycorrhizal hyphae: Effect of soil temperature and atmospheric CO_2 enrichment. *Global Change Biology* 9:106-116.

Gavito, M.E., P.S. Curtis, T.N. Mikkelsen, and I. Jakobsen (2001). Neither mycorrhizal colonization nor atmospheric CO_2 has strong effects on pea root production and root loss. *New Phytologist* 149:283-290.

Goicoechea, N., M. Sanchez-Diaz, R. Saez, and J. Iraneta (2004). The association of barley with AM fungi can result in similar yield and grain quality as a long term application of P or P-K fertilizers by enhancing root phosphatase activity and sugars in leaves at tillering. *Biological Agriculture and Horticulture* 22:69-80.

Grace, J. (2004). Understanding and managing the global carbon cycle. *Journal of Ecology* 92:189-202.

Heinemeyer, A. and A.H. Fitter (2004). Impact of temperature on the arbuscular mycorrhizal (AM) symbiosis: Growth responses of the host plant and it's AM fungal partner. *Journal of Experimental Botany* 55:525-534.

Heinemeyer, A., K.P. Ridgway, E.J. Edwards, D.G. Benham, P.W. Young, and A.H. Fitter (2003). Impact of soil warming and shading on colonization and community structure of arbuscular mycorrhizal fungi in roots of a native grassland community. *Global Change Biology* 10:52-64.

IPCC (2001a). *Climate Change 2001: Synthesis Report. A contribution of Working Groups I, II, and III to the Third Assessment Report of the Intergovernmental Panel on Climate Change.* R.T. Watson and the Core Writing Team (eds.), Cambridge, UK and New York, NY: Cambridge University Press. 398 pp.

IPCC (2001b). *Climate Change 2001: The Scientific Basis. Summary for Policymakers (SPM) and Technical Summary (TS)*, Geneva, Switzerland: IPCC. 98 pp.

Janos, D.P. (2004). The Fourth International Conference on Mycorrhizae from four perspectives. *Mycorrhiza* 14:143-144.

Jifon, J.L., J.H. Graham, D.L. Drouillard, and J.P. Syvertsen (2002). Growth depression of mycorrhizal *Citrus* seedlings grown at high phosphorus supply is mitigated by elevated CO_2. *New Phytologist* 153:133-142.

Liu, A., B. Wang, and C. Hamel (2004). Arbuscular mycorrhizal colonization and development at suboptimal root zone temperature. *Mycorrhiza* 14:93-101.

Luscher, A., U.A. Hartwig, D. Suter, and J. Nosberger (2000). Direct evidence that symbiotic N-2 fixation in fertile grassland is an important trait for a strong response of plants to elevated atmospheric CO_2. *Global Change Biology* 6:655-662.

Monz, C.A., H.W. Hunt, F.B. Reeves, and E.T. Elliot (1994). The response of mycorrhizal colonization to elevated CO_2 and climate change in *Pascopyrum smithii* and *Bouteloua gracilis*. *Plant and Soil* 165:75-80.

Norby, R. and Y. Luo (2004). Evaluating ecosystem responses to rising atmospheric CO_2 and global warming in a multi-factor world. *New Phytologist* 162:281-293.

Rillig, M.C. and M.F. Allen (1999). What is the role of arbuscular mycorrhizal fungi in plant-to-ecosystem responses to elevated atmospheric CO_2? *Mycorrhiza* 9:1-8.
Rillig, M.C., S.F. Wright, M.F. Allen, and C.B. Field (1999). Rise in carbon dioxide changes soil structure. *Nature* 400:628.
Rillig, M.C., S.F. Wright, B.A. Kimball, P.J. Pinter, G.W. Wall, M.J. Ottman, and S.W. Leavitt (2001). Elevated carbon dioxide and irrigation effects on water stable aggregates in a *Sorghum* field: A possible role for arbuscular mycorrhizal fungi. *Global Change Biology* 7:333-337.
Rogers, H.H., G.B. Runion, and S.V. Krupa (1994). Plant responses to atmospheric CO_2 enrichment with emphasis on roots and the rhizosphere. *Environmental Pollution* 83:155-189.
Ryan, M.H. and J.H. Graham (2002). Is there a role for arbuscular mycorrhizal fungi in production agriculture? *Plant and Soil* 244:263-271.
Smith, S.E., F.A. Smith, and I. Jakobsen (2004). Functional diversity in arbuscular mycorrhizal (AM) symbioses: The contribution of the mycorrhizal P uptake pathway is not correlated with mycorrhizal responses in growth or total P uptake. *New Phytologist* 162:511-524.
Staddon, P.L. (2004). Carbon isotopes in functional soil ecology. *Trends in Ecology and Evolution* 19:148-154.
Staddon, P.L. and A.H. Fitter (1998). Does elevated atmospheric carbon dioxide affect arbuscular mycorrhizae? *Trends in Ecology and Evolution* 13:455-458.
Staddon, P.L., A.H. Fitter, and D. Robinson (1999). Effect of mycorrhizal colonization and elevated atmospheric CO_2 on carbon fixation and below-ground partitioning in *Plantago lanceolata. Journal of Experimental Botany* 50:853-860.
Staddon, P.L., J.D. Graves, and A.H. Fitter (1999). Effect of enhanced atmospheric CO_2 on mycorrhizal colonization and phosphorus inflow in 10 herbaceous species of contrasting growth strategies. *Functional Ecology* 13:190-199.
Staddon, P.L., A. Heinemeyer, and A.H. Fitter (2002). Mycorrhizae and global environmental change: Research at different scales. *Plant and Soil* 244:253-261.
Staddon, P.L., K. Thompson, I. Jakobsen, J.P. Grime, A.P. Askew, and A.H. Fitter (2003). Mycorrhizal fungal abundance is affected by long-term climatic manipulations in the field. *Global Change Biology* 9:186-194.
Syvertsen, J.P. and J.H. Graham (1999). Phosphorus supply and arbuscular mycorrhizae increase growth and net gas exchange responses of two *Citrus* spp. grown at elevated [CO_2]. *Plant and Soil* 208:209-219.
Treseder, K.K. and M.F. Allen (2000). Mycorrhizal fungi have a potential role in soil carbon storage under elevated CO_2 and nitrogen deposition. *New Phytologist* 147:189-200.
Treseder, K.K., L.M. Egerton-Waburton, M.F. Allen, Y.F. Cheng, and W.C. Oechel (2003). Alteration of soil carbon pools and communities of mycorrhizal fungi in chaparral exposed to elevated carbon dioxide. *Ecosystems* 6:786-796.
Van den Staaij, J., J. Rozema, A. van Beem, and R. Aerts (2001). Increased solar UV-B radiation may reduce infection by arbuscular mycorrhizal fungi (AMF) in dune grassland plants: Evidence from five years of field exposure. *Plant Ecology* 154:169-177.

Wang, B., D.M. Funakoshi, Y. Dalpé, and C. Hamel (2002). Phosphorus-32 absorption and translocation to host plants by arbuscular mycorrhizal fungi at low root-zone temperature. *Mycorrhiza* 12:93-96.
Woodward, F.I. and C.P. Osborne (2000). The representation of root processes in models addressing the responses of vegetation to global change. *New Phytologist* 147:223-232.

Index

© 2007 by The Haworth Press, Inc. All rights reserved.
doi:10.1300/5425_09

Printed in the United States
by Baker & Taylor Publisher Services